Living in a Material World

Inside Technology
edited by Wiebe E. Bijker, W. Bernard Carlson, and Trevor Pinch

For a list of the series, see pp. 401–403.

Living in a Material World

Economic Sociology Meets Science and Technology Studies

edited by Trevor Pinch and Richard Swedberg

The MIT Press
Cambridge, Massachusetts
London, England

For information on quantity discounts, email special_sales@mitpress.mit.edu.

Set in Stone Serif and Stone Sans on 3B2 by Asco Typesetters, Hong Kong. Printed and bound in the United States of America.

Library of Congress Cataloging-in-Publication Data

Living in a material world: economic sociology meets science and technology studies / edited by Trevor Pinch and Richard Swedberg.
    p.   cm.—(Inside technology)
Includes bibliographical references and index.
ISBN 978-0-262-16252-4 (hbk. : alk. paper) — ISBN 978-0-262-66207-9 (pbk. : alk. paper)
1. Economics—Sociological aspects. 2. Technology—Economic aspects. I. Pinch, Trevor. II. Swedberg, Richard.
HM548.L59   2008
306.301—dc22   2008018948

10 9 8 7 6 5 4 3 2 1

Contents

Living in a Material World

Introduction

Although it is generally agreed in the social sciences that technology plays an important role in the economy, it is also recognized that it is difficult to understand what this role is and how to conceptualize it. Economists have traditionally treated technology as an exogenous factor and a black box. So-called growth theory has succeeded in endogenizing technology but has made little progress in developing a concrete and empirical type of analysis. Though economists inspired by Joseph Schumpeter have, in recent decades, developed the influential "economics of innovation" (Rosenberg 1994; Dosi 2000; Freeman 1982; Nelson and Winter 1982; McKelvey 2000), this approach does not, by and large, open the black box of technology, and it fails to engage with the increasingly sophisticated analyses of technology coming from history and sociology of technology. Bringing economy and technology together in one coherent analysis that is both analytically interesting and empirically oriented is, therefore, still very much on the agenda of the social sciences.

In this book, an attempt is made to reconceptualize the meeting between the economy and technology with the help of Science and Technology Studies (STS) and economic sociology.[1] Both of these approaches are relatively young and have developed new sets of ideas and concepts that have not yet been assimilated into mainstream social science.[2] The theoretical point at which we suggest that economic and technological analysis may come together is in the idea of *materiality*, or the notion that social existence involves not only actors and social relations but also objects. This is an approach that has been developed in STS and which we think constitutes a useful point of departure.[3] We have titled the volume *Living in a Material World* because the word 'material' also has another meaning than objects and materia, as the 1984 song by Madonna reminds us: it can also refer to something economic.[4]

A number of disciplines have recently embraced the "material turn." For instance, in communications in the 1980s there were calls to analyze the material dimension, led by the German scholars Hans Ulrich Gumbrecht and Karl Ludwig Pfeiffer.[5] But it is within anthropology and archeology, with the revival of the notion of "material culture," that one finds sustained attempts to develop a notion of materiality. Arjun Appadurai's 1986 collection *The Social Life of Things* marked a renewed anthropological interest in things and in particular commodities, and in how these might be tied to concerns with culture. Things and how they circulate and are exchanged in different historical and social milieus can be thought of as the low-hanging fruit of materiality. Ever since Georg Simmel's investigations into the role of money, conducted around the turn of the twentieth century (see Simmel 1978), it has been clear that objects can take on value in exchange, and, as Appadurai argues, this insight can usefully be extended to different regimes of value in space and time. That material forms matter to social science was increasingly recognized in the 1980s and the 1990s. It is now obvious that the social world is partly constituted by things, including the built environment of the cities we inhabit, the clothes we wear, the restaurant menus we peruse, and the food we eat.[6]

This new interest in material forms and in what they mean for humans has been marked by the emerging subdiscipline of Material Culture Studies along with a new *Journal of Material Culture*. The notion of "material cultures" has been of particular significance to the group of anthropologists around Daniel Miller at University College London, and Miller has edited two collections (1998, 2005) on materiality and material cultures.[7] Miller and his collaborators share with us a desire to theorize materiality without falling into the usual dichotomies raised by treating signification as separate from materiality *per se*, as in familiar tropes of subject versus object. This means moving beyond treating materiality and the world of things as passive objects that gain meaning only in symbolic terms in regard to the signification work that humans alone do. Most of the work on material culture, however, does not yet examine the technical working of technologies (as is done in STS) or explore the workings of the economy (as is done in economic sociology).

If things and commodities are the low-hanging fruit for social scientists, things that "bite back," or things that themselves have emergent powers or to which some form of agency may be ascribed, are much trickier to deal with. The classic example is, of course, technology. The pitfalls of examining technology from the perspective of hermeneutic social science were noted by Michael Mulkay (1979) at the dawn of the emergence of the field of Science and Technology Studies. Mulkay argued that there is a world of difference between the sociological analysis of a television that is working and one that is sitting in a room broken. At stake is what it is materially that such an object comprises. The non-working television can certainly be invested with human meaning; it can mark boundaries in a house, it can even serve as an object for exchange, and in the spirit of the literature on cargo cults it might take on all sorts of properties assigned to supernatural beings. But a functioning television is a very different sort of object because materially it has different capabilities. For one thing, a functioning television is embedded within a complex socio-legal-technical network—what sociologists of technology call a *sociotechnical ensemble*. Just try to write down the list of things (and actors) that are involved with a working television in the United States—obvious items might include electricity, cables, plugs, television studios, advertisements, actors and presenters, Hollywood, Fox, Rupert Murdoch, and the Federal Communications Commission, but this is only a start. As soon as one thinks about particular television programs, such as the popular *American Idol*, the list becomes even larger and would include other socio-technical ensembles, such as the telephones whereby watchers send in their votes for contestants. The signification of the different genres of programs for the viewers is itself a whole field of cultural analysis.[8] But the analysis of a technology such as television becomes even more complex if one takes up Mulkay's challenge and examines the material technology that enables a television to work at all. This means delving into the different ways televisions work—cathode-ray tubes versus flat screens, LEDs versus plasma. It also means delving into the struggles of engineers and television manufacturers as they develop the new standards, and into the visual and sonic technologies that are hidden within the box we users operate. The rallying cry within the sociology of technology and STS in general is to "open the black box" of technology; to see that the social does not start or stop with processes of signification produced by programs but that televisions are social all the way down (Bijker, Hughes, and Pinch 1987; Latour 1987; MacKenzie 1991; Bijker 1995a; Pinch and Trocco 2002). This richer notion of materiality, which encompasses technology, the social practices that constitute it, and the myriad ways we interact with it, is at the heart of STS.

STS offers a series of concepts that, we suggest, may be of help in developing a better understanding of technology and economy. The terms 'actor network' (Latour 1999) and 'sociotechnical ensemble' (Bijker 1995b) are used in STS to suggest that objects and humans should be understood to always exist together. Material objects and humans mutually constitute each other and should not be separated for analytical purposes. Analysis must start from the fact that people and objects always come together and that it would be artificial to draw a sharp line between the two. Objects and people are always entangled to various degrees. One may even argue that, as technology develops, this quality of entanglement—which is material as well as symbolic—becomes increasingly complex and important.

As one would expect of a lively new field, STS has not reached unanimity on how to analyze technical objects. Where there is unanimity, it is on the requirement that the analysis of materiality should not shy away from treating the same technical entities that engineers deal with. For example, Diane Vaughan's 1996 sociological analysis of why the Space Shuttle *Challenger* crashed (see also Collins and Pinch 1998) involves looking at how the testing of the O-rings of the solid-fuel booster rockets was carried out, because the social analysis of the accident rests in part on understanding how technical uncertainty was dealt with by different groups of engineers working within different organizational contexts. Where analysts part company is on how to treat the powers, emergent properties, or affordances that make technology so interesting to examine. It is obvious that technologies can do new things and that technologies are better than humans at doing some things. An electronic synthesizer can make a range of sounds of which no human is capable, an airplane can fly in a way no human has ever mastered, and a tractor can quickly beat the strongest "tug of war" team. But this way of phrasing the issue is not quite correct. In setting up some sort of opposition between technologies and humans, we tend to play down or forget that technologies gain their powers through the often hidden work of humans. Airplanes may fly, but they cannot fly without flight controllers and pilots. Indeed, it is in looking at the detailed embedding of humans with machines, as Edwin Hutchins (1995) does in examining the "distributed cognition" required to land a modern airplane, that one see the complexity involved in the coordination of humans and machines and the embedding of each with the other. A pilot may be talking to the flight controller one moment, manually adjusting a dial in the cockpit (a routine skill) the next moment, and assigning control of the airfoils to a computer the next. Much of today's STS research is concerned with the "plans and situated actions" (Suchman 1987) and "communities of practice" (Lave and Wenger 1991) that exist in the liminal space between machines and humans. Here one finds that materiality means examining not only the affordances enabled by machines (see chapter 11 of this volume) but also the material social and cognitive practices whereby humans interact with technology.

No easy separation between human and technological agency is possible when the thick description of technologies is concerned (Alder 2007; Bijker 2007). Furthermore, we fully agree with Bruno Latour (2007) that an idealistic notion of materialism, where some geometrized property of machines, as is found in engineering diagrams that delineate technologies in terms of their functions, must be resisted. Nevertheless, legitimate differences exist in the field of STS as to how best to treat the agency given to technology. Some analysts try, with ever-increasing complexity (Collins and Kusch 1998), to keep ledgers of the kinds of actions that can properly be assigned to humans and to non-humans. Others (e.g., Latour and Callon in their development of Actor Network Theory[9]) level the playing field, refuse to make any analytical distinction between human and non-humans, and talk in general about "actants." Some of these debates over humans versus non-humans surface in the present volume, most obviously in Mirowski and Nik-Khah's chapter on the allocation of the FCC spectra, which is in part a polemic against Latour and Callon's actor-network approach and against what Mirowski and Nik-Khah interpret as neglect of some good old-fashioned human political influences in the Federal Communications Commission's allocation of spectra.

Two other concepts from STS can usefully be employed to relate economy and technology to one another: *interpretive flexibility* and *closure* (Bijker 1995; Pinch 2006a). The former refers to the fact that actors are capable of interpreting a technology differently or investing it with different meanings. For example, a bicycle that old people consider dangerously unsafe may be perfectly acceptable to sporty young men (Bijker and Pinch 1987). These meanings of a technology are highly consequential for agency because they lead to different uses and different design trajectories. For example, some early bicycle companies, in responding to the meaning of the "ordinary" or "penny farthing" bicycle as "macho," built bicycles with larger and larger front wheels to make them even faster and more thrilling to ride.[10] Closure means that a novel technology will eventually stabilize, at which point it acquires a generally accepted meaning. Thus, the significantly named "safety bicycle," which emerged from a variety of design possibilities in the period 1880–1890, remained a remarkably stable technology until the appearance of the "mountain bike" in the early 1970s (Bijker 1995a; Rosen 2002). The links between objects and actors, in short, can be drawn differently depending on the meaning structure involved—but there is also a tendency for the meaning to stabilize in the sense of becoming general and accepted by large numbers of social groups.[11] The roles of *users* and *intermediaries* in how these meanings are generated and stabilized are also important in the STS account of technology (Oudshoorn and Pinch 2003).

Economic sociology, in its turn, has also developed concepts and ideas that can be useful in the attempt to relate the economy and technology to each other.[12] The term 'embeddedness', introduced by Karl Polanyi, is often used in sociological circles in connection with networks. (See e.g. Granovetter 1985.) Whereas most sociologists (including economic sociologists) look only at social relations and ignore the role of objects, here we suggest that the term 'embeddedness' should be used together with 'materiality', in the sense that objects and people are indissolubly embedded in each other. This new type of material embeddedness has its own distinct structure, which it is up to the analyst to try to outline. This structure can be described as a configuration of objects and social relations. Using 'embeddedness' in this sense is close to Actor-Network Theory in STS (Latour 2005). Actor-network approaches, however, have not been used to analyze economic topics until recently (Callon 1998; Latour forthcoming).

Economic sociologists have used networks to describe and explain a huge number of economic phenomena, often with a high degree of technical skill. Among the topics that have yielded quite nicely to this approach are markets, industries, consumption, entrepreneurship, business groups, and relationships inside as well as between firms (Baker 1984; Powell 1990; Burt 1992; DiMaggio and Louch 1998; McGuire and Granovetter 1998; Granovetter 2005). Insights from studies of this type would benefit STS— which, in turn, would add its insights about technology and materiality, with new and interesting insights as a result.

The concept of *field*, as used in economic sociology, may also be of assistance in further developing the idea of materiality in dealing with technology and economy. A field is usually understood as a type of social space or social structure that assigns a place to each actor (Powell and DiMaggio 1991: 64–65; Wacquant and Bourdieu 1992: 94–115). Power is part of a field; the actors may also constitute the field either through interaction or through orientation to other actors. These actors can be individuals as well as organizations; and both of these are typically perceived by sociologists as purely social entities, devoid of any materiality. But even if the concept of field does not take materiality into account, it nonetheless complements the idea of collectivity in a useful manner, not least in drawing attention to the structures of inequality and hierarchy that tend to develop between individuals as well as organizations. Fligstein's 1990 study of the evolution of the huge firm in the United States since the late 1800s is an example of this.

If a field is thought of as constituted by both social and material entities, hierarchies and inequalities can be conceived of in a new way. Material technologies allow old hierarchies to be reconfigured. Think, for instance, how the hierarchy in the nuclear family can be subverted by the introduction of the cell phone. Teenagers can now arrange meetings with their friends without parental control. The introduction of the telephone itself played a dramatic role in reconfiguring social relationships, particularly between women in rural communities (Martin 1991; Fischer 1994).

The concern with materiality in STS has an old ancestry in the philosophical school of materialism. This deserves to be highlighted, not least because this older type of materialism influenced Karl Marx, who is considered one of the founding fathers of economic sociology. Marx reacted as strongly against the abstract nature of Hegel's philosophy as Latour and others are reacting today against a generation of social scientists who see only social relations when they look at reality. Marx rejected the mechanical and old-fashioned type of materialism and strongly advocated the introduction of history and social relations into material analysis. Proceeding in a similar way may be useful for advocates of the modern approach to materiality too, not least since Marx's view of materiality is wedded to a sharp insight into the nature of capitalism and the centrality in modern life of economic power.

In *Capital*, Marx's version of materiality comes out most clearly in two ways. First, according to Marx, workers have to be reproduced if surplus value is to be produced, and Marx carefully describes the costs for reproducing the body of a worker and how these costs differ between countries. Here 'materiality' refers primarily to the body and its needs. Marx's concern with materiality in *Capital* is also evident from the attention he pays to the everyday life of the workers, such as the physical constitution of the factories in which they spend most of the day and the machines with which they work. Drawing on factory reports, Marx pointed to a number of material circumstances that wore down and tormented English workers as they carried out their tasks.

Marx also initiated the tradition in economic sociology of trying to analyze the systemic dimension of capitalism and connect it to what happens at the micro level in the factory. What drives this type of economic system, according to Marx, is the need for accumulation. *"Accumulate! Accumulate!* That is the Moses and the Prophets [of the modern economy]," we read in *Capital* (Marx 1976: 742). Similarly, Max Weber analyzed capitalism at a later stage of its development and emphasized its "pursuit of profit, and forever renewed profit, by means of continuous, rational, capitalistic enterprise" (1988: 17). Profit's role in driving the modern economy forward is central to Marx and Weber, and it must be taken into account in modern analyses of materiality (including technology).

The mentions of Marx and Weber remind us that modern economic sociology has a long classical tradition to draw on—a tradition that is centered around a strongly realistic and social structural approach. Issues such as the relationship between the economy and (say) religion, law, or politics were worked out more than a century ago (Swedberg 1998; Wright 2005).

To the Marxian-Weberian model of capitalism, Schumpeter added attention to the role of the entrepreneur, and Polanyi added the notion of modern capitalism as a radical utopian project. Schumpeter's (1934) idea that entrepreneurship can be conceptualized as a new combination of already existing elements is easily wedded to the idea that materiality is important and to changing relations between people and objects. Similarly, Polanyi's (1944) notion of modern capitalism as a peculiarly utopian project that is against nature through its peculiarly abstract and radical quality can be recast in material terms.

Contemporary economic sociology has continued the tradition from Marx onwards of trying to theorize the capitalist machine, and it has done so in a way that can be linked to the insights of STS. In theories of contemporary capitalism, the point of departure for economic sociology is in the conventional definition of the economy as consisting of production, distribution, and consumption (Swedberg 2005). It is usually also assumed in mainstream economics that what is being produced is distributed via the market, and then consumed. It is, however, clear that the assumption that distribution takes the form of exchange in the market is by no means obvious, even in a society with a market or capitalist economy. Different forms of distribution besides that of exchange exist, such as distribution via the state (what Polanyi termed 'redistribution') and distribution according to certain norms of reciprocity (Polanyi 1957).

The link between production and consumption may, by way of summing up the argument so far, be constituted in different ways—via exchange, redistribution, or reciprocity, to use Polanyi's terminology. Early preindustrial societies organized their economies with the help of redistribution and reciprocity, according to Polanyi, while capitalist society relies mainly on exchange. It is also clear that the capitalist type of society uses not only exchange but also redistribution and reciprocity. In today's OECD countries, for example, the state channels some 30–50 percent of GNP. And reciprocity (in other words, exchanges based on preexisting social roles, such as between family members, with no attempt to gain a direct advantage) remains the main principle governing individual households.

Finally, there exists an interesting difference between economies in which distribution mainly takes place via the market (exchange), on the one hand, and those where it mainly takes place via the state (redistribution) or the household (reciprocity). While the former tends to be dynamic, economies organized on the basis of redistribution and reciprocity do not. The reason for this has to do with the fact that exchange is entered into not only because of a desire to consume, but also to make a profit. This is not the case with redistribution and reciprocity; both of which are modeled on the idea of a household and the desirability of being able to satisfy the needs of its members rather than to make a profit (as in the individual household, the medieval manor, or the socialist state). This means that, although there definitely exists a necessity for reproduction in redistributive systems and systems based on reciprocity, there is usually not a push to constantly expand. Exchange is also a form of distribution that tends to mobilize both parties to a transaction, while redistribution and reciprocity typically only mobilize the actor who initiates the distributive process, not the one who receives the service or the good. In the former case, both parties have to go to the market, whereas in the latter case one party assumes a passive role. (See figure I.1.)

What roles do technology and materiality play in these different types of the economic process? It is clear that materiality is absolutely central to all of them, for the simple reason that all individuals have to reproduce themselves, and this is also a need that each of our three processes can accommodate. Also, in each of the three systems of distribution materiality is present at each of the three stages (production, distribution, and exchange).

The last statement deserves to be explicated, since we have now reached the point in our account of economic sociology where it is appropriate to bring STS and its concern with materiality back into the discussion. The apparently stable boxes in figure I.1 are perhaps better understood as networks making up the collectivities of individuals and objects to which we referred earlier. These networks exist not only in production (the part of the economic process to which technology is usually assigned) but also in distribution and consumption. The market, for example, is not just some abstract structure of social relations or an institution consisting of rules and regulations; it also involves material objects, be it in the form of balances, coins, tickers, telephones, or computers. Similarly, consumption involves objects, and not only objects to be consumed, but also other

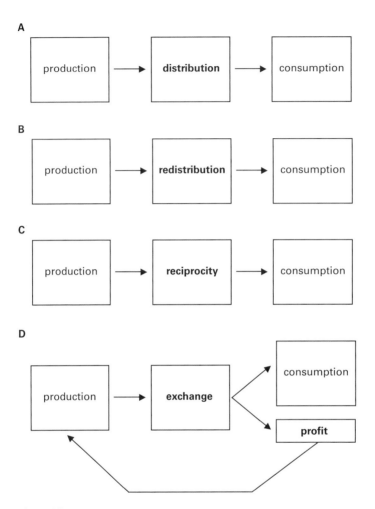

Figure I.1

The economic process and how it can be organized. (A) The economic process in general. (B) The economic process when "redistribution" (Polanyi) is predominant. (C) The economic process when "reciprocity" (Polanyi) is predominant. (D) The economic process when "exchange" (Polanyi) is predominant. The economic process in any society can be defined as consisting of production, distribution, and consumption. Exchange characterizes the capitalist organization of the economy; this type of economy derives its dynamic from the fact that the end goal of the economic process is not exclusively consumption, but also profit. The more this profit is reinvested into production, the more dynamic the economy will be. The two main mechanisms in capitalism, in other words, are organized exchange (the market) and the feedback loop of profit into production. It is the use of these two that makes the organization of economic interests in the form of capitalism an effective machine for transforming economic reality.

objects—including such "means of consumption" as advertisement, pack-aging, stores, malls, and parking lots (Ritzer 1999). Consider frozen orange juice. The widespread consumption of orange juice, and indeed much of the food-distribution system in the United States, depends crucially on the development of the technology of mobile refrigeration units on trucks, the existence of interstate highways, and the establishment since World War II of an independent trucking industry (Hamilton 2003).

There is also, again, the fact that consumers often use the objects they buy in ways that are unforeseen by the producers, and in this sense become "co-constructors" of many new objects (Oudshoorn and Pinch 2003). Even if it can be analytically instructive in some cases to assume that there are distinct boundaries between production, distribution, and consumption, these boundaries are often ambiguous in reality and shot through with links to objects and people (in the other boxes in figure I.1, so to speak).

If technology and materiality have been neglected thus far, how come economics is such a successful science without them? One way to answer this question is to simply point out that, although neoclassical economics no doubt is a great academic success and has its very own Nobel Prize, it has also shown itself to be singularly unable to produce the kind of knowledge that translates into predictions of what will happen in reality (McCloskey 1985, 1990). Macroeconomic forecasting, as Evans (1997) points out, has a particularly dismal record. Nevertheless, economists still offer a veneer of understanding, some nuggets of real wisdom, and legiti-mation for policies. Economic techniques such as cost-benefit analysis, as Porter (1995) has shown, originated in the world of politics as a numerical way of displacing and resolving political disputes. The application of such techniques involves a constant set of translations back and forth between calculative practices and politics (Ashmore, Mulkay, and Pinch 1989). "Trust in numbers" enables the state to use suites of calculative tools, prac-tices, and techniques to help maintain political and social order and stabil-ity (Porter 1995). These calculative practices are, as Mirowski (2002) has shown, increasingly part and parcel of the technologies (e.g., computers) that are indispensable to forecasting. Economic assumptions are them-selves sometimes built into the "laws of the market," as Callon has persua-sively demonstrated, thereby providing a reflexive circle of verification that enables the market to be understood as an economic phenomenon. (For Callon, economic activity is not strictly that of economists but would in-clude auditors, accountants, and policy makers.) Materiality and technol-ogy are everywhere, and once technologies move from interpretative flexibility to stability they enable assemblages of humans and non-humans to work together in the choreography of any modern economy. It is this "glue," this stability in the entanglements, that provides the illusion of the iron hand of the market.

A classical question is whether technology drives economic development or whether the profit motive is the driving force. In capitalism it is clearly the latter, even if technology is an integral part of all three ways in which an economy can be organized. But this does not mean that the technology that exists in a capitalist society is directly shaped by the profit motive alone. All that can be said at this general level is that technology can be used to accelerate profit. This is a very broad formulation that allows for a multitude of ways of conceptualizing how technology develops.

One may, however, also ask whether the materiality perspective that we are advocating does not invite a new and different conceptualization of the problem of how profit and economic development are related. At some level it makes sense to conceptualize technology in terms of the entanglement of objects and people, and this is true for all three forms of organizing the economy (reciprocity, redistribution, and exchange). The more technology there is, it would appear, the more entanglement; and as objects are shaped by people, these are transformed in their turn. A dynamic capitalist system could, from this perspective, be understood as a system that has the capacity to speed up this interaction considerably, and with this comes an increased capacity for the reproduction of people. Entanglement of this type would seem to be related to the concept of productivity in economics, even if it is broader in nature and even if it also takes human meanings and social relations into account.

These ideas also bear on how we understand the specific role of information technologies in the economy. Though information may seem "non-material," in reality this type of technology permits new forms of entanglements between people and objects and can crucially change the material circumstances whereby exchange of goods and knowledge occurs and where things and ideas circulate. The changing material arrangements and social organization of exchange are highlighted in chapters 5, 7, and 8 of this book. Similarly, several chapters deal with the material and social changes brought about by computer software in the arena of consumption (chapter 10), in new "open-source" systems for allocating value to items during exchange (chapter 11), and in new online systems for managing human resources (chapter 12). In all these chapters we see that the details of changing material arrangements underpin the social and economic understanding of the new information economy.

This book is a first exploration of how the idea of materiality may be used as a bridge between STS and economic sociology for two-way traffic between these two fields. The individual chapters can be grouped according to whether they raise general questions in this respect (part I), questions relating to infrastructure (part II), questions about the material arrangements of the market (part III), or questions about the use of technology (part IV).

In chapter 1, Michel Callon ponders one of the fundamental issues in economics: the role of the individual and the reality of the famed *homo economicus*. Many commentators agree that the actor assumed in *homo economicus* is a pale shadow of real human actors, but Callon accepts neither of the standard responses to this recognition. Whereas economic sociology argues that the way forward is to embed the too thinly endowed actor with human characteristics (in short, to embed the actor in a real social and cultural context), neoclassical economists try to free *homo economicus* from the institutional ties and make the actor even more rational. Callon's position is to reject the debate and to instead focus on how and under what circumstances individuals can have agency. Drawing on the notion of agency earlier developed by Callon and Latour within the well-known Actor-Network Theory, Callon argues that agency is configured within a network of both human and non-human actors. Drawing on examples such as buying decisions by consumers at supermarkets and studies of people with disabilities, Callon investigates what social policy options there are for dealing with individual actors in the changed world of network economies. He argues that with "prosthetic" policies the individual is endowed with new competencies to deal with the network, in the same way that disabled people are equipped with prosthetic devices to help them manage their handicap. "Habilitation" policies, on the other hand, construct new socio-technical arrangements, such that that they facilitate new forms of individual agency within networks, rather in the way that the world can be restructured better to suit the needs of a handicapped person by allowing handicapped access to all buildings. In terms of economic networks, by changing labor regulations firms can tailor training directly to their employees' career plans and thus enable employees to engage in the running of organizations in new ways. It is clear that Callon favors the latter sorts of policies and thus offers a normative prescription for dealing with some of the socio-economic issues raised by the new economy.

In chapter 2, Richard Swedberg brings the discussion firmly back to the material realm—from *homo economicus* to home economics. He notes that, although people live in houses, eat food, interact with machines, and produce and use objects, economics at best acknowledges this materiality in an indirect way. He argues that the ambitious goal the authors in this volume have set themselves—the goal of building a new form of economic analysis that attempts to theorize the economy in terms of entanglement and inter-penetration of things and people—will also require us to mine the history of economic thought for leads. Swedberg makes a start as to how to do this by recovering the concern of early economics with the material theory of the household. He locates the origins of this approach in ancient Greece in the often-neglected economic writings of Aristotle and Xenophon. He then shows how in the classical period of political economy (1600s–late 1800s) the material realm of the household began to vanish in the writings of Adam Smith and Karl Marx, who, although paying attention to the body and technology, focused more on production and exchange outside the household. For example, women's work, emphasized by the ancient Greeks, is almost completely absent from the accounts of Smith and Marx, where women (and children) are mentioned only when they enter the labor market. Swedberg concludes his survey by looking at one modern institutional context for economics: Cornell University in the early twenti-eth century, where in 1916 Frank Knight famously gave the theory of *homo economicus* a classical formulation. Interestingly, the discipline of "home economics," with its material vision of the household, emerged at Cornell at the same time—a development largely ignored by and outside the main-stream of economic thought.

In chapter 3, Phillip Mirowski and Edward Nik-Khah use the case of the 1994 public auction by the Federal Communications Commission of spec-trum licenses to the highest bidder as a way of examining Callon's ideas about the performativity of the economy. Callon (1998: 30) argues in his well-known book *The Laws of the Market* that "the economy is embedded not in society but in economics." By this Callon means that economists and economic thinking and methods have played a role in the way that markets are built. This shifts the traditional economic sociology problem of studying the embedding of economy in society or the social explana-tion of economics to studying how the economy in a sense must work according to economic laws because economists have built an "economic machine" that made it that way. *Prima facie*, the case of FCC spectrum auc-tions seems to fit Callon's performativity argument, because many leading game theorists were involved in the design of the auction and have claimed credit for it. Mirowski and Nik-Khah provide a much more detailed account of the different groups of game-theory and experimental economists in-volved in the auctions. They show a diversity of aims and understandings amongst the different players in the auction and the power played by the telecom corporations in redefining the government's goals for the auction.

In the end they claim that a more traditional science studies analysis which includes power among social groups better explains the outcome—an outcome that was deliberately masked such that one influential group of economists—game theorists—could "bask in the limelight and take the credit." Mirowski and Nik-Khah offer a different view of how STS analysis can be taken into the realm of economics. Their view is "not at all isomorphic to the performativity thesis," but it does grant more rigid and hence constraining roles to nature and society than Callon and Latour are prepared to grant.

Part II of the volume is titled Infrastructure, a term that is sometimes used in STS and by which is meant not only the traditional types of infrastructure (such as electrical networks and railroad systems; see Chandler 1977) but also laws, organizational forms, systems of filing, administration classification, and the like (Star 1999; Star and Bowker 1999; Yates 2005). This part opens with a chapter on one particular form of infrastructure that is very important in the modern economy, namely accounting. David Hatherly, David Leung, and Donald MacKenzie argue that some areas of "economic reality" are constituted via the classification of economic transactions in the form of accounting. This is especially the case with profit and loss, which are both of utmost importance to investors, governments (e.g. for taxation purposes), and employees (e.g. for the determination of bonuses). Accounting is often manipulated, as evidenced by recent corporate scandals in the United States involving Enron and WorldCom. New financial instruments, such as derivatives, also present special problems when it comes to accounting, sometimes because of their complexity. Hatherly, Leung, and MacKenzie argue that accounting rules can be interpreted and used in a potentially unlimited number of ways. The reason for this is that rules of any kind cannot be locked into place once and for all. According to finitism (a Wittgensteinian approach developed within the sociology of scientific knowledge), rules can always be interpreted in new and unpredictable ways, even if they cannot be used in just any way. According to Hatherly et al., this quality makes the approach of finitism well suited to deal with the many complexities that an empirical study of accounting entails.

The next chapter moves on to the topic of global configurations of trading and information technologies. Karin Knorr Cetina and Barbara Grimpe focus their attention on two such systems: FOREX (used in the foreign exchange market) and DMFAS (a debt management and financial analysis system) first developed in 1979 after a United Nations Conference on Trade and Development and now used internationally to monitor and control countries' debt levels and debt repayments. Knorr Cetina and Grimpe argue that these technologies not only network the world together, as Manuel Castells and others have argued, but introduce a specific "scopic mechanism" of coordination. By this they mean a mechanism that collects and focuses activities, interests, and events on one surface (in the case of foreign exchange markets, a computer screen), from whence the results may then be projected again in different directions. They identify what they call a Global Scoping System (GSS) as the configuration of screens, capabilities, and contents that traders in financial markets confront all over the world. The goal of the chapter is to examine how the material architecture of global financial markets leads to this new form of "scoping" coordination. They trace the origins and development of FOREX and DMFAS and show that, although both infrastructures provide scoping, FOREX provides a continual instantaneous scoping of the market whereas DMFAS depends on International Monetary Fund and World Bank schedules for producing reports and inputs on fixed time scales. Despite these temporal differences, both technological systems provide for reflexive scoping in that traders (in the case of FOREX) and national debt officers (in the case of DMFAS) participate reflexively in creating and participating in a standard global representation to which they and other actors in turn respond. Knorr Cetina and Grimpe conclude that FOREX and DMFAS embody different strategies of globalization. FOREX follows the strategy of global exclusivity and maintains a separate province of the global world, co-existing with the rest of the planet rather than integrating it. DMFAS, on the other hand, is tied in with the nation state and thus helps sustain the global character of world economic institutions and an emerging global governance.

In chapter 6, Elizabeth Popp Berman analyzes a different type of infrastructure: that of law and the various institutions in which legal rules are embedded—courts, police, lawyers, and so on. Patent law, Popp Berman argues, is an institution that mediates between technology and economic forces in modern society, and it can do so in a number of ways. One alternative would be for the state to pay for the research and make it available for free; another would be to assign to the inventor the exclusive right to an invention for a specified period of time. Which way is preferable is largely unknown, according to Popp Berman, and some of the confusion involved is mirrored in the largely accidental passage of the Bayh-Dole Act of 1980, which opened up the rapid commercialization of academic inventions in the United States. While economists have very firm opinions about what type of patent law is needed and why, Popp Berman suggests that economic sociologists as well as people working in the STS tradition are needed to disentangle the complex social, technological, and economic issues involved. Economic sociologists may, for example, explore the different ways in which scientists and investors react to different types of patent law. The issues may also differ depending on what scientific field and what industry are involved. Input from STS scholars is also needed in this effort, since their expertise is in scientific and technical issues. The fact that STS scholars have a very broad and social approach to science and technology is a particularly important asset in this regard.

The chapters in part III (Technology and the Material Arrangements of the Market) examine particular technologies and their roles in financial markets. Alex Preda examines the history of the first technology specifically designed to be used in financial markets: the stock ticker. This device, invented in 1867, was initially a printing telegraph that printed out a security's name, price quote, and traded volume. It soon became ubiquitous on stock markets, and it remained so until it was replaced in 1960 with an electronic ticker. Preda examines the context from which the ticker emerged, how it was first used, and how it affected stock markets. He shows that the stock ticker was adopted because it enabled official stockbrokers to maintain their monopoly over credible, authoritative price data.

There is much concern in economic sociology with how prices in markets are determined not only by economic efficiency and computational rules but also by social networks, interests, and status. Preda shows that what must be added to the economic sociology account is the part played by technological systems in how price data are generated and observed. In capturing the new forms of agency which the stock ticker enables in financial markets, Preda introduces the notion of a "generator"—a concept that captures temporal structures, representational languages, and cognitive tools. The price ticker made market exchanges visible as they happened, transforming them into more abstract and visible forms available to everybody at once. The paper strips of the stock ticker were a forerunner of financial charts that today make market exchanges visible in real time. In short, the stock ticker enabled the transformation of multiple, unsystematic, discontinuous, and unrecorded heterogeneous price information into the single, continuous, homogeneous nearly real-time price variations on which traders have come to rely.

In chapter 8, Daniel Beunza and David Stark take us into the heart of the modern trading room. They present the results of an extensive ethnography of a Wall Street trading room in a major international investment bank. The traders they examine are concerned with arbitrage, which involves essentially the construction of comparability across different assets.

In arbitrage small differences in these comparative valuations are exploited to make a profit on the deal. In the process of calculating these values equivalencies are established and the process of recognizing these equivalencies and how they offer opportunities depends crucially on the "tools of the trade." Like Preda, Beunza and Stark demonstrate the importance of instrumentation (in this case, assemblages of instrumentation) in understanding how the socio-technical, the socio-cognitive, and the socio-economic are intertwined. Just as Latour and others have argued that the scientific laboratory gains its strength as a place where diverse instruments are gathered together, Beunza and Stark argue that traders turn their trading rooms into laboratories in which they experiment and deploy an array of instruments, including networked computers, mathematical formulas, and "robots" capable of automated trading. Their ethnography shows also the importance of spatial layout and physical proximity—in terms of our earlier discussion, the materiality of local arrangements really counts. They succeed in showing that in order to understand the sophisticated instruments of quantitative finance we need to analyze the entanglements of actors and instruments in the "sociotechnology of the trading room laboratory."

In chapter 9, Fabian Muniesa focuses on the role of a rather old technology—the telephone—in the trading room. Although computers and electronic networks were expected to make the telephone redundant in financial markets, Muniesa shows that this is far from the case. Rather as Beunza and Stark found that face-to-face communication between proximal actors in a trading room was crucial, so too it turns out to be the case that telephone communications can still facilitate trading operations. Just as the stock ticker has been updated, the technology of the telephone has changed such that current market operators obtain greater diversity and flexibility than ever before with the use of a special box that permits as many as 24 telephone conversations to be accessed in a variety of ways at the touch of a button. A microphone even permits some conversations to be relayed on and over heard by a variety of other listeners. All conversations are recorded "back room" and if need be can serve as a legal record of transactions. Muniesa's research, rather than treating the telephone as a passive device, an instrument serving human interaction, shows how the materiality of the device (its technical features) shapes action and enables its users to perform functions crucial to the operation of markets. He examines the use of the telephone in three different empirical locations: in market making, in a stockbroking environment where orders are largely being filled, and in a sales environment where the particular needs of clients must be met. He shows how the telephone serves a crucial function in all these environments, enabling counterparts (defined by Muniesa as a "client," a "trader," and a "broker") for trades to be identified and providing a means of negotiating and trust building amongst actors in social networks— aspects not available with anonymous electronic trades. He shows how in some circumstances telephony practices are transformed by the other new market technologies (such as computer screens) and how these enable reconfigurations of social networks. The overall goal is to explore the correspondence between "social networks" in the traditional economic sociology sense and material networks of communication that allows for "ties" to be articulated and expressed in a particular code and manner.

In the first chapter in part IV (Technology, Economy, Use), Christian Licoppe addresses how e-commerce is changing modern life. Licoppe argues that before e-commerce existed, consumers were used to viewing shopping as either a kind of planned activity (as when one brings a list to a shop) or as a form of impulsive buying. E-commerce lends itself much more easily to planned shopping, according to Licoppe, while browsing and spontaneous purchases are harder. One type of purchasing behavior, in brief, is typical for stores and supermarkets, and another for the consumer sitting in front of his or her computer.

Electronic shopping also differs from traditional shopping in the way its sequences are timed. When you buy on the Internet there is especially a significant time gap that does not exist when you shop in a store or a supermarket, namely between the expressed intention to buy and the actual delivery of the goods. From an economic viewpoint, e-commerce adds little to the traditional view of exchange, since you still have to first hand over the money before you get what you want. From the perspective of buying as a social type of activity, in contrast, buying on the Internet represents a novel experience; and the customer often reacts with anxiety to the time gap between the display of intention to buy and the delivery of the goods.

How the technology that is used in e-commerce has led to many changes in the ways that people interact with one another is also at the center of the second chapter in part IV. What interests Shay David and Trevor Pinch is that, whereas a few years ago it was common for books and other cultural goods to be reviewed only by expert reviewers, the new technology has led to the emergence of a new type of reviewers, who may be called amateur reviewers or user reviewers. Their activities, according to David and Pinch, are reshaping the operation of "the reputation economy." Taking as their point of departure the fact that some amateur or user reviewers duplicate the product reviews that they post on websites such as Amazon.com, David and Pinch raise a series of general issues concerning new technology and how it is used in predictable and unpredictable ways. "Interpretive flexibility" and "technological affordance" are seen as particularly helpful in explaining why electronic product reviews have come to serve a number of purposes besides providing information about the item that is being sold, such as the construction of identity and the promotion of one's own books and CDs. They do so, according to David and Pinch, by emphasizing that the use of an item is not somehow inherent in its essence but instead is decided by the social, historical, and economic context of the item.

That mainstream economic analysis is not particularly well suited to explain what happens when technology is transferred, and that alternative types of explanations from the STS literature are better at this, is the main theme of chapter 12. By 'transfer' Nicholas Rowland and Tom Gieryn do not mean the process of moving some piece of technology from one place to another, but something that involves the social setting—more precisely, the organizational processes that accompany the transfer of some technology from one organizational setting to another. The example Rowland and Gieryn use to make their point is the trouble that has resulted from the recent decision by the Indiana University Business School to switch to a standardized IT system produced by PeopleSoft. If one analyzes this situation with the transaction-cost approach that is associated with the work of Oliver Williamson, Rowland and Gieryn argue, it is very difficult to make sense of the difficulties that ensued at the Indiana University Business School. If instead one uses the STS idea that it is not possible simply to replicate an experiment, since this does not take tacit knowledge into account, one is in a much better position to address "transfer problems."

Notes

1. We are not the first to suggest this linkage. Michel Callon's 1998 book *The Laws of the Market* has been highly influential. Callon's approach has also been taken forward (and critiqued) in a special issue of *Economy and Society* edited by Andrew Barry and Don Slater (2002). Important studies also have been conducted by Knorr Cetina and Bruegger (2002), MacKenzie and Millo (2003), and Callon and Muniesa (2003). See also MacKenzie, Muniesa, and Siu 2007.

2. Both fields have produced handbooks of their main themes and approaches. For STS, see Jasanoff, Markel, Petersen, and Pinch 1996 and Hackett, Amsterdamska, Lynch, and Wajcman 2007; for economic sociology, see Smelser and Swedberg 1994, 2005.

3. There are other ways of bringing STS and economics together. For example, the role of numbers, figures and calculative practices and their use in economics has been investigated from an STS perspective (Ashmore, Mulkay, and Pinch 1989; Porter 1995; Kalthoff, Rottenburg, and Wagener 2000). Calculative practices are at the heart

of Callon's (1998) approach, which we discuss in more detail in the text and which Callon elaborates upon in his own contribution. A related body of work draws parallels between accountancy practices and the increasing prevalence of what might be called an "audit society" (Miller 2001; Power 1990; Strathern 2000). McCloskey's (1985) influential work on the rhetoric of economics offers another approach to the practice of economics as an academic discipline (see also McCloskey 1990). The history and philosophy of economics is also a rapidly developing area; see, e.g., Morgan 1990; Mirowski 1998; Weintraub 2002.

4. "[T]he boy with the cold hard cash / is always mister right / 'cause we are living in a material world / and I am a material girl"—Madonna, "Material Girl" (1984). Andrew Pickering (1989) deserves credit for being the first person to use Madonna's catchy song title in an academic context.

5. Their German reader was eventually published in English as *Materialities of Communication* (Gumbrecht and Pfeiffer 1994).

6. For another attempt to take materiality seriously within science studies, see Mukerji 1997.

7. For another such collection, see Buchli 2002.

8. The cultural analysis of television was famously discussed by the cultural critic Raymond Williams in *Television, Technology and Cultural Forms* (1974).

9. See also Akrich 1992.

10. The feminist approach to technology has also been extremely influential in STS. The focus upon bodies and the gendered meanings and practices built around technologies provides another important route into the issue of materiality. See, e.g., Cockburn and Ormrod 1993; Oudshoorn 1994; Wajcman 1991; Haraway 1991; Oldenzeil 1999. For a very recent bringing together of feminism with materiality, see Alaimo and Hekman, 2008.

11. "Closure" and "stabilization" are given slightly different meanings in Bijker 1995a.

12. For overviews of economic sociology, see Smelser and Swedberg 2005; for some often used anthologies Granovetter and Swedberg 2001; Biggart 2002; Dobbin 2004. Michel Callon, Karin Knorr Cetina, Donald MacKenzie, Alex Preda, David Stark, and a few other scholars span the two areas of STS and economic sociology.

References

Akrich, M. 1992. The de-scription of technological objects. In *Shaping Technology/ Building Society*, ed. W. Bijker and J. Law. MIT Press.

Alaimo, S. and S. Hekman eds. 2008. *Material Feminisms*. Indiana University Press.

Alder, K. 2007. Thick things: Introduction. *Isis* 98: 80–83.

Appadurai, A., ed. 1986. *The Social Life of Things: Commodities in Cultural Perspective*. Cambridge University Press.

Ashmore, M., M. Mulkay, and T. Pinch. 1989. *Health and Efficiency: A Sociology of Health Economics*. Open University Press.

Baker, W. 1984. The social structure of a national securities market. *American Journal of Sociology* 89: 775–811.

Barry, A., and D. Slater. 2002. Introduction: The technological economy. *Economy and Society* 31, no. 2: 175–193.

Biggart, N., ed. 2002. *Readings in Economic Sociology*. Blackwell.

Bijker, W. 1995a. *Of Bicycles, Bakelites and Bulbs: Toward a Theory of Sociotechnical Change*. MIT Press.

Bijker, W. 1995b. Sociohistorical Technical Studies. In *Handbook of Science and Technology Studies*, ed. S. Jasanoff et al. Sage.

Bijker, W. 2007. Dikes and dams, thick with politics. *Isis* 98: 109–123.

Bijker, W., T. Hughes, and T. Pinch, eds. 1987. *The Social Construction of Technological Systems: New Directions in the Sociology and History of Technology*. MIT Press.

Bourdieu, P., and L. Wacquant. 1992. *An Invitation to Reflexive Sociology*. University of Chicago Press.

Buchli, V., ed. 2002. *The Material Culture Reader*. Berg.

Burt, R. 1992. *Structural Holes: The Social Structure of Competition*. Harvard University Press.

Callon, M., ed. 1998. *The Laws of the Markets*. Blackwell.

Castells, M. 2000. *The Rise of the Network Society*, second edition. Blackwell.

Chandler, Alfred D. 1977. *The Visible Hand: The Managerial Revolution in American Business*. Belknap.

Cockburn, C., and S. Ormrod. 1993. *Gender and Technology in the Making*. Sage.

Collins, H., and M. Kusch. 1998. *The Shape of Actions: What Humans and Machines Can Do*. MIT Press.

Collins, H., and T. Pinch. 1998. *The Golem at Large: What You Should Know about Technology*. Canto.

DiMaggio, P. 1988. Interest and agency in institutional theory. In *Institutional Patterns and Organizations*, ed. L. Zucker. Ballinger.

DiMaggio, P., and H. Louch. 1998. Socially embedded consumer transactions: For what kind of purchases do people most often use networks? *American Sociological Review* 63: 619–637.

Dobbin, F., ed. 2004. *The New Economic Sociology: A Reader*. Princeton University Press.

Dosi, G. 2000. *Innovation, Organization and Economic Dynamics: Selected Essays*. Elgar.

Fischer, C. 1994. *America Calling: A Social History of the Telephone to 1940*. University of California Press.

Fligstein, N. 1990. *The Transformation of Corporate Control*. Harvard University Press.

Freeman, C. 1982. *The Economics of Industrial Innovation*. MIT Press.

Granovetter, M. 1985. Economic action and social structure: The problem of embeddedness. *American Journal of Sociology* 91: 481–510.

Granovetter, M. 2005. Business groups and social organization. In *The Handbook of Economic Sociology*, second edition, ed. N. Smelser and R. Swedberg. Princeton University Press.

Granovetter, M., and P. McGuire. 1998. The making of an industry: Electricity in the United States. In *The Laws of the Market*, ed. M. Callon. Blackwell.

Granovetter, M., and R. Swedberg, eds. 2001. *The Sociology of Economic Life*. Westview.

Gumbrecht, H., and K. Pfeiffer, eds. 1994. *Materialities of Communication*. Stanford University Press.

Hackett, E. J., O. Amsterdamska, M. Lynch, and J. Wajcman, eds. 2008. *The Handbook of Science and Technology Studies*, third edition. MIT Press.

Hamilton, S. 2003. Cold capitalism: The political economy of frozen orange juice. *Agricultural History* 77: 557–581.

Haraway, D. 1991. *Simians, Cyborgs and Women: The Reinvention of Nature*. Routledge.

Hutchins, E. 1995. *Cognition in the Wild*. MIT Press.

Jasanoff, S., G. Markel, J. Petersen, and T. Pinch, eds. 2004. *The Handbook of Science and Technology Studies*. Sage.

Kalthoff, H., R. Rottenburg, and H.-J. Wagner, eds. 2000. Facts and Figures: Economic Representations and Practices. *Oekonomie und Gesellschaft Jahrbuch* 16. Metropolis.

Kline, R., and T. Pinch. 1996. Users as agents of technological change: The social construction of the automobile in the rural United States. *Technology and Culture* 37: 763–795.

Knorr Cetina, K., and U. Bruegger. 2002. Global microstructures: The virtual societies of financial markets. *American Journal of Sociology* 107, no. 4: 905–950.

Knorr Cetina, K., and A. Preda, eds. 2004. *The Sociology of Financial Markets*. Oxford University Press.

Latour, B. 1996. *Aramis: or, The Love of Technology*. Harvard University Press.

Latour, B. 1999. *Pandora's Hope: Essays on the Reality of Science Studies*. Harvard University Press.

Latour, B. 2005. *Reassembling the Social: An Introduction to Actor-Network Theory*. Oxford University Press.

Latour, B. 2007. Can we get our materialism back, please? *Isis* 98: 138–142.

Latour, B. Forthcoming. *Gabriel Tarde and Economics*.

Lave, G., and E. Wegner. 1991. *Situated Learning: Legitimate Peripheral Participation*. Cambridge University Press.

MacKenzie, D. 1991. *Inventing Accuracy: A Historical Sociology of Nuclear Missile Guidance*. MIT Press.

MacKenzie, D., and Y. Millo. 2003. Negotiating a market, performing theory: The historical sociology of a financial derivatives exchange. *American Journal of Sociology* 109, no. 1: 107–146.

MacKenzie, D., F. Muniesa, and L. Siu, eds. 2007. *Do Economists Make Markets? On the Performativity of Economics*. Princeton University Press.

Martin, M. 1991. *"Hello Central?" Gender, Technology and Culture in the Formation of Telephone Systems*. McGill–Queens University Press.

Marx, K. 1999. *Capital: A Critique of Political Economy*. Penguin.

McCloskey, D. 1985. *The Rhetoric of Economics*. University of Wisconsin Press.

McCloskey, D. 1990. *If You're So Smart: The Narrative of Economic Expertise*. University of Chicago Press.

McKelvey, M. 2000. *Evolutionary Innovations: The Business of Biotechnology*. Oxford University Press.

Miller, D., ed. 1998. *Material Cultures: Why Some Things Matter*. University of Chicago Press.

Miller, D., ed. 2005. *Materiality*. Duke University Press.

Miller, P. 2001. Governing by numbers: Why calculative practices matter. *Social Research* 68, no. 2: 379–395.

Mirowski, P. 1998. *Against Mechanism: Protecting Economics from Science*. Rowman and Littlefield.

Mirowksi, P. 2002. *Machine Dreams: Economics Becomes a Cyborg Science*. Cambridge University Press.

Morgan, M. 1990. *The History of Econometric Ideas*. Cambridge University Press.

Mukerji, C. 1997. *Territorial Ambitions and the Gardens of Versailles*. Cambridge University Press.

Muniesa, F. 2003. Des Marchés commes algorithmes: Sociologie de la cotation électronique à la Bourse de Paris. Ph.D. dissertation, Ecole des Mines, Paris.

Nelson, R., and S. Winter. 1982. *An Evolutionary Theory of Economic Change*. Harvard University Press.

Oldenzeil, R. 1999. *Making Technology Masculine: Men, Women and Modern Machines in America, 1870–1945*. Amsterdam University Press.

Oudshoorn, N. 1994. *Beyond the Natural Body: An Archaeology of Sex Hormones*. Routledge.

Oudshoorn, N., and T. Pinch. 2003. *How Users Matter: The Co-Construction of Users and Technologies*. MIT Press.

Pickering, A. 1989. Living in the material world: On realism and experimental practice. In *The Uses of Experiment*, ed. D. Gooding et al. Cambridge University Press.

Pinch, T. 2006. The sociology of science and technology: a review." In *21st Century Sociology: A Reference Handbook*, volume 2, ed. C. Bryant and D. Peck. Sage.

Pinch, T. 2008. Technology and institutions: Living in a material world. *Theory and Society*.

Pinch, T., and W. Bijker. [1984] 1987. The social construction of facts and artifacts: or how the sociology of science and the sociology of technology might benefit each other. In *The Social Construction of Technological Systems*, ed. W. Bijker et al. MIT Press.

Polanyi, K. 1944. *The Great Transformation*. Farrar & Rinehart.

Polanyi, K. 1957. The economy as instituted process. In *Trade and Market in the Early Empires*, ed. K. Polanyi et al. Regnery.

Porter, T. 1995. *Trust in Numbers: The Pursuit of Objectivity in Science and Public Life*. Princeton University Press.

Powell, W. 1990. Neither market nor hierarchy: Network forms of organization. *Research in Organizational Behavior* 12: 295–336.

Power, M. 1990. *The Audit Society: Rituals of Verification*. Oxford University Press.

Ritzer, G. 1999. *Enchanting a Disenchanted World: Revolutionizing the Means of Consumption*. Pine Forge.

Rosen, P. 1993. The social construction of mountain bikes: Technology and postmodernity in the cycle industry. *Social Studies of Science* 23: 479–513.

Rosen, P. 2002. *Framing Production: Technology, Culture, and Change in the British Bicycle Industry*. MIT Press.

Rosenberg, N. 1994. *Exploring the Blackbox: Technology, Economics, History*. Cambridge University Press.

Schumpeter, J. 1934. *The Theory of Economic Development*. Harvard University Press.

Simmel, G. 1978. *The Philosophy of Money*. Routledge.

Smelser, N., and R. Swedberg, eds. 1994, 2005. *The Handbook of Economic Sociology*. 1st and 2nd rev ed. Russell Sage Foundation and Princeton University Press.

Star, S. 1999. The ethnography of infrastructure. *American Behavioral Scientist* 43: 377–391.

Star, S., and G. Bowker. 1999. *Classification and Its Consequences*. MIT Press.

Strathern, M., ed. 2000. *Audit Cultures: Anthropological Studies in Accountability, Ethics and Academy*. Routledge.

Suchman, L. 1998. *Plans and Situated Actions*. Cambridge University Press.

Swedberg, R. 1998. *Max Weber and the Idea of Economic Sociology*. Princeton University Press.

Swedberg, R. 2005. The economic sociology of capitalism: An introduction and agenda. In *The Economic Sociology of Capitalism*, ed. V. Nee and R. Swedberg. Princeton University Press.

Vaughan, D. 1996. *The* Challenger *Launch Decision: Risky Technology, Culture and Deviance at NASA*. University of Chicago Press.

Wajcman, J. 1991. *Feminism Confronts Technology*. Pennsylvania State University Press.

Weber, M. 1988. *The Protestant Ethic and the Spirit of Capitalism*. Peter Smith.

Weintraub, E. 2002. *How Economics Became a Mathematical Science*. Duke University Press.

Williams, R. 1974. *Television, Technology and Cultural Forms*. Technosphere Books.

Woolgar, S. 1991. Configuring the user: The case of usability trials. In *A Sociology of Monsters*, ed. J. Law. Routledge.

Wright, E., ed. 2005. *Approaches to Class Analysis*. Cambridge University Press.

Yates, J. 2005. *Structuring the Information Age: Life Insurance and Information Technology in the 20th Century*. Johns Hopkins University Press.

I General Concerns: Economy, Materiality, Power

1 Economic Markets and the Rise of Interactive *Agencements*: From Prosthetic Agencies to Habilitated Agencies

Michel Callon

Many analysts contend that the continuous extension of the market sphere contributes powerfully to the lasting process of individualization character-izing Western societies (Beck and Beck-Gernsheim 2002). The new forms of organization of markets and the importance of innovation in their structur-ing strengthen this movement by combining two trends. The first of these trends is associated with the growing importance of networks (of coalition and cooperation between heterogeneous actors and organizations) as a new form of coordination of economic activities. This networking of markets stems primarily from the centrality of innovation and, consequently, the increasing singularization of goods and services (Powell and Smith-Doerr 1994). The second trend corresponds to the upsurge of individual autono-mous actors (in design, production, or distribution, as well as in consump-tion), sometimes compared to self-entrepreneurs capable of developing their own projects and therefore responsible for the consequences of their acts.[1] Networks and individual agents can be seen as two complementary realities. The former mobilize individuals capable of ensuring their devel-opment and functioning; the latter opt for those networks that allow them to realize their projects. Even though these two transformations are complementary and difficult to disentangle, in this chapter I would like to focus on the second one and, more particularly, on the modalities and effects of the contribution of markets as networks[2] to the process of individualization.

The scope and reality of this overall process of individualization to which economic activities contribute, as much as the consequences it generates, are a subject of debate. There is nothing new about this controversial situa-tion; it was born with the appearance of *homo economicus* on the intellec-tual scene. Critics, mainly but not only from sociology and anthropology, have applied themselves—with some degree of success—to challenging the very existence of this original being. For some he is a caricature of reality.

Nowhere does this type of human being—capable of defining objectives and calculating the appropriate means to achieve them, sometimes to maximize or optimize his own interests—exist. To make him resemble beings in the real world, critics say, he needs to be enriched with the emotions, passions, and values without which his behaviors and choices are meaningless, or to be put back into his institutional and social context. It was Durkheim who expressed such criticism most forcefully. "The human being that we know, real man," he wrote, "is far more complex; he belongs to a period and a country, he lives somewhere, has a family, religious beliefs and political ideas." (Durkheim 1970). This critique shows that *homo economicus* is not self-sufficient. It is the anthropological monster—to use Pierre Bourdieu's striking term—that is denounced. *Homo economicus*, irrespective of his material richness, cannot exhaust the richness of what constitutes our common humanity; he is a fiction that should not cause us to forget that only *homo sociologicus* exists in reality. The opposite position, found mostly among economists, holds that the capacity of *homo economicus* to calculate his interests and optimize his actions corresponds to an anthropological reality that appears as soon as the conditions favorable to its expression are met.

This opposition between disciplines has been and still is intense. It is more than a family quarrel between *homo economicus* and *homo sociologicus*, for it concerns the interpretation of the individualization process itself. It seems, however, that it should not be overemphasized. Since Durkheim, points of view have evolved and moved closer together. The idea is gaining currency that, under certain institutional conditions, individual agents who behave willy-nilly as *homo economicus* can exist. Of course this does not prevent some from thinking that this configuration is the best and most efficient one possible, whereas others consider that there is no one best way and that *homo economicus* is simply one possible form of existence—and not necessarily the most prevalent one at that. Yet in both cases there is recognition that, to exist, *homo economicus* needs a favorable environment and institutional affiliations.[3]

There are several reasons for this convergence, including, above all, interaction between disciplines that have given up their trench warfare and become more attentive toward one another. But from my point of view, the appearance of a new species of *homo economicus* is the main reason for the convergence and for the resulting relative pacification. Until recently, *homo economicus*, which I will now qualify as version 1.0, was the only known species. *Homo economicus 1.0* was highly introverted and relied only on himself and his own resources. He had to be able to unambiguously determine his interests, have preferences and classify them, and then choose the most economic means to concretize them. He was miles away from *homo sociologicus*, with whom there was no way of cohabiting. By contrast, when it comes to *homo economicus 2.0*, the individual agent who arrives with markets as networks, sociologists, anthropologists, and economists tend to have similar points of view. This is a considerably upgraded and enhanced version. *Homo economicus 2.0* engages in strategic activities and interacts with other agents. To survive and prosper as an individual, he draws on diverse material and emotional resources and relies on interpersonal networks. As he is fleshed out and gains substance, he starts resembling *homo sociologicus*. Sociologists are prepared to accept him, while economists readily recognize that to live and survive he needs the assistance and institutional support that provide him with critical mental and cognitive resources. In short, when he exists, *homo economicus 2.0* is caught up in the institutional devices constituting the ecological niche on which he depends. He is assisted, helped, surrounded, relayed. No one would dream of denying that he is "embedded" in assemblages endowing him with the resources, competencies, and assistance (financial, social, cultural, emotional) needed for his existence. It becomes clear that, without these assemblages, he is destined to extinction or, worse, to a very poor existence.

This new approach alters our view of the process of individualization and its different stages. The question is no longer so much one of the anthropological realism of this evolution and especially of the change to *homo economicus 2.0*; it now concerns the conditions allowing for the evolution, the difficulties it encounters and its consequences. From a normative perspective, we might even wonder whether it is good to encourage it and, if so, how and on what terms. We might also reflect on conceivable alternatives. This inquiry leads to a revival of the social question or, more precisely, the question raised since the middle of the nineteenth century by the identification and political management of what is seen as the (potentially) negative effects of the irresistible expansion of economic markets. In this case, the issue is the capacity that some people have (or do not have) to turn into *homo economicus 2.0*, that is, to participate in their own right in economic life and the welfare that it can create. Through the arrangements that they imply and impose, do markets not produce new forms of exclusion by brutally eliminating all those who cannot adjust to their conditions of functioning?

Existing approaches contain partial answers to this question since they emphasize the importance of institutional "niches," without which *homo economicus 2.0* collapses. But one of their limits stems from the fact that they tend to overlook materialities and technologies in the description and analysis of these "niches." Economists, anthropologists, jurists, and sociologists all recognize that, as Beck and Beck-Gernsheim put it, the individual exists only when he is institutionalized and therefore tied up, entangled, caught in mechanisms of coordination. However, they take into account only regular social institutions. Some recent studies on the sociology of markets, inspired by STS, have shown the crucial importance of technologies and materialities for understanding not only the functioning of markets but also the shaping of agencies and their competencies. Market arrangements are, of course, made up of rules, routines, incorporated skills, incentives, norms, interpersonal relations, access to resources, and so on, but also of heterogeneous material devices (Callon, Muniesa, and Millo 2007). In this chapter I wish to show the advantage of analyzing these materialities, mainly by highlighting the concept of socio-technical *agencement* for understanding the emergence of *homo economicus 2.0.* This approach enables us to address the social question and policies concerning it in new terms.

In the first section, based on studies of distributed action, I introduce the notion of individual agency, which leads me, in the second section, to the notion of socio-technical *agencement.* These two concepts enrich the notion of the institutionalized individual and consequently enable us to consider the diversity of forms of individual agency. In the third section I rely on this analytical frame, and I relate it to the interesting but hitherto largely unexploited sociological literature on disabilities in order to examine the conditions enabling *homo economicus 2.0* to emerge and thrive. This leads me to distinguish two strategic configurations: one that enacts *homo economicus 2.0* by an arrangement of prostheses, and one that enacts him by habilitation. I then show, in the fourth section, how these two configurations pave the way for a distinction between two types of policy. In the conclusion I briefly discuss this approach. I highlight the fact that the question of the viability and the conditions of dissemination of *homo economicus 2.0* concern only certain (social) problems raised by the development of markets as networks of innovation. I also emphasize the fact that this approach includes material elements in social policies and consequently provides a more realistic view of the latter. As the reader will gather, this is a tentative study, one of the aims of which is to suggest the interest of this general analytical frame for understanding markets and their effects. A considerable amount of work is still required to adapt it to markets.

Distributed Action: From Individual (Institutionalized) Agents to Individual Agencies

To study the variety of conditions in which *homo economicus* exists, survives and thrives, I propose that we give up the idea of implicitely or explicitely comparing his poor equipment (in terms of relations, rules to follow, cognitive and bodily competencies, values, motivations, and will) to the infinitely richer equipment of *homo sociologicus*. As I aim to show, it seems more appropriate to take into consideration the modes of distribution of individual action (especially if we want to integrate the role of materialities into the analysis) and give up the study of individual agents. My assumption is that we should replace the notion of an individual by that of individual agencies, and to add that the latter are as diverse as the configurations shaping them. To analyze this diversity we have to start with the concept of a distributed agency which can in no way be reduced to that of a "dressed" (or embedded) agency.[4]

The reason I introduce the notion of agency rather than social science's more traditional concepts of actors or agents is because the analysis of action poses two distinct problems. The first concerns the effects it produces, which make it possible to affirm that an action did actually take place. The action, and this is its minimal definition, is what produces an observable effect. The second problem concerns its origins: Who or what caused the action?

By employing the now well-established concepts of actors or agents, the social sciences tend to dangerously limit the range of possible answers and to underestimate the uncertainties surrounding action, its content, its effects, and its origins. They generally grant more importance to the role of humans and thus impose a very narrow definition of action. Their approach causes them to contrast actions which are determined entirely by causal chains (behaviors) with those which introduce a solution of continuity between cause and effect (that can be described as autonomous, strategic, intentional, etc.). A course of action will thus alternate between behavioral episodes, based on a mechanical type of rationale, and episodes in which the human agent takes over to steer the course of events in new directions (Collins 1995). Explaining these changes is one of the most difficult tasks the social sciences have to deal with. It generally revolves around the opposition between forces said to be external to actors (or agents) but running through them (structures, interests, the subconscious, technological development, etc.) and internal forces that explain why the course of action changes suddenly (intentionality, cognitive skills, strategic calculation, etc.). With this distinction, action is contrasted with its context and, in its most extreme forms, the individual agent with the social structures in which he is plunged (sometimes drowned). Hence, the social sciences are confronted with the interminable task of explaining relations between these two orders of reality. The most common solution consists in evoking the existence of dialectical relations between the two (Giddens 1991; Bourdieu 2000).

The notion of agency is a way out of this dialectic and offers a glimpse of a solution. First, it leaves the uncertainties concerning sources of action open. Who is acting? Is it an individual? A collective? Are things (or nonhumans) involved in the action, and, if so, in what forms? There is no general answer to these questions; only the particular circumstances of the action count. Agency—and this is the first advantage of the concept—can be attributed to heterogeneous and unexpected entities which are not necessarily human beings ("the French economy" that creates unemployment; "biotechnology" which generates ethical problems; "genes" which are said to cause severe diseases and/or impairments). Second, the content, nature, and effects of the actions that the agency triggers off are also widely diverse. What differences are produced? How can they be characterized? What evidence is there of the existence of these differences? The answers are open, since with this perspective the electron acts as much as the engineer interacting with it, albeit differently. Third, by restoring the richness and diversity of action and leaving its characterization open, the notion of agency modifies the respective contributions that social scientists and participants in action make towards the analysis of that action. The origins, effects, and modalities of action are of interest to all those participating in it, and their points of view, conceptions, and theoretical elaborations influence the course of action (Beck et al. 1994).[5]

Thus defined, the notion of agency substantially enhances the descriptions of action that the social sciences can give. In particular, it enables us to grant the place it deserves to the concept of distributed action which leads us to talk no longer of individual action but of individual agency.

To clarify what we mean by distributed action, the following two examples seem to be relevant illustrations of situations in which economic agents find themselves. The first is piloting (in this case an airplane, but the action applies equally to a firm or to a project); the second is the making of a choice.[6]

Take, first, the simple case of an airline pilot, a good example of individual agency.[7] Depending on the situation, and especially in cases of serious crisis, she can play on and adjust to the circumstances with the aim of safe-guarding her passengers' lives (or, on the contrary, with the deliberate in-tention of dying with the passengers, as was recently the case). This ability to set (arbitrary) goals and to develop a course of action that may make it possible to achieve them, while remaining responsible for her acts, is pos-sible (a) because the pilot is not acting alone and (b) because collective action is configured in such a way as to make her play an important role. She is caught in a socio-technical *agencement* that performs and organizes the actions of a large number of entities. These are called on as needed, and they "propose" solutions to which the pilot would not otherwise have access and of which she would not even have thought had she been isolated. The action denoted by the verb "to pilot" is a collective action in which a host of heterogeneous entities (or actants) participate (air-traffic controllers, radars, gyroscopes, control levers, pilots and co-pilots, landing strips, international regulations, etc.), all cooperating to enable the Air-bus 420 to travel from Paris to Marseilles. In fact, 'cooperating' is not the most appropriate word. Saying that the action is distributed is more ac-curate, first because the action (of piloting) is spread among several actants (human and non-human, individual and collective), second because that collective action consists of sequences whose order can vary depending on the events (distributed action is organized but cannot be reduced to a pre-established plan), and third because none of the participants in the action can be considered independently of the others (a pilot without dials and screens to read, suggesting what ought to be done, is not a pilot, and vice versa). Specialists of distributed action say that each of the actors makes proposals to the other actors in the course of the action, and that it is these proposals that act as affordances or what I propose to call *promissions*.[8]

We could go further in the description and contend that not only action but also cognition is distributed. The pilot could be taken again to illustrate this point,[9] but in order to show the generality of the phenomenon we will consider an ordinary economic agent: a customer in a supermarket. How can a customer choose between two packets of sliced ham which seem identical? Franck Cochoy (2002) has shown that in order to make this choice the supermarket customer is helped by a host of "assistants": quality labels, *appellations contrôlées*, data on the composition and origin of the ham, advice from other consumers, consumer magazines, advertising, and, obviously, the price. The work of Kjellberg (2001), Barrey (2001), and Grandclément (2006) has enabled us to pursue this exploration and to highlight the involvement of a heterogeneous population of market profes-sionals who belong to either the world of production or that of distribution.

These professionals design software, tools for making comparisons, and presentation devices that preformat the space of consumers' choices in close interaction with them. This space of distributed calculation (Callon and Muniesa 2005) encompasses not only human beings but also a set of material devices in which the shopping cart plays an essential part (Grand-clément and Cochoy 2006).

This example calls for two observations. First, all this equipment does not belong to the consumer; it comes from the outside, so to speak, and constitutes her calculation and decision-making competencies. Second, talking of equipment can be misleading because in reality these are entities participating in the calculation, not suppliers of information that the consumer is content simply to take into account. The data at her disposal are the result of prior calculations by those who transmit them. The customer's ability to calculate is therefore distributed among (human and non human) assisting entities, in the same way that the pilot's action was distributed (we could, moreover, show that piloting or steering is also calculating, and that any distributed action is also distributed cognition) (Callon 1998; Callon and Muniesa 2005; Beunza and Stark, this volume; Beunza et al. 2006).

The two examples of the pilot and the supermarket customer show us what a distributed agency is. In both cases an action is underway, with observable effects (the choice of the airplane's trajectory; the decision to purchase); in both cases the action is collective, in the precise sense of mobilizing a large number of (human and non-human) entities taking part in the action; in both cases the participants in the action (pilot, navigation instruments, radars, quality labels, friends' advice, consumer organization, shelves, shopping carts, packaging, etc.) take turns to further the action, with each entity acting in its own particular way; in both cases the origins of the action thus conceived are not easily assignable or are at least subject to debate: is it the pilot or the designer of the radar, the joystick or the Airbus, the air-traffic controller or the builder of the runway that constitutes the main source of the action? Each of these participants contributes toward the action, and there is no reason *a priori* to give any one of them special treatment. Yet in both cases, due to the very configuration of the *agencement*, in which formal procedures, practical and technical knowledge, software, skills, and rules of action must be included, we impute responsibility for the action to the pilot, in the first case, and to the customer, in the second. This responsibility, enacted by law, can be re-opened at any time, especially during a crisis or a catastrophe resulting in legal proceedings.[10]

We see the advantage of the concept of agency. In no way does it impede interpretation. In both examples it makes a description possible in which the capacity to act is granted to all the participants of the action. That is what the term 'distributed agency' means. At the same time, this (distributed) agency can be assigned to a particular figure or form which, in the examples proposed, is that of individual agency. It is to the pilot that the action of piloting is ascribed because it is on her that the sequences of action undertaken by the participating entities converge. It is to the customer that the action of buying is imputed because she is the destination of all the calculations constituting the framework of her own calculation. In both cases the law confirms and legitimates this imputation.[11] We could thus agree that the notion of individual agency denotes a course of action (engaging a large number of heterogeneous entities participating in the action) which is imputable to an individual since the actors themselves are led (or channelled by the arrangement) to consider the individual to be the source of the action. To the question "Who pilots the airplane?" the passengers, the pilot herself, and her crew all answer "The pilot!" To the question "Who chooses the sliced ham?" the consumer herself, her friends and family, and the manager of the supermarket all answer "The consumer!"

Individual agency is simply one possible form of agency, one that encompasses a wide variety of possible forms. Individual agencies can indeed be altruistic and non-calculating or selfish and calculating (Callon and Law 2005). But we could mention others: collective agency ("the company's employees" decided to reject the proposed retirement plan), anonymous agency ("market forces" ended up causing prices to drop substantially). Irrespective of the form, recognizing that an agency (whether it is related to economic activities or not) is distributed and simultaneously that it is embodied in a variety of possible forms leads to the notion of socio-technical *agencement.*

Socio-Technical *Agencement*

Any action is distributed. Through it, humans as well as procedures, calculation tools, instruments, and technical devices collaborate and participate in a coordinated manner. All these entities contribute in their own way to the collective action that consequently consists of a series of ordered acts. The analysis of this collective action (Who participates? With what effects?) is not only in the hands of the observer, for the participants have their say in characterizing the motivating forces of the action and the identities of those involved.

To describe the action, one has to be able to describe the strange assemblages that we could agree to call *socio-technical agencements*. The word *agencement* has the advantage of being close to the notion of agency: an *agencement* acts, that is, it transforms a situation by producing differences. The modifier 'socio-technical' underscores the fact that the entities which are included in the *agencement* and participate in the actions undertaken are both humans and non-humans. The set pilot + air-traffic-controllers + instruments-of-flight + radars + runways + airplane-manufacturers + communication-protocols + jurisprudence-concerning-the-responsibility-of-navigation-staff etc. is just such a socio-technical *agencement*.

As Callon, Muniesa, and Millo (2007) emphasize, the notion of agencement is close to notions of ordinary language that foster a similar intuition (display, assemblage, arrangement),. But it is also a philosophical concept whose proponents, Gilles Deleuze and Félix Guattari, can be considered representatives of a French pragmatist tradition. (See Deleuze and Guattari 1980.) In his discussion of Foucault's notion of 'device' (*dispositif* in French), Deleuze (1989) develops an account that is closer to the idea of *agencement*. For Deleuze, the subject is not external to the device. In other words, subjectivity is enacted in a device—an aspect, I think, that is better rendered through the notion of *agencement*. In Deleuze's phrasing, a device "is a tangle, a multi-linear ensemble. It is composed of different sorts of lines. And these lines do not frame systems that are homogeneous as such (e.g. the object, the subject, the language). Instead, they follow directions, they trace processes that are always at disequilibrium, sometimes getting close to one another and sometimes becoming distant from one another. Each line is broken, is subjected to *variations in direction*, diverging and splitting, subjected to *derivations*." (Deleuze 1989: 185, Muniesa and Callon translation) In actor-network theory, a perspective always attentive to the distributed character of action, the notion of 'socio-technical device' (*dispositif socio-technique* in French) is also close to this idea of *agencement*—an idea which emphasizes the distribution of agency and with which materiality comes to the forefront. An *agencement* is constituted by fixtures and furnishings, by elements that allow lines to be drawn and territories to be constituted. It is only when devices are understood as *agencements* that the evolving intricacies of agency can be tackled by the sociologist or the anthropologist (otherwise she may need to conform to the great agency divides that so often characterize the sociological tradition).

Why introduce this notion of *agencement* rather than being content to talk of singular distributed agencies whose functioning would have to be described case by case, depending on the particular socio-technical configurations shaping them? The answer to this question relates to the formulation of a hypothesis that seems to be supported by empirical observations. These suggest that the arrangements organizing action and deciding on prevailing forms of agency can be grouped together in a limited number of families. Each of these families ought to be able to be associated with a particular form of distributed agency.[12] In the following I describe one of these configurations corresponding to what Andrew Barry, himself inspired by Foucault and Deleuze, calls the interactive diagram (Barry 2001).[13] The agency associated with this *agencement* is what I suggest calling an *interactive individual agency*. My thesis is that, because of the important part that it gets *homo economicus 2.0* to play, the network economy tends to mobilize the interactive diagram on a massive scale.

The interactive diagram is a socio-technical *agencement* configured in such a way that at the center of the collective action we find an individual who is capable of developing projects and is endowed with a will to accomplish them, and who holds herself (because she is held) responsible for her acts and their effects. This diagram constitutes a particular answer to questions concerning the modalities of action. To the question "Who is at the source of the action?" the diagram answers "The individual and her projects." To the question "What is the status of the different participants in the action?" it answers "On the one hand the individual defining and undertaking projects, whose identity changes and adjusts in relation to feedback and results, and on the other hand the technical devices with which she interacts and which constantly suggest original courses of action." To the question "What does the action produce?" it answers "The discovery of possible new worlds, the unexpected, constant experimentation."

How is the interactive diagram able to produce this responsible individual, capable of projects and innovations and at the same time condemned to govern her own affairs? The answer to this question warrants subtle analysis and needs further investigation. I will settle here for a brief characterization of these *agencements*, based on the discursive elements, procedures, and technical devices that constitute them.[14]

• Discursive elements produce indisputably real effects. They may be very general discourses, such as those which praise the market economy, the new economy, or the service economy, or, more directly, all the discourses that argue for individuals' responsibility and/or that see the ability to define and realize projects as essential for any human being in his own right. Luc Boltanski and Eve Chiapello (2006) thus highlight the significance of managerial discourses in the creation of a spirit that serves as a reference for action: employees must be capable of defining projects, bearing responsibility for them, and agreeing to be judged on their results. Jean Gadrey (2002) shows that the "new economy" consists more in a discourse aimed at enacting and supporting a certain form of network economy than in an "objective" description of reality. These discursive elements contribute to describing, shaping, and supporting the autonomous and responsible individual. They also include references to teamwork, pluri-disciplinarity, hybridization, and cross-fertilization that emphasize the need to be open, mobile, and interactive in order to generate opportunities and be able to grasp them (Barry 2001).

• Procedures, forms of organization, and incentives are also a powerful motivating force in the configuration of the interactive diagram. The quasi-contractual practices developed in the corporate world produce employees who have to define objectives and engage in their realization. The same applies to employee incentives, competency reviews (in French, *bilan de competences*), and recruitment techniques (Eymard-Duvernay and Marchal 1997). These management techniques and the heterarchical organization they imply give substance to the responsible and innovative individual (Stark 1999). The law furthermore endorses this procedural construction every time it decides in favor of individual responsibilities. Preferred forms of organization (project groups, task forces) also encourage employees to embark on collaborative projects in which roles are poorly defined and mutually influenced. More generally, what Nigel Thrift (2005) calls the knowing capitalism, or what Michael Power (1997) and Marylin Strathern (2000) call the audit society, contributes to enacting this singular form of agency.

• The technologies that facilitate the creation of individual agency are usually termed 'information technologies' and 'communication technologies'. They are interactive, they generate connections and networks, and they facilitate unplanned encountering (Thrift 2004). Interactivity forces the user to be reactive, to use initiative, to be imaginative, and constantly to boost the action in order to test new possibilities and take observable results into account. The creation of connections makes it possible to mobilize, at the same place, the different instruments, equipment, data, and information constituting and multiplying the individual's cognitive capacities.[15] New possibilities of encounters constantly emerge and facilitate an uninterrupted process of exploration and investigation. The individual is thus endowed with intentions, calculation, reflection, projects, and imagination. Because they facilitate the traceability of actions, these technologies reinforce the demand for coherence and along with it rational action.

They likewise facilitate imputation of actions. Note that these technical devices include all software used for project management, monitoring actions, compiling customer files, tracking and tracing goods and people, and individualizing supply and demand, as well as all tools used for calculation and evaluation which provide algorithms for ranking decisions.[16]

Socio-technical *agencements* corresponding to the configuration of the interactive diagram construct a form of agency that implies an informed individual capable of intention, anticipation, reactivity, calculation, and control of her own actions. But the above discussion shows that we must avoid confusing the individual (a pilot, a customer of a supermarket, or an industrial manager) with (interactive) individual agency, since:

• Alone the individual cannot be an individual agency. She is a stakeholder in an *agencement* which (like any socio-technical *agencement*) configures distributed actions involving a large number of humans and non-humans. The figure of interactive individual agency associated with it is inseparable from the (interactive) socio-technical *agencements* that make it exist.

• Through the extension of network economies it is not individualism that is spreading but interactive *agencements* which multiply to become a dominant form of organization of collective action. Barry suggests that this configuration, fairly close to the post-Fordian model (to use the terminology proposed by the *École de la régulation*), is found in the worlds of services and industry but also in that of culture. The reasons for this diffusion are multiple and are obviously not directly relevant here. They end up mutually reinforcing one another to produce a society which seems to be inhabited by active and enterprising individuals but which, in reality, consists of a multitude of closely connected interactive socio-technical *agencements*.

• With the generalization of these *agencements* and the responsible and proactive individuals that go along with them, a new form of subjectivity is spreading as well. Not only does the individual (autonomous, strong-willed, responsible, and creative) impose herself as the dominant figure of agency; her experience of herself also coincides with that view: she senses that she is required to be autonomous, responsible, and interactive (Rose 1999). In short, the individual experiences and accepts herself as an active and interactive individual.

• This subjectivity leads to the construction of plastic, open identities that are shaped, circulated, and exchanged along with the interactions and experiments in which individual agency is engaged.

The two notions of distributed agency and socio-technical *agencement* enable us to grasp the diversity of forms of agency without being hampered by the hypothesis of a common anthropological base. They also allow us to identify and characterize a particular form: interactive individual agency. Other forms of individual agency are possible, such as the disciplined individual agency described by Barry. I will now use the latter to contrast two forms of social policy: prosthetic policies and habilitation policies.

From Prosthesis and Discipline to Habilitation and Interaction

The notion of socio-technical *agencement* enables us to build a complete picture of the elements that need to be present and mutually adjusted if interactive individual agency is to appear. It is therefore crucial for an understanding of the conditions of existence and development of network economies, whose functioning implies the mobilization and coordination of such agencies. It is also essential for explaining the shortcomings of these agencies and describing the policies designed to prevent them. This is the issue to which I will now turn. To describe these shortcomings and their effects, I propose to use the repertoire of handicaps and the political elaborations it has spawned. I will draw on the highly innovative work of Ingunn Moser (2003) and Myriam Winance (2007).

The notion of 'handicap' has a long history. It has generated and continues to generate rich debate, some of which, particularly interesting from a political point of view, concerns the terminology to use.[17]

The forms and nature of handicaps are multiple. Some are qualified as mental and others as physical; some are accidental and others congenital; some are evolving and others stable. Agreement has gradually been reached on the idea that handicaps are not (only) linked to individual themselves: they also stem from the relationship between individuals and society. A handicap is not located exclusively in the handicapped person, even if the two cannot be entirely dissociated; it is synonymous with maladjustment.

The point of introducing the idea of a handicapped person into the analysis of effects produced by network economies relates above all to the fact that one of the models shared by a vast majority of those striving to define what a social handicap policy could be is precisely that of the individual agent. Using my terminology, I would say that in the field of handicap a very common challenge is the transformation of people living with handicaps into individual agencies, that is, autonomous and responsible individuals, in some cases capable of developing projects and implementing them.[18]

Strategies that have been imagined and enforced for achieving this transformation are multiple. We could say, very briefly, that the history of the handicap field is that of the gradual shift from an individual-focused model to a society-focused model, with the two models not mutually exclusive and frequently overlapping. Each of the models has a different distribution of the sources of the handicap and hence of the origins of the weakness it causes. In the individual model, society is seen as constituting a normative frame, so that the adjustment primarily concerns the person who must be either repaired or re-equipped in order to be integrated into the collective. In the social model, the environment is considered as the essential source of the maladjustment. In the latter case, the recommended strategy is either to make the environment accessible or to transform it in such a way as to overcome the observed maladjustments. These two strategies choose the same starting point: the individual/society couple (with the word 'society' signifying the social, cultural, or technical environment). They have the same objective: performing a readjustment so that the handicapped individual becomes an autonomous individual. But in order to do so they choose different means, for they pose the question of relations between the members of this twosome in different terms. As a result, the individual agency models that they implicitly have in view are also different. To describe these two strategies I suggest using the two concepts of prosthesis and what I propose to call *habilitation*, which will enable me to define the modalities of political interventions and to identify the one that leads to the production of interactive individual agencies.[19]

In the individual model, the option is taken either to repair the individual concerned or to equip her so that she (again) becomes an autonomous agency in her own right. She is given (human or technical) prostheses to restore the lacking competencies, thus making her capable of the same actions as any non-handicapped (disciplined) person, defined by the "normal" performance of which that person is capable. This prosthetic conception aims to equip the person with tools, competencies, and resources that will enable her to overcome some of her limits. The prostheses, as extensions equipping the individual, are the way in which her capacities are restored and her disabilities compensated for. They are designed to complete and prolong the person's body: relations between the individual and her prostheses are at the same time relations of intimacy, proximity, and familiarity that rely on fine adjustments and tinkering of the different elements (Winance 1996). Identities are transformed, but with the aim of stabilizing them. The wheelchair that ends up being part of the handicapped person's body is an emblematic prosthesis. The animal (monkey or dog) or human being mobilized to assist a handicapped person can also act as a prosthesis, that is, an instrument or tool that is attached or articulated rigidly with the individual in order to extend her, to be incorporated into her. Prostheses, irrespective of what they are, equip individuals in such a way as to give them a capacity to act and move in society. This capacity for action (which is, as any action, distributed) imposes a very specific model in which the individual is autonomous to the precise extent that, in a disciplined way, she follows the course of action allowed by the prostheses and inscribed in them (Akrich 1992).

Habilitation is a quite different approach. It is based on the idea that there is no reason to act exclusively on handicapped persons to reduce the maladjustments they are suffering from. Instead of focusing on the extension of the individual by successive articulations and integrations of prostheses, as in the case of prosthetic adjustment, habilitation is also directed at the individual, but starting from the outside environment. As it goes along, it shapes devices, procedures, and forms of organization, aiming for the inclusion of the handicapped person in an interactive diagram. Here we are moving away from "disciplinarization" and toward the interaction, encounters, and initiatives that transformation of the environment is bound to allow. The word 'habilitation' aptly describes this sought-after configuration, for the aim is to put the handicapped person in a position to define her own projects by constructing the *agencements* enabling her both to conceive them and to accomplish them.[20] Here again a very simple example demonstrates the rationale of the approach. In the mid 1980s, the AFM (Association Française contre les Myopathies, the French muscular dystrophy association) set up regional services (SRAI) for providing daily assistance to people with serious neuromuscular diseases. Generally these are evolving diseases which in some cases result in situations of extensive disability. The AFM decided to consider these people as capable of having projects (called "life projects") whose definition and elaboration ought to be in the hands of the person concerned (Callon and Rabeharisoa 2007). It was with the aim of stimulating the development of projects and allowing their realization that the SRAI were created. These services consist of pluridisciplinary teams whose mission is not to adapt the patients but to set up interactions enabling them gradually to discover what they want and to try out solutions. They help also to organize the technical, economic, and administrative environment (or niches) in such a way that these projects can evolve in relation to the evolution of the disease and the feedback from experience. In its most extreme form this approach consists in leaving aside the specific characteristics of the flesh-and-blood individual and working only on the *agencement* in which she is a stakeholder, thus making her an individual agency capable of having projects and being responsible for her acts.

Prosthesis and habilitation are two symmetrical approaches. Both aim to compose an individual agent: the former by acting primarily on the person, the latter by striving to transform the environment. If we agree to use the proposed terminology, we can say that both compose individual agencies, but according to radically different models. Habilitation shapes an interactive agency and at the same time endows the individual with the capacity to define projects and realize them. By contrast, the addition of prostheses extends the individual to enable her to conform to common norms. The aim is to grant the individual an individual agency, but one that is disciplined. Habilitation constantly allows for the appearance of new possibilities, whereas the prosthesis limits the possible fields of action. The distinction proposed by Barry (2001) between a disciplinary diagram and an interactive diagram perfectly describes these two options.[21]

Prostheses and habilitation are not mutually exclusive. In the above example, the muscular dystrophy patient interacting with the SRAI does not envisage being deprived of her wheelchair. On the contrary, she needs one, even several, and configured in a certain way, personalized, in order to be able to pursue the projects she has been encouraged to propose.[22]

The notions of distribution and socio-technical *agencement* are crucial in distinguishing these two approaches. The handicapped person's agency is constructed differently, depending on the option chosen. In the prosthetic conception, individual agency is embedded in devices which define possible scenarios of action in a fairly rigid and restrictive way. In the habilitating conception, the entities involved (be they humans or non-humans) are more numerous, more diversified, more autonomous, and less disciplined, while the individual is put in a position to be able to interact with them in order to define projects. In one case, the injunction is ''Be like the others!'' In the other, it is ''Be what you are!'' On the one hand the identity is given; on the other it is a result of an open range of associations, ties, and bonds, defined along with interactions and experimentations. We would be wrong to say that in one case the individual is dependent and in the other free. Such notions are irrelevant. What changes is the content of the injunctions and the conditions of their felicity.

Prosthetic Policies and Habilitation Policies

My argument is that the figure of the handicapped person is crucial for understanding the difficulties encountered by the individual who is unable to fit into the mold of the Western neo-liberal subject (capable of defining and undertaking projects) and, more particularly, of adjusting to the habits and habitus of *homo economicus 2.0*. Such an individual is comparable to the handicapped person faced with situations that impede the will and autonomy with which she is credited and which she is told to exercise. This thesis leads me to distinguish two different approaches to social policy aimed at compensating for maladjustments encountered by individuals in their professional and private lives.

The first approach, as we have seen, consists in considering that maladjustment is related to the person and her shortcomings, without the environment itself being responsible. In this case we can talk of a prosthetic policy aimed at "repairing" the person concerned and/or at restoring the functionalities of which she is deprived. Interventions are primarily focused on the person, not her environment. Whether one looks at the world of work or that of consumption, it could be said, for instance, that some individuals have neither the material resources nor the symbolic or social resources (i.e. membership of networks) that might enable them to exercise their individual agency. The resulting prosthetic policies aim at restoring access to all these resources. They translate into upgrading which is basically integration (or readjustment), for the implicit assumption is that the individual suffers from a lack of resources which can (and must) be remedied. The existence of an unquestionable model with a quasi-universal definition of what a human being in society is is implied. This model serves as a reference in designing the necessary equipment for readjustment.

The second approach is diametrically opposed to the first. It rejects the assumption of a common anthropological base, and it (more or less) implicitly recognizes the constructed and contingent nature of forms of agency. Instead of positing the existence of concrete individuals who want the competencies they lack in order to cope with the resulting maladjustments, it takes these maladjustments as the starting point to argue for an adaptation of the world and particular situations to these individuals— and not vice versa. The slogan is no longer "Adjust, with the help of prostheses, to finally be self-entrepreneurial individual agents"; it becomes "Let us produce socio-technical *agencements* that are flexible, adjustable, and robust and that allow different individuals to fit into the interactive rationale characterizing neo-liberal individual agency, irrespective of where they are and the period of their lives."[23] This policy can be qualified as a habilitation policy.

My assumption is that, when applied to a network economy, prosthetic policies may produce strong maladjustments because they overlook the role of interactive socio-technical *agencements* in the emergence of individuals able to define projects and undertake them. Instead of promoting the development and diffusion of interactive niches, these prosthetic policies assume that appropriately equipped individuals can survive in any situation: *homo economicus 2.0* is seen as an upgrading of *homo economicus 1.0* rather than being considered as the emergent outcome of specific *agencements*. In contrast, I suggest that a habilitation policy is more in phase with network economies. In my view this distinction (prosthesis vs. habilitation policy), although borrowed from the handicap world, nevertheless has a general value because it is based (albeit implicitly) on the two general notions of socio-technical *agencements* and individual agencies; this distinction makes clearer and more manageable the (also general) distinction between disciplined and interactive diagrams. Finally, this distinction could be used to characterize (and possibly design) the different social policies devised to deal with the effects produced on individuals by the network economy (including for example employment, educational or "family" policies). Simply to suggest the operational nature of this analysis, here are two brief illustrations.

▪ Let us revert to the supermarket customer studied by Cochoy, Barrey, and Grandclément, and to his ability to make calculated choices. The above description shows the distributed nature of his agency: the consumer's calculative capacities are mostly delegated to a set of devices that were designed and arranged in the supermarket, by professionals of distribution. He behaves like those consumers who are just strolling about, letting themselves be lost in the maze of alleys, pushing their carts, without any particular idea in mind.[24] The socio-technical *agencements* in which the customer is plunged are for the most part defined, constructed, and evaluated by marketers, packagers, and other professionals of distribution. The individual agency thus formatted is obviously a disciplined agency. The customer calculates, but according to a program designed in relation to the distributor's clearly understood (and calculated) interests. This is reflected in the existence of a deep asymmetry between consumer and seller.

This asymmetry is not too problematic in the area of mass consumption, where consumers' calculative capacities can fairly easily become routine. It becomes problematic when, as in the case of the network economy, the innovation regime is a singularization of the available services and of their co-design. This asymmetry can be remedied (first strategy) by retaining the assumption of a detached consumer whom the seller must constantly supply with more information: the latter continues to design and organize distributed calculation in a way that reproduces the disciplined configuration. Such a prosthetic policy does not consider consumers as proactive agents struggling to define what they want. They are seen as processors of information.

But a balance can also be obtained (second strategy) by rearranging the socio-technical *agencement* itself. The consumer is no longer put in an environment fitted out with prostheses transformed into black boxes, which calculate for her and with which it is out of the question for her to interact and dialogue, in case non-programmed scripts emerged. In and around the supermarket a rearrangement aimed at promoting interactivity would revive the possibility of changing modes of calculation, of exploring other options. The consumer, instead of seeing a disciplined individual agency imposed on her (especially by market professionals), could "slide" into an interactive agency. Considering the importance and strength of the disciplinary model, prevailing almost everywhere, a fictive scene might be helpful to give an idea of what this type of consumer might be like.[25] Imagine that the customer is put into a position of directly resorting to the services of a consumer union, that access to websites is possible in the supermarket itself, or that the spokespersons of small traders can be consulted on the spot. Imagine that she has with her a list discussed and negotiated with her companion, that she promised to call friends she has invited for a week to know how they could react to the choices she is ready to make, that she is invited to taste different sorts of cheese on the spot, etc. The customer, instead of acting as a simple part of a script that makes her hesitate between a slice of ham and ... another slice of ham, starts to act as an individual agency engaged in a series of interactions which from the beginning invite her to set up plans (think of the list) and to revise her plan, and finally pushes her to go elsewhere and buy something different.[26] Such habilitation practices could, can, be completed by legal mechanisms which grant time for reflection and consultation before execution of a contract and/or which involve the supplier's responsibility far beyond the transaction *per se* (guarantees and insurance). The world in which the consumer lived and moved about would then be larger and inhabited by new agencies cooperating more willingly with her, proposing new options to her, and putting her in charge of steering the device. The most elaborate form of this interactive (consumptive) agency is found not in mass consumption, which was designed essentially for *homo economicus 1.0* and is difficult to upgrade so as to welcome *homo economicus 2.0*, but in sectors where consumers are actively called upon to participate in the design of goods or services intended for them. (See Von Hippel 2004.) These very rudimentary indications are intended simply to suggest the relevance of the proposed distinction with regard to evaluating consumption policies or designing new ones.

· As has often been pointed out, an employee in the network economy is required to be mobile and flexible, to be capable of cooperating and then disengaging himself, and to take the initiative to move dfrom one position in the network to another. As many observers have noted, maladjustments abound and create obvious social injustices (Boltanski and Chiapello 1999). When this question of maladjustment is seen as a social issue, calling for political treatment, the two approaches distinguished above could be considered as constituting two contrasting political strategies.

A prosthetic policy would consist in providing the employee with the resources he needs to meet the demands of mobility and flexibility, without considering the socio-technical *agencements* in which he acts, that is to say, without taking into account the socio-technical organization of work settings (in French: *postes de travail*). The employee is seen as an individual in his own right with deficiencies which are compensated for as necessary. For example, additional training will be proposed to help him in his mobility and allow any required reskilling; financial compensation will be given to facilitate readjustments and cover the costs involved in his projects; in some cases psychological assistances will be provided, tranquilizers will be refunded by social security and access to networks of social relations (career guidance, psychological advice, etc.) will be facilitated to enable him to find a new position.

A habilitation policy would aim more to arrange the world, that is, to construct socio-technical *agencements* which allow the deployment of individual agencies, i.e. the transformation of (more or less well-adjusted) individuals into interactive individual agencies. In his book on the new economy, Gadrey clearly illustrates the realism of this habilitation strategy. First, he emphasizes the fact that a way of adjusting the world better to the worker (handicapped person) who has difficulty moving is to reduce the need for mobility.[27] He shows that if labor regulations are amended, a firm, without compromising its efficiency, can nurture employees' loyalty, involve them in training directly related to the elaboration of their career plans, and organize their participation in the running of the organization in such a way that they are involved in a wide range of interactions, and therefore of suggested actions to take. This type of policy, which relies on adjustments outside of employees (instead of searching for solutions on their side, by transforming them or, worse, by firing them), produces socio-technical environments which locally can facilitate the emergence of interactive individual agencies. Gadrey's second suggestion is complementary to the first. Attaching an employee to the firm by changing it in such a way that ties and relations proliferate within it, and thus opening up the field of possibilities, is one thing. But as employees' movements are inevitable, he emphasizes that such *agencements* need to be generalized rather than being noteworthy exceptions. This implies, in particular, new forms of professional trajectories and careers. For example, it has been proposed to organize *tirages sociaux* (social drawing rights) which give employees the right (and the resources) guaranteed by law to take leave for training courses of their own choice, in relation to their professional and existential project (which can be discussed with counsellors and advisers). As we can see, these different measures are not intended to equip the individual with ever more prostheses so that he can adjust better to a non-adjustable environment. On the contrary, it is the *agencements* in which the individual acts that are considered to be transformed and widely diffused.[28] Such policies do not abolish networks. They take their existence seriously, but instead of considering them as superficial devices, without substance binding individuals to one another, they recognize the existence of socio-technical *agencements* which have to be established, configured, maintained and linked to one another.[29] As Gadrey shows, a supermarket designed to promote the emergence and action of individual (interactive) agencies, implies different forms of socio-technical organization to those chosen by supermarkets which consider that they employ individual agents capable of individual actions (even if this means equipping them, when necessary, with compensatory prostheses).[30]

These brief illustrations show that social policies formulated in response to the failings of individuals who have difficulty fitting into economic networks can (and must?) be designed not only as prosthetic policies (aimed at producing disciplined individual agencies) but also (and above all?) as habilitation policies (production of interactive individual agencies). This proposition does not exhaust all the (social) issues raised by the new economy. As stated in the introduction, network economies trigger collective agencies which require other types of social policies.

Concluding Remarks

The analytical framework proposed in this chapter should serve to enhance and complete studies of the process of individualization, and to better define the contribution of economic markets to that process. It departs from denunciatory positions (*homo economicus 2.0* is a manipulated being) without taking us into the camp of the unconditional supporters of the network economy (*homo economicus 2.0* represents decisive progress in the history of human beings' emancipation). By relativizing the question of the realism or non-realism of *homo economicus 2.0*, and putting aside the idea of a human nature that is just waiting to be revealed, it concentrates, more modestly and pragmatically, on the conditions needed for *homo economicus 2.0*'s appearance and possibly the pursuit of a pleasant existence.

To understand how *homo economicus 2.0* can exist, I have introduced the notion of socio-technical *agencement*, which adds the materialities and particularly technical elements to those usually taken into account by the social sciences. This opens onto a more flexible and richer interpretation of individual agency, as well as a more precise analysis of the conditions under which different types of individual agency can appear and prosper. In this approach the notion of distributed agency and certain studies devoted to disabilities have been a major resource. They have the advantage of helping us to break out of the traditional and paralyzing opposition between *homo sociologicus* and *homo economicus*. Socio-technical *agencements* and the abilities they enact change the definition of the word 'social' and result, for example, in a distinction between prosthetic social policies and habilitation social policies.

In this chapter I set out to propose ways for studying the transformation of markets and some of its effects on the constitution of individual agencies. We now need to go further and to characterize more precisely the functioning and constitution of the socio-technical *agencements* that shape *homo economicus 2.0*, whether in the field of design, production, distribution, or consumption. It should thus be possible to compare the different conceivable social policies better. This reflection would, moreover, have to be accompanied by research on the different matters of concern produced by markets as networks of innovation. As we have seen, some affect individual agency but others are at the origin of movements that lead to the discussion of new collective identities (Callon and Rabeharisoa 2007; Callon 2007).

Notes

1. For a synthesis of this literature, see Boltanski and Chiapello 2006.

2. I will use "markets as networks of innovation," "markets as networks," or "network economies" to denote this new form of economic organization. As many authors (including White, Granovetter, and Burt) claim, economic markets might

be described as (social) networks I refer here to another (common) meaning of the notion of network defined as a modality of economic co-ordination characterized by the importance of flexible alliances between heterogeneous actors and organizations (firms, laboratories, users etc.) and project forms of innovation management.

3. On this convergence between sociological and economic neo-institutionalism, see Callon and Caliskan forthcoming.

4. Saying of these individuals that they are equipped, dressed up or embedded is, in my view, equivalent. It asserts both the existence of individuals as such, and the necessity of their (variable) upgrading.

5. Should eternal France be granted a capacity for action? The question does not call for a univocal answer. What is a gene or electron capable of? The different points of view developed at different times by scientists are obviously crucial in determining possible answers.

6. On distributed action, see Hutchins 1995.

7. I could have chosen surgeons instead of pilots as an example of how "individual decisions" are made. For a striking analysis based on distributed agency, see Moreira 2004. By focusing on socio-technical agencies we highlight categories of action cutting across different fields of activity: one pilots a plane just as one pilots a business; one performs surgical operations on a human body or on an organization, for instance to amputate an unprofitable subsidiary. It would be interesting to compare the agencements performed by the same actions, to determine what they have in common.

8. 'Promission' is a concatenation of two words: permission and promise. An affordance allows (permission) and suggests (promise) some courses of action.

9. Hutchins's (1995) example of the large naval vessel is even more convincing, because it shows that faced with an unexpected event, a collective action emerges and is deployed, so that an emergency solution can be found.

10. For an analysis of this process, see Law and Mol 2002.

11. Attribution or imputation are not arbitrary or, as phrased by some sociologists, socially constructed: the material configuration and distribution of the action powerfully contributes to shape them.

12. Deleuze (1987) calls the different observable forms of *agencement* diagrams.

13. That is why I will also talk of interactive individual agencies.

14. To characterize socio-technical *agencements* and distinguish between them, I suggest positing that they are a combination of these three components.

15. The individual strategist, consistent with Burt's intuition, is an individual whose profile of connections is "star"-shaped.

16. Professional of the markets play an important role in the design and implementation of these devices (Barrey, Cochoy, and Dubuisson 2000).

17. See Winance 2001. I nevertheless keep on using 'handicap' or 'handicapped' as pointing out social issues that are pretty well structured.

18. This statement obviously should be qualified. Some handicapped people demand the right to remain as they are, in their being and identity, as attested by the existence of organizations of deaf persons who refuse cochlear implants (Blume 1997).

19. Winance gives another meaning to the notion of habilitation. She uses it to designate the general process whereby persons are endowed with abilities.

20. In French the word 'habilitation' refers in particular to the *recognized* capacity of an academic to undertake research projects and to run them.

21. Suchman's analyses are also illuminating since they contrast forms of agency in which non-humans are instrumentalized and made docile, with those in which they are endowed with autonomy and initiative. In the latter case, the environment which is, we dare say, intelligent, is nevertheless profiled to participate in a distributed action imputable to individuals (Suchman 1987).

22. The notion of accessibility is halfway between prosthesis and habilitation. By lowering a pavement, equipping a bus, adapting an apartment, or making autistic children's access to ordinary schools mandatory, both repertoires of prosthesis and habilitation are played on. *Agencements* are created which give handicapped persons the means to formulate projects without entirely predetermined content. At the same time, they are equipped with prostheses giving them access to those *agencements*.

23. One can understand why information and communication technologies are at the heart of these *agencements*.

24. Following Benveniste, we proposed to define calculation as a three-step process: a) sorting out entities, detaching them, putting them in a single space/time; b) relating them to one another; c) summa(rizing) them. This definition which has the advantage of creating a continuum between mere calculation and mere judgement, allowed us to describe a supermarket as a particular form of calculation device (Callon and Muniesa 2005). This device can welcome consumers strolling about as a complementary element of the calculative action. In addition, it is possible to relate this configuration to the discipline diagram, as presented by Barry, in which bodies are central actants: they are used as a calculation tool which is rigidly articulated to the calculation device proposed by the supermarket arrangement.

25. For realistic and detailed analyses of consumers as interactive individual agencies, in particular in e-commerce or exchanges, see Licoppe 2001.

26. Market professionals are aware of this danger and accordingly display such possibilities of interactivity in supermarkets, but ones that are closely controlled and disciplined: "Taste X and Y salami and if you prefer, just try this ham!," "Talk to our wine specialist, he'll help you to choose the vintage you need!" For a detailed analysis of these interactive devices see Clark and Pinch (1995).

27. As for handicapped persons who have difficulty moving, a convenient solution is to transform their environment so as to decrease the need to move.

28. It is interesting to note, in the case of handicapped persons—at least in France—that habilitation policies consist in setting up what the new legislation has called *Dispositifs pour une vie autonome* (devices for an autonomous life), and that one of the associations' demands is for the generalization and diffusion of these "*dispositifs.*" They are a kind of French counterpart of "center for independent life" (US) or "centre for integrated living" (UK).

29. We could show that, in order to transform and adapt these *agencements*, it is necessary to play on their different legal, discursive, and technical components.

30. It would be interesting to apply the distinction between prosthetic and habilitation policies in order to classify the different modes of intervention regarding unemployment.

References

Akrich, M. 1992. The de-scription of technological objects. In *Shaping Technology/ Building Society*, ed. W. Bijker and J. Law. MIT Press.

Barrey, S. 2001. On ne choisit jamais seul: la grande distribution des choix. *Consommations et Sociétés* 1: 25–36.

Barrey, S., F. Cochoy, and S. Dubuisson. 2000. Designer, packager et merchandiser: trois professionnels pour une même scène marchande. *Sociologie du Travail* 42.

Barry, A. 2001. *Political Machines: Governing a Technological Society*. Athlone.

Beck, U., and E. Beck-Gernsheim. 2002. *Individualization: Institutionalized Individualism and Its Social and Political Consequences*. Sage.

Beck, U., A. Giddens, and S. Lash. 1994. *Reflexive Modernization: Politics, Tradition and Aesthetics in the Modern Social Order*. Polity.

Beunza, D., and D. Stark. 2004. Tools of the trade: The socio-technology of arbitrage in a Wall Street trading room. *Industrial and Corporate Change* 13: 369–401.

Beunza, D., I. Hardie, and D. MacKenzie. 2006. A price is a social thing: Towards a material sociology of arbitrage. *Organization Studies* 27: 721–745.

Blume, S. 1997. The rhetorics and counter rhetorics of a "bionic" technology. *Science, Technology and Human Values* 32: 51–56.

Boltanski, L., and E. Chiapello. 2006. *The New Spirit of Capitalism*: Verso.

Bourdieu, P. 2000. *Pascalian Meditations*. Polity.

Durkheim, E. 1970. *Leçons de sociologie*. Presses Universitaires de France.

Callon, M., ed. 1998. *The Laws of the Markets*. Blackwell.

Callon, M. 2007. An essay on the growing contribution of economic markets to the proliferation of the social. *Theory, Culture & Society* 31: 499–536.

Callon, M., and J. Law. 2005. On qualculation, agency and otherness. *Environment and Planning D: Society and Space* 23: 717–733.

Callon, M., and F. Muniesa. 2005. Economic markets as calculative collective devices. *Organization Studies* 26: 1129–1250.

Callon, M., F. Muniesa, and Y. Millo, eds. 2007. *Market Devices*. Blackwell.

Callon, M., and V. Rabeharisoa. 2008. The growing engagement of emergent concerned groups in political and economic life lessons from the French association of neuromuscular disease patients. *Science, Technology, and Human Values* 33: 230–261.

Callon, M., and K. Caliskan. (In preparation). New directions in the social studies of markets: Towards an anthropology of economization.

Clark, C., and T. Pinch. 1995. *The Hard Sell: The Language and Lessons of Street-wise Marketing*. HarperCollins.

Collins, H. 1995. Humans, machines, and the structure of knowledge. *Stanford Humanities Review* 4: 67–83.

Deleuze, G. 1987. *Foucault*. Athlone.

Deleuze, G. 1989. Qu'est-ce qu'un dispositif? In *Michel Foucault philosophe: rencontre internationale Paris 9, 10, 11, janvier 1988*. Seuil.

Eymard-Duvernay, F., and E. Marchal. 1997. *Façons de recruter. Le jugement des compétences sur le marché du travail*. CEE-Métailié.

Gadrey, J. 2002. *New Economy, New Myth*. Routledge.

Giddens, A. 1991. *Modernity and Self-Identity: Self and Society in the Late Modern Age*. Polity.

Grandclément, C., and F. Cochoy. 2006. Histoire du chariot de supermarché. Ou comment emboîter le pas de la consommation de masse. *Vingtième siècle*, Juillet-Septembre: 77–93.

Hutchins, E. 1995. *Cognition in the Wild*. MIT Press.

Kjellberg, H. 2001. *Organising Distribution: Hakonbolaget and the Efforts to Rationalise Food Distribution, 1940–1960*. Stockholm: EFI.

Law, J., and A. Mol. 2002. Local entanglements or utopian moves: An inquiry into train accidents. In *Utopia and Organization*, ed. M. Parker. Blockwell.

Licoppe, C. 2001. Pratiques et trajectoires de la grande distribution dans le commerce alimentaire sur Internet. Vers un modèle de coordination pour le commerce électronique? *Revue économique* 52: 191–211.

Moreira, Tiago. 2004. Self, agency and the surgical collective: Detachment. *Sociology of Health and Illness* 26: 32–49.

Moser, I. 2003. Road Traffic Accidents: The Ordering of Subjects, Bodies and Disabilities. PhD thesis, Universitetet i Oslo.

Powell, W., and L. Smith-Doerr. 1994. Networks and economic life. In *The Handbook of Economic Sociology*, ed. N. Smelser and R. Swedberg. Princeton University Press.

Power, M. 1997. *The Audit Society: Rituals of Verification*. Oxford University Press.

Rose, N. 1999. *Powers of Freedom: Reframing Political Thoughts*. Cambridge University Press.

Stark, D. 1999. Heterarchy: Distributing intelligence and organizing diversity. In *The Biology of Business*, ed. J. Clippinger. Jossey-Bass.

Strathern, M., ed. 2000. *Audit Cultures*. Routledge.

Suchman, L. 1987. *Plans and Situated Actions: The Problem of Human Machine*. Cambridge University Press.

Thrift, N. 2004. Remembering the technological unconscious by foregrounding knowledges of position. *Environment and Planning D: Society and Space* 22: 175–190.

Thrift, N. 2005. *Knowing Capitalism*. Sage.

Von Hippel, E. 2004. *Democratizing Innovation*. MIT Press.

Winance, M. 2001. Thèse et prothèse. Le processus d'habilitation comme fabrication de la personne. In *Centre de Sociologie de l'Innovation*. Ecole des mines de Paris.

Winance, M. 2006. Trying out the wheelchair: The mutual shaping of people and devices through adjustment. *Science, Technology, and Human Values* 31: 52–72.

Winance, M. 2007. Being normally different? Changes to normalisation processes: From alignment to work on the norm. *Disability and Society* 22: 625–638.

2 The Centrality of Materiality: Economic Theorizing from Xenophon to Home Economics and Beyond

Richard Swedberg

During the last few years an attempt has been made to look at economic theory from the perspective of performativity. (For an overview as well as an introduction, see MacKenzie et al. 2007.) The basic idea is that, rather than assuming that economists do a fine job in analyzing reality with their sophisticated models and categories, the economic world has been constructed on the basis of economic theory—and it is this that explains why economists are so successful in their analyses. The most important theoretician in performativity theory is Michel Callon, but the two most famous applications have been made by two other social scientists. One is Marie-France Garcia-Parpet, who has shown how a local marketplace for strawberries was physically changed in order to better accommodate the neo-classical theory with its vision of the market as an auction in which a group of buyers meets a group of sellers (Garcia-Parpet [1986] 2007). The second empirical example comes from the work of Donald MacKenzie, and here (MacKenzie 2006) the case consists of a market in options constructed on the basis of the theory for the pricing of options as elaborated by Black, Scholes, and Merton.

At the heart of performativity theory is the notion that economic ideas can materialize, be it into a building (Garcia-Parpet) or into specific forms of behavior of market actors (MacKenzie). While much remains to be done in performativity theory, I will take a different approach to materiality and economic theory in this chapter. My point of departure will be that the natural focus of any economic theory is in many ways materiality, and that this is something that economic theory must take into account if it wants to be able to explain what is going on in the economic realm.

When I say that materiality constitutes the natural focus of economics, I mean that economic life is anchored in materiality: people live in houses, they eat food, they interact with machines, they produce objects, and they use objects. Modern economic analysis, on the other hand, distances itself from objects and at best acknowledges them in an indirect way, say in the form of measures for poverty, consumption, or economic growth. The modern science of economics is typically presented as a perspective or as an abstract way of looking at reality, rather than as a type of analysis that is concerned with objects or people anchored in material life. Take, for example, Gary Becker's famous definition of economic theory (1976: 5): "The combined assumptions of maximizing behaviour, market equilibrium, and stable preferences, used relentlessly and unflinchingly, form the heart of the economic approach as I see it."[1] Material objects are conspicuously absent from this definition, and in referring to the early definition of economics as a science concerned with wealth Becker writes that "the definition of economics in terms of material goods is the narrowest and the least satisfactory" (ibid.: 4).

From a social science perspective that takes materiality seriously, it is clear that this abstract type of economic analysis is not very satisfying. What is needed is a type of analysis that attempts to theorize the economy in terms of relations, objects, and the interpenetration of objects and human meanings. To construct a novel type of economics along these lines represents a daunting task, and it will not be undertaken here. It is also a task that would have to entail a recasting of the history of economic theory, including an attempt to see what can be salvaged from the past and what cannot.

What I shall try to do in this chapter is instead to make a modest contribution to the task of reinterpreting the history of economic thought, in an effort to better understand what an economics that takes materiality into account might look like. I will mainly be concerned with the way that Western economics started out as a material theory of the household, and with how it then gradually eliminated this focus from its concerns and replaced it with an abstract, non-materialistic theory of the economy centered around a very abstract notion of what constitutes market activities. Eventually, an abstract and non-materialistic theory of the household also emerged. It was known as "new household economics."

In the rest of this chapter I try to capture the main features of this development rather than follow its evolution over the centuries. The chapter is divided into three sections.

In the first section, I present the beginnings of Western economics in ancient Greece. At this stage of history, economic life was firmly centered around the household. Economic theory also began as a theory of the household which was material in nature.

I then move on to the time when economics termed itself political economy (1600s to late 1800s). The two examples I discuss are the works of Adam Smith and Karl Marx. During this period of economic life, an increasing amount of economic activities had begun to take place outside of the household and this was reflected in economic thought. During this stage of economic theory, a concern with the individual household began to disappear, and the element of materiality started to vanish.

In the third section the emphasis is moved to the United States and in particular to Cornell University in the early twentieth century. The reason for singling out Cornell University is that it was there that the fully abstract theory of economics emerged—the theory of *homo economicus*, in the work of Frank Knight. At the same time at Cornell University, interestingly enough, a new type of economics of the household also emerged. This was *home economics*, which was centered around a material vision of the modern household. It has largely been ignored in histories of economic thought.

A note on terminology is in order. Following Max Weber in *Economy and Society*, I will argue that, from a theoretical viewpoint, all economic activities are centered either around households or around profit-making enterprises (Weber 1978: 86–100). Households are essentially oriented toward the needs of their members, while profit-making enterprises attempt to exploit opportunities of gain in order to make a profit. While increasing consumption and wealth represent the goals of households, seizing opportunities and increasing capital represent the goals of profit-making enterprises. Historical examples of households include the individual household of a family, the medieval manor, and the socialist state. Profit-making enterprises include the commenda, the modern firm, and the capitalist economy. Economic reality typically contains a mix of households and profit-making enterprises. (See figure 2.1.)

The Economy as a Material Household in Antiquity

Western economic theory was born in ancient Greece; and the two most important texts on were written by Aristotle (384–322 B.C.) and Xenophon (c. 430–c. 356 B.C.).[2] Aristotle's analysis of the economy is concentrated in Book I of *Politics*, while Xenophon's text is a full work in its own right, called *Oeconomicus*. As opposed to the works of such people as Adam Smith, John Stuart Mill, and Alfred Marshall, these two texts are rarely read today and have a low status in the history of economics. In this part of his work,

	Householding	Profit making
Short-term goal	Consumption	Seize opportunities for profit
Long-term goal	Wealth	Capital
Form of calculation	Budget	Capital accounting
Institutional expression	Individual household, *oikos* in Antiquity, manorial economy socialist economic system	Profit-making enterprise, capitalist economic system

Figure 2.1
Max Weber's distinction between householding and profit making. According to Weber, economic actions and orders fall either into the category of householding (*Haushalt*), the category of profit making (*Erwerben*), or a mixture of both. Source: Weber, *Economy and Society*, pp. 86–100.

Aristotle is typically only remembered for his attack on money-making, and Xenophon for having written a pedestrian treatise on agriculture.

This view of Aristotle and Xenophon misses most of what they actually say on the economy. If one follows their arguments closely instead of trying to cast these two thinkers as predecessors to modern economics, a very different picture emerges. From the perspective of materiality à la Latour, Callon, and Pinch, for example, Aristotle and Xenophon become very interesting since they both explicitly introduce objects into their analyses and discuss how people use them in order to prosper. Xenophon was especially concerned with agriculture: the nature of the soil, how to sow, how to harvest, and so on. Some of the existing technology in ancient Greece, such as tools, are also discussed in the analysis that one can find in Aristotle and Xenophon, as is a deep concern with the human body, including sexuality. The family is of central importance as well, both the relationship between husband and wife and that between parents and children. Knowledge of how to prepare food and of various ways of keeping the house in order were also considered crucial to a good economy. Trade and commerce were held in less high regard.

That the opening chapter of *Politics* is devoted to the economy has to do with the fact that Aristotle viewed economic life as part of the life of the polis. The economy was "embedded" in the rest of society (Polanyi 1971: 81). The goal of the polis was autarchy, or self-sufficiency, which ruled out extensive commercial contacts with merchants outside the polis. While Xenophon does not say much on the issue of the role of the polis in the economy, it would seem that he agrees with Aristotle on this point. Both had as their ideal a polis that was self-sufficient and in which the citizens were good warriors as well as prosperous in their peacetime activities.

Aristotle's analysis of economic life contains praise for what he terms the art of household management (*oeconomic*), on the one hand, and a sharp condemnation of what he terms the art of acquisition (*chrematistic*), on the other.[3] While the former is natural, he says, the latter is unnatural. The reason for this is that the resources of the household come from "plants and animals," while the latter is "made at the expense of man" (Aristotle 1946: 28). Aristotle's famous distinction between use value and exchange value is also related to his argument about what is natural and unnatural in the economy:

All articles of property have two possible uses. Both of these uses belong to the article as such, but they do not belong to it in the same manner, or to the same extent. The one use is proper and peculiar to the article concerned; the other is not. We may take a shoe as an example. It can both be used for wearing and for exchange. Both of these uses are uses of the shoe as such. (Aristotle 1946: 23)

The art of acquisition, the reader is told, comes from the act of exchange, and its goal is to make money. What is further characteristic of this type of economic activity is that it is infinite in nature; one can never get enough money. What drives economic behavior of this type is "anxiety about livelihood, rather than about well-being," and this anxiety can never be fully satisfied (Aristotle 1946: 26). The desire to make money also has a tendency to overtake areas of human life that have nothing to do with the economy. One's concern with courage, for example, should be directed at warfare, and not at making money; one's concern with medical knowledge should be used to create health; and so on. "But those of whom we are speaking turn all such capacities into forms of art of acquisition, as though to make money were the one aim and everything else must contribute to that aim." (ibid.: 27)

The art of the household, as opposed to the art of acquisition, is primarily concerned with the direct use of resources and not with exchange. It has to do with the use of what has been produced; and its importance derives from the fact that "it is impossible to live without means of subsistence" (ibid.: 19). Reproduction is also essential, and it takes place within the household. Ultimately, according to Aristotle, "true wealth" has more to do with "human beings than with inanimate property" (ibid.: 21, 33).

True wealth means a concern with the full and moral development of the citizens of the polis or with "the good conditions of human beings" (ibid.: 34). "The art of household management is a moral art, aiming at the moral goodness of the members of the household," to cite one of Ernest Barker's comments on *Politics* (ibid.: 33).

At the core of the art of the household, according to Aristotle, are three relationships of authority: between the free man and his slaves, between the free man and his wife, and between parents and children. The art of command is crucial to the operations of the household and differs among these three cases. While authority over slaves means command over non-free subjects, this is not the case with command over one's children or wife. Slaves also lack the capacity of deliberation, while women have this faculty to some extent.

The slave is a natural part of the free man's household; "a complete household consists of slaves and freemen" (ibid.: 8). Slaves constitute animate objects, just as oxen and various tame animals. Aristotle infamously states that "just as some are by nature free, so others are by nature slaves and for these latter the condition of slavery is both beneficial and just" (ibid.: 14).

If we now leave Aristotle and turn to Xenophon, it should first of all be noted that *Oeconomicus* is cast in the form of a Socratic dialogue. The knowledge about the economy that one can find in Xenophon's work is, in other words, generated through the questions of Socrates, and we may see this work as an account of Socrates' view of the economy. Most of the volume deals with the household; only a few lines are devoted to the market and the art of acquisition. As opposed to *Politics*, *Oeconomicus* provides a wealth of details, both when it comes to the running a household and when it comes to attending to land. Of great importance to Xenophon's account is further the division of labor between men and women. His fascination with leadership and its role in directing the work of others must also be mentioned. The ethical dimension of economic life is much more complex and fully developed in *Oeconomicus* than in *Politics*. Since Socrates lived before Aristotle, one may well argue that it was Socrates, and not Aristotle, who discovered the economy (Polanyi 1971). As I shall try to show, Socrates' view of the economy is also considerably more interesting than Aristotle's.

The first part of *Oeconomicus* takes the form of a dialogue between Socrates and a wealthy young Athenian named Critobulus. Socrates argues that economics is an art, just as medicine or carpentry, and thus can be taught. What Socrates has in mind is not so much economics as we today understand this topic, but a practical type of knowledge that can be of use in economic life. To Socrates, economics is both a noble and a necessary type of knowledge. The economy, we read in *Oeconomicus*, is as important as war, and is also a useful complement to warfare. The art of war and the art of economics constitute "the noblest and most necessary pursuits" (Xenophon 1923: 399).

The heart of the economy is agriculture, and Socrates sings the praise of husbandry:

The land also stimulates armed protection of the country on the part of the husbandmen, by nourishing her crops in the open for the strongest to take. And what art produces better runners, throwers and jumpers than husbandry? What art rewards the labourer more generously? What art welcomes her follower more gladly, inviting him to come and take whatever he wants? What art entertains strangers more generously? . . . What other art yields more seemly first-fruits for the gods, or gives occasion for more crowded festivals? What art is dearer to servants, or pleasanter to a wife, or more delightful to children, or more agreeable to friends? To me indeed it seems strange, if any free man has come by a possession pleasanter than this, or found out an occupation pleasanter than this or more useful for winning a livelihood? (Xenophon 1923: 400)

In economic affairs one aims to produce a surplus or a balance, and this will come about when what goes out is less than what comes in. Women are typically in charge of what goes out and men of what comes in. "If both do their part well, the estate is increased," and the art of economics teaches how to accomplish precisely this (Xenophon 1923: 389). What often prevents a positive balance from developing are laziness, gluttony, lechery, and the like. Xenophon calls these the "unseen rulers" of men and says that they often destroy their wealth (ibid.: 371).

Wealth, Socrates makes Critobulus realize, does not so much consist of objects as of the way in which these are used. The category of objects is used in a very broad sense by Socrates. One can, for example, increase one's wealth through the use of one's friends as well as through the use of one's enemies. Socrates also argues that although Critobulus owns a hundred times more than Socrates, Critobulus is not as rich. The reasons for this are that many of Critobulus' resources are committed to various obligations and that his overall balance is low. Socrates adds that if Critobulus would ever be in need of money, his friends would not help him out. If Socrates, on the other hand, was in trouble, he would get assistance from his friends.

Socrates states at one point in his dialogue with Critobulus that since he himself has never been rich, he lacks important knowledge in this matter. And when one lacks knowledge, he says, one should consult someone who has it. It is also important to "watch people" and to carry out "investigations" (Xenophon 1923: 385, 389). While Aristotle seems happy to argue from principles, when it comes to economic matters Socrates is much more open to fresh experience and willing to learn from others.

The rest of *Oeconomicus* contains a dialogue between Socrates and Ischomachus (one of the wealthiest citizens in Athens). This dialogue is the heart of the work. Socrates is first told how Ischomachus has educated and instructed his wife about her tasks inside the house and how these are related to his own tasks, which are located outside the house. The goal of the relationship between husband and wife, Ischomachus says, is to create "a perfect partnership in mutual service" (Xenophon 1923: 419). Ischomachus informs his wife that she was chosen by him and her parents in the hope that she would become "the best partner of [his] home and children" (ibid.: 419).

If the husband or the wife fails in his or her duties, Ischomachus explains, the household will be like a "leaky jar" (ibid.: 427). The long-time goal of both of them is "that their possessions shall be in the best conditions possible, and that as much as possible shall be added to them by fair and honourable means" (ibid.: 419). Children are important, and they will provide for the parents when they are old.

Ischomachus explains in great detail which duties belong to the husband and which belong to the wife. Most of what the husband does takes place on the outside: sowing, plowing, harvesting, and so on. The husband is in charge of production and of the defense of the estate. The wife is responsible for what takes place on the inside; this includes tasks such as storing what has been produced, caring for the children, and being in charge of food and clothing. According to *Oeconomicus*, the minds and the bodies of men and women suit their respective tasks very well. Men are stronger and more courageous than women, who are weaker and more fearful. Both, however, have the same capacity for memory, attention, self-control, and authority.

Another important task of women inside the house is to train and oversee the domestic servants. The emphasis on the importance of the wife's duties led a latter-day commentator on *Oeconomicus* to note that Xenophon is "the first Greek author to give full recognition to the use-value of women's work, and to understand that domestic labour has economic value even if it lacks exchange-value. This idea was radical in the formal literature of classical Greece, and has yet to gain acceptance in modern times." (Pomeroy 1994: 59; see also cf. 36, 87ff.)

Ischomachus also emphasizes the importance of order in the household. It is absolutely crucial, he says, that everything is in its place, so that one can easily find it and so that it does not get wasted. A household, just as an army, must be in order:

> How good it is to keep one's stock of utensils in order, and how easy to find a suitable place in a house to put each set in, I have already said. And what a beautiful sight is afforded by boots of all sorts and conditions ranged in rows! How beautiful it is to see cloaks of all sorts and conditions kept separate, or blankets, or brazen vessels, or table furniture! Yes, no serious man will smile when I claim that there is beauty in the order even of pots and pans set out in neat array, however much it may move the laughter of a wit. There is nothing, in short, that does not gain in beauty when set out in order. (Xenophon 1923: 437)

Order also means that the right object is placed in the right room. The most valuable blankets and utensils, for example, belong in the storeroom, and the corn in the dry covered rooms. Wine should be placed in the cool room, and art and vessels that need light in the well-lit rooms. If the house has been properly built, the decorated living rooms will be cool in the summer and warm in the winter. The rooms in which male and female slaves sleep should be separated by "a bolted door" so that they cannot breed without permission (Xenophon 1923: 441).

The successful art of the household also includes wise management of the emotional and sexual relationship of husband and wife. As Michel Foucault has argued (1985, 1986), *Oeconomicus* and similar texts exemplify a trend in Greek ethics toward "care of the self," according to which husband and wife have a moral (but unequal) obligation to one another. If the wife carries out her duties well, according to this ethic, the husband should repay this with respect, including sexual respect when she grows old and becomes physically less attractive. The husband should also consider that, while a slave does not have intercourse of free will, a wife may do so if the husband acts well toward her. As the Athenians in his days, Ischomachus was also against the idea that the wife should use makeup. Husband and wife know each other's bodies in great physical detail, so it would be false to present an exterior that does not answer to reality. By performing her household duties, the wife will keep her figure beautiful.

Socrates is very curious in *Oeconomicus* to find out how Ischomachus has become so successful in economic affairs, and he asks him a number of questions on this theme. Ischomachus answers by first describing how he keeps himself in good physical form in order to manage his household and his military duties. He also explains in great detail how an agricultural estate should be run: how and when to plant, how and when to sow, and so on. An important part of managing an agricultural estate, he emphasizes, has to do with training some slaves to oversee the other slaves. This should be done, according to Ischomachus, by developing virtue and loyalty through rewards. Bailiffs may also get part of the gain. "The power to win willing obedience" is of great importance to Ischomachus, who says that the gift of leadership is "divine" (Xenophon 1923: 525).

While Socrates challenged Critobulus' claim that he was richer than Socrates, it is different with Ischomachus. When the latter is asked "do you really want to be rich and have much, along with much trouble," Socrates receives an answer that he had not expected:

Yes, I do indeed. For I would fain honour the gods without counting the cost, Socrates, help friends in need, and look to it that the city lacks no adornment that my means can supply. (Xenophon 1923: 455)

It is clear that Socrates not only respects Ischomachus' answer, but also that he wants to know how Ischomachus has been able to gather his wealth. Ischomachus continues:

I will tell you what principles I try my best to follow consistently in life. For I seem to realise that, while the gods have made it impossible for men to prosper without knowing and attending to the things they ought to do, to some of the wise and careful they grant prosperity, and to some deny it; and therefore I begin by worshipping the gods, and try to conduct myself in such a way that I may have health and strength in answer to my prayers, the respect of my fellow-citizens, the affection of my friends, safety with honour in war, and wealth increased by honest means. (ibid.: 455)

The only point at which Socrates takes Ischomachus to task and directly challenges his ideas about the successful art of household management is when Ischomachus explains how his father has taught him how to buy, fix up, and sell landed properties. One way to create wealth, Ischomachus explains to Socrates, is to locate good landed properties that are mismanaged, develop them, sell them, and then start the whole process over again. Socrates rejects this way of acting. While he approved of Ischomachus' way of managing his estates, Socrates does not accept the idea of trading estates for profit.

It can be said that the material dimension of economic life was well understood in Antiquity. The body (including sexuality), physical objects, and the soil itself were all included in the art of the household. There is also an emphasis on the interaction of people and objects; and that only by taking this into account can wealth be produced and ensured. This is particularly

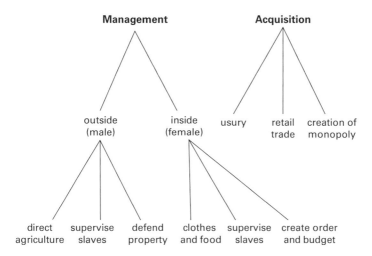

Figure 2.2
The art of the management of the household versus the art of acquisition in ancient Greek thought. In ancient Greece, economic activities were seen as either part of the art of householding or as part of the art of acquisition. Use value was associated with the former and exchange value with the latter. Since Xenophon is the foremost source for the art of householding, his work has been used to present *oeconomic*. Aristotle has similarly been used to portray *chrematistic*, or the art of acquisition. Sources: Aristotle, *Politics* (1948), pp. 1–38, 324–31; Xenophon, *Oeconomicus* (1923).

the case with the analysis of Socrates, as portrayed in *Oeconomicus* by Xenophon. The famous analysis of the economy that can be found in *Politics* by Aristotle is, in contrast, much more concerned with abstract conceptualizations, not only of agriculture and the domestic economy but also of the market. (See figure 2.2.)

Political Economy and Materiality (Adam Smith and Karl Marx)

The two works by Adam Smith and Karl Marx that will be discussed in this section—*The Wealth of Nations* (1776) and *Capital* (1867)—were produced more than 2,000 years after *Politics* by Aristotle and *Oeconomicus* by Xenophon. They were also published during a particularly dramatic and dynamic period in England's economic history, namely the hundred or so years during which the Industrial Revolution took place and England became the world's first truly capitalist nation. *The Wealth of Nations* appeared at the beginning of this period and *Capital* toward its end, but both represent attempts to conceptualize the new economic reality that confronted their authors. Smith as well as Marx essentially tried to argue that the emphasis in economic analysis had to be shifted away from the household to production and exchange outside the household. Indeed, both Smith and Marx attributed so little importance to the household in the working of the modern economy that for all practical purposes it disappeared from their analyses.

It is clear that one can find much less of a concern with materiality in the works of Smith and Marx than in the Greek classics. The turn toward nonmateriality that is characteristic of modern university economics had now begun, even if it is clear that Smith and Marx still tried to anchor their analysis in the human body and also to incorporate material objects and technology into their analyses. To account for the forces that create complex intellectual works such as *The Wealth of Nations* and *Capital* is of course impossible, but it does seem possible to at least single out some of the factors that were involved in this evolution away from the household and materiality. The development in England during 1770–1870 away from an economy centered around the household and use value and toward an economy centered around exchange value and the market was one of the factors that operated against an emphasis on materiality. The reason for this is made clear by Smith as well as by Marx: use value is concrete, while exchange value, which is the only way to coordinate buyer and seller in the market, is abstract (Smith 2000: 31; Marx 1976: 125–26). On the other hand, the fact that both Smith and Marx saw economics in a very practical way—as a guide for the statesman (Smith) and as a tool for the proletariat (Marx)—may have operated against this loss of materiality.

Today *The Wealth of Nations* is often read as an homage to the liberal market economy. The metaphor of the market operating as an invisible hand is typically cited, as well as the fact that the butcher, the brewer, and the baker all do their work because of individual interest and not because they have any particular desire to serve the public interest. It is the mechanism of competition, we are told, that makes the meat of the butcher, the beer of the brewer, and the bread of the baker to be of such high quality. Smith's skepticism against government intervention is also singled out from today's perspective, and so is his argument for free trade.

One may, however, attempt a different reading of Adam Smith, not least if one is interested in the issue of materiality. While it, for example, is clear that the concept of wealth is central to the work of Smith since it appears in its title, it also seems clear that Smith had some difficulty in handling it, and that these difficulties were to have important consequences for the role of materiality in his analysis. On the one hand, as Max Weber reminds us of, the concept of wealth belongs to the vocabulary of householding (*Haushalt*), as opposed to that of profit-making (*Erwerben*), where the equivalent term is 'capital'. *The Wealth of Nations* does, for example, contain an effort to spell out in physical detail exactly what wealth consists of. On the other hand, Smith was primarily trying to understand and conceptualize the new reality of markets, and this pushed him in a different and much more abstract direction than the one associated with the traditional concept of wealth.

In the spirit of householding, Adam Smith often refers to the fact that the aim of the economy is to produce "the necessaries and conveniences of life." We find, for example, this expression already in the opening sentence of *The Wealth of Nations* (Smith 2000: xxii). At one point Smith also spells out in detail what he means by it. There is first of all different types of food, such as grain, turnips, carrots, cabbages, potatoes, onions, and apples (Smith 2000: 89). There is also the material for clothes, such as linen and woollen cloth, as well as (unspecified) furniture and tools. Finally, what is today considered a necessity, Smith notes, may not always have been one; and this can be exemplified by what is seen as necessary in two different countries (ibid.: 938–39). In England, for example, everybody has to have a linen shirt, while this was not the case in ancient Greece and Rome. Similarly, each and every person in England has to have leather shoes, while this is not true in France.

At another point in *The Wealth of Nations*, an attempt is made, again in the spirit of householding, to establish "the stock" of a country (Smith 2000: 302ff.). According to Smith, the stock of a country falls into two categories: what is available for immediate consumption and "capital." The former consists of the necessary conveniences just mentioned, plus houses and other places to live in. The latter consists of two types of capital: fixed and circulating. Fixed capital means machines and buildings to be used for business; circulating capital covers raw material, items that have been produced but not yet sold, and whatever else is in stock.

Everything related to householding, in brief, tends to be concrete and easily specified. What relates to profit-making in the Weberian sense, pulls, however, in the other direction. At one point in *The Wealth of Nations* one can read that "wealth is the power of purchasing" (Smith 2000: 34). Power of purchase, of course, is of interest only if there is something to buy or if what is being produced passes through the market. To this should be added that while labor is what creates wealth, according to Smith, there is productive as well as unproductive labor. Productive labor, we are told, results in a commodity, which is not the case with unproductive labor, as exemplified by the activities of servants, soldiers, and lawyers. The effect of using labor as a unit of measure therefore means two things, both of which detract from materiality: a homogenization of different types of labor, and that certain types of labor are eliminated from the concept of wealth, namely those that do not result in commodities.

The consequences for economic analysis of equating wealth with commodities that go on the market are very important, especially when it comes to the household. The reason for this is that all that is done inside the home is now eliminated from the analysis, including such activities as ordering things, cleaning the house, making food, budgeting, and caring for the children. While women's work was central to Xenophon's analysis of the aristocratic and slave-owning economy, it has disappeared from the democratic economy of Adam Smith. When women's work is mentioned at all, it is only in connection with manufacturing—that is, with paid work or market work. *The Wealth of Nations* is more than a thousand pages long; the space devoted to the economic role of women takes up only about half a page. If one adds what Smith has to say about the economic role of the home and the family, including children, the half page becomes only a few lines longer.

What is said in *The Wealth of Nations* about agriculture is also of interest in this context since this is the type of economic activity that comes closest to the ancient concept of householding, apart from what goes on inside the house. According to *The Wealth of Nations*, there exist "two different systems of political economy": "the system of agriculture" and "the system of commerce" or "the modern system" (Smith 2000: 455). Only the latter can bring about true growth in wealth. While *Oeconomicus* by Xenophon contains detailed instructions for how to plant, how to harvest, and so on, in *The Wealth of Nations* one only finds the repeated cliché that "soil and climate" are of much importance (ibid.: 7, 109). As one would expect, Smith also rejects the physiocrats' argument that agriculture is the one and only source of wealth.

While it is often said that *The Wealth of Nations* contains a confrontation between two radically different economic systems (the modern market economy and the mercantilist system), a closer reading shows that Adam Smith confronts what he terms the system of commerce with different types of householding. One of these types of householding, we read in *The Wealth of Nations*, can be found in Antiquity, a period when agriculture was "honored" while commerce and foreign trade were discouraged (Smith 2000: 741ff.). Smith also mentions that according to the Greeks, engaging in manufacturing and crafts would make the male body less suitable for warfare. The system of slave labor, he notes, was unproductive. Slaves, for example, were discouraged from working hard and never took any initiatives since this was punished.

Not only life in Antiquity, but also in feudalism, as Adam Smith saw it, was based primarily on agriculture and the logic of householding in a way that prevented wealth from growing. As long as the feudal estates did not engage in commerce, everything that was produced had to be consumed on the spot, something that resulted in "servile dependency" among the local population (Smith 2000: 440). Once commerce came into the picture, on the other hand, the feudal lords could sell their surplus, with the result that their subjects grew less dependent on their masters and eventually became free and secure in their own property.

In mercantilism—a version of "the modern system" that Adam Smith criticized with great energy—the nation was cast as a household and the state as the main administrator of this household (e.g. Smith 2000: 273). The central idea was that the nation should increase its wealth, understood as its holdings of precious metals. A series of measures to increase these holdings were typically introduced and policed by the state. Imports of whatever could be produced at home should be discouraged, while domestic manufacturers and exports from these should be encouraged. Colonies were useful, according to the same logic of householding, since one could extract raw materials from these and also use them as markets for manufactured goods.

Adam Smith says over and over again that it makes no sense from the perspective of profit-making to have a mercantilist policy. A country is much better off, for example, if it buys from abroad what it can only be produced at a higher price at home. Money is not the same as wealth; it is a *measure* of wealth. It is also at this point of his argument that Smith gets to introduce what he himself sees as the main force that creates wealth, namely labor under the condition of an advanced division of labor. Being the practical man that he was, Smith becomes quite materialistic when he sings the praise of the division of labor. A few pages into *The Wealth of Nations*, he uses the democratic example of the woollen coat of a worker to show how important the division of labor is:

The woollen coat, for example, which covers the day-labourer, as coarse and rough as it may appear, is the produce of the joint labour of a great multitude of workmen. The shepherd, the sorter of the wool, the wool-comber or carder, the dyer, the scribbler, the spinner, the weaver, the fuller, the dresser, with many others, must all join their different arts in order to complete even this homely production. How many merchants and carriers, besides, must have been employed in transporting the materials from some of those workmen to others who often live in a very distant part of the country! How much commerce and navigation in particular, how many ship-builders, sailors, sail-makers, rope-makers, must have been employed in order to bring together the different drugs made use of by the dyer, which often come from the remotest corners of the world! (Smith 2000: 12)

In discussing the link between wealth and the division of labor, we also come to the topic of Adam Smith and technology. When he discusses what makes the division of labor so important, Smith sometimes points to its relationship to technology ("machines"); at other times, technology seems to be more of an independent factor (e.g. Smith 2000: 7, 279). Regardless of this, technology, according to Smith, is what allows a person to produce more than he or she otherwise could. That this, however, is not always positive, is also clear from Smith's famous remarks about the pin-making industry. Workers in this type of enterprise soon get in bad physical shape and become, thanks to the division of labor, "as stupid and ignorant as it is possible for a human creature to become" (ibid.: 4–5, 840).

In discussions of Adam Smith and technology (e.g. Koebner 1959) it is often mentioned that there is no awareness whatsoever in *The Wealth of Nations* that England was undergoing the Industrial Revolution at the time the work was written. This is true, and also that references in this work are more to manufactures than to factories. But even if Smith did not understand the importance that modern machinery, such as the Spinning Jenny and the steam engine, would have for the English economy, he nonetheless had a good material sense for technology. This comes out in the following passage, in which Smith discusses division of labor and technology:

What a variety of labour too is necessary in order to produce the tools of the meanest of … workmen! To say nothing of such complicated machines as the ship of the sailor, the mill of the fuller, or even the loom of the weaver, let us consider only what a variety of labour is requisite in order to form that very simple machine, the shears with which the shepherd clips the wool. The miner, the builder of the furnace for smelting the ore, the feller of the timber, the burner of the charcoal to be made use of in the smelting-house, the brick-maker, the brick-layer, the workmen who attend the furnace, the mill-wright, the forger, the smith, must all of them join their different arts to produce them. (Smith 2000: 12)

How machines, raw material, and labor meld into each other and become a product also comes out very nicely in some passages in *The Wealth of Nations*, such as the following:

A piece of fine cloth, for example, which weighs only eighty pounds, contains in it, the price, not only of eighty pounds weight of wool, but sometimes of several thousand weight of corn, the maintenance of the different working people, and of their immediate employers. (ibid.: 437)

When one moves from the analysis in *The Wealth of Nations* to that of *Capital*, a different picture of the economy emerges. This is not surprising: Smith, in the terminology of Marx, was addressing issues relating to the manufacturing period (c. 1550–c. 1775), while Marx was concerned with the period of large-scale industry. And while Smith battled the semi-capitalist, semi-feudal economic system that he famously termed 'mercantilism', Marx had a different target. From the very first lines of *Capital*, it is clear that Marx analyzed a world where production for the market had replaced production in the household:

The wealth of societies in which the capitalist mode of production prevails appears as an "immense collection of commodities"; the individual commodity appears as its elementary form. Our investigation therefore starts with the analysis of the commodity. (Marx 1974: 125)

Everything in capitalism, according to Marx, is drawn into the need for ever more profit: "Accumulate! Accumulate!" (Marx 1974: 742) This emphasis on accumulation also tends to shift the focus of the analysis away from materiality since what constitutes exchange value, for Marx as well as for Smith, is "abstract human labor" (e.g. ibid.: 142). Prices are based on labor, something that becomes possible only if all types of labor are seen as fundamentally alike in some respect. While this is true, Marx's analysis nonetheless succeeds in avoiding the worst dangers of being too abstract and non-materialistic. One reason for this is that Marx sees production as much more important than the market, something which has to do with the role that he assigns to surplus value in his analysis. What capitalists fight about is not profit generated in the market through the act of exchange, say by buying cheaply and selling expensively, but surplus value generated through production by the workers in the factory. To look only at the prices of commodities and compare these to one another, Marx says, is to mystify what goes on in the economy. Prices express relations between people and not between objects ("fetishism of commodities"; Marx 1974: 164–165).

That Marx's analysis of capitalism pulls in a materialist direction has also another and very obvious explanation, namely that Marx had been a materialist since early on. As a young and radical Hegelian, it was precisely the abstract and non-materialistic quality of Hegel's thought that he rebelled against. The well-known expression "der Mensch ist was er isst" ("you are what you eat") had been coined by another Hegelian, Ludwig Feuerbach.

And in Marx's attempt to go beyond Feuerbach's type of materialism in his "Theses on Feuerbach" we find an argument for a "new materialism" (Marx 1978: 145). The new materialism, as opposed to "the old materialism," understands for example that there also exist material reasons why people think in non-material or religious terms. People suffer for a number of very material reasons—and having a God alleviates their suffering.

Attention to the physical world, including the human body, is characteristic of all of Marx's writings. In *Capital*, the human body is discussed in primarily two contexts. There is first of all a need in capitalism to reproduce the body of the worker. Secondly, the body of the worker is severely abused in this type of economic system. The need to reproduce the body of the worker comes from the fact that labor is the only commodity that can produce surplus value, and to pay for this commodity means to pay for its physical reproduction. Or, to cite *Capital*:

If the owner of labour-power works today, tomorrow he must again be able to repeat the same process in the same conditions as regards health and strength. His means of subsistence must therefore be sufficient to maintain him in his normal state as a working individual. His natural needs, such as food, clothing, fuel and housing vary according to the climatic and other physical peculiarities of his country. (Marx 1974: 275)

Marx's concern with the body of the workers can also be seen in his attempt to assess how many calories a worker needs per day, as measured in the terminology of the time: "nutrive elements," consisting of "carbon" and "nitrogen" (Marx 1974: 808 ff.). He notes in addition that since workers one day will die and have to be replaced, the price of labor (= the price for the reproduction of a worker) must also include the cost of his children. "The labour-power withdrawn from the market by wear and tear, and by death, must be continually replaced by, at the very least, an equal amount of fresh labour-power." (ibid.)

In their eagerness to make a profit, Marx argues, the capitalists typically exploit the workers and hurt them physically and psychologically:

In its blind and measureless drive, its insatiable appetite for surplus labour, capital oversteps not only the moral but even the merely physical limits of the working day. It usurps the time for growth, development and healthy maintenance of the body. It steals the time required for the consumption of fresh air and sunlight. It haggles over the meal-times, where possible incorporating them into the production process, so that food is added to the worker as a mere means of production, as coal is supplied to the boiler, and grease and oil to the machinery. It reduces the sound sleep needed for the restoration, renewal and refreshment of the vital forces to the exact amount of torpor essential to the revival of an absolutely exhausted organism. (Marx 1974: 375–376)

Despite his sensitivity to the material dimension of labor, Marx—just as Adam Smith—eliminated everything that happens in the home or the household from his analysis of the economy, including the domestic work of women. The importance of this last type of work at the time of Marx has been well established in scholarship (e.g. Tilly and Scott 1978; see Folbre 1991 for a discussion of the absence of domestic work from economic thought in the 1800s). Women (and children) are mentioned in *Capital* only when they enter the labor market, something which they did when large-scale industry was introduced in England. In brief, Marx ignored the household.

One can on the other hand find a significant attempt in *Capital* to introduce science and technology into the analysis of the economy (cf. MacKenzie 1996). Technology is conceptualized as the practical application of science (e.g. Marx 1974: 775, 929). Science and technology, the reader is told, also set man apart from animals since human beings can think about the different ways in which they interact with nature *before* they do so. They also use instruments that they have constructed, when doing so. While human beings have made instruments throughout history, in order to accomplish various tasks, the role of these instruments changes dramatically with capitalism. From now on, human beings have to adjust to their instruments rather than the other way around. Man becomes, as Marx puts it, a "living appendage to the machine" (ibid.: 548). The workers also need little skill to run the machines that are now being used; they become de-skilled.

The main reason for this situation is not so much technology or science *per se*, according to Marx, but capitalism and the fact that the workers have nothing to do with the decision of what is to be produced. The person who does the conceptualization is the capitalist, and the reason for his interest in science and technology and to decide on what is being produced, has primarily to do with the profit motive. The way to beat your competitors is by being able to lower the price, and this can be accomplished through the introduction of new machines. "The battle of competition is fought by the cheapening of commodities" (Marx 1974: 777). This is why science has to be "pressed . . . into the service of capital" (ibid.: 482). And the introduction of science and technology into a type of economic system that has ever more profit as its goal, means that the technology—and people working with this technology—will always be in a process of change:

Modern industry never views or treats the existing form of a production process as the definitive one. Its technical basis is therefore revolutionary, whereas all earlier modes of production were essentially conservative. By means of machinery, chemical processes and other methods, it is continually transforming not only the technical basis of production but also the functions of the worker and the social combinations of the labour process. At the same time, it thereby also revolutionizes the division of labour within society, and incessantly throws masses of capital and of workers from one branch of production to another. (ibid.: 617)

To summarize my section on political economy: It is clear that compared to Xenophon and Aristotle, the household has disappeared and so has some of the materiality of early economic analysis. There is, for example, no concern in the works of Smith and Marx with women's work in the household, the task of bringing up children, and the sexual relationship between husband and wife. Materiality is still present in both Smith and Marx, but there are also clear signs that it is on its way out. This is especially clear when it comes to the discussion of the market, exchange value (price) and labor (abstract labor).

There is also the fact that, while the actors in the works of Smith and Marx still live in a material world, the material dimension is little theorized and mainly taken for granted. While this may be a weakness from the perspective of the new materiality of Science and Technology Studies, it is nonetheless easy to feel nostalgia for the high days of political economy. Why this is the case should become obvious in the next section of this chapter which is devoted to the phase of economics in which it cut its very last links to materiality and became a highly abstract science.

The Immateriality of *Homo economicus* and the Materiality of Home Economics

In presenting the art of household management I have looked at ancient Greece, and in presenting political economy, nineteenth century England. In now turning to *homo economicus*, the main focus will shift to the United States in the twentieth century. This change of scenery to the United States also reflects the fact that mainstream economics has developed the most forcefully in this country since the early twentieth century, just as political economy came to its classical expression in England during the 1800s. I shall in particular turn the spotlight on Cornell University around 1900, because of two very interesting developments that took place there. First, this is where the theory of *homo economicus* came to its classical expression, in the doctoral dissertation of Frank Knight. And second, around the same time Cornell University also helped to create a type of economics that is usually ignored in the histories of economic thought, but which is very important for the concerns of this chapter, namely a material theory of the household that went under the name of home economics.

Let us start out by taking a look at what according to Kenneth Arrow constitutes the pioneer formulation of *homo economicus*, namely the section devoted to this topic in *Risk, Uncertainty and Profit* by Frank Knight (Arrow 1987: 203; Knight 1979: 76–81). Knight's book was published in New York in 1921 and is based on his dissertation in the Department of Economics at Cornell University (Knight 1916). According to Knight and Arrow, the discussion of the theory of *homo economicus* in *Risk, Uncertainty and Profit* only makes explicit what is already present in "a large part of the economic literature" (Knight 1971: 81). It is usually agreed that it was John Stuart Mill who make the first attempt to introduce the idea of *homo economicus* in the mid-nineteenth century, so one can perhaps phrase it so that Knight in his dissertation gave voice to ideas on this theme that had developed in mainstream economics over something like half a century (e.g. Persky 1995).

Knight prefaces his presentation of the nine central assumptions of *homo economicus* with a statement which can also be found in John Stuart Mill and several other economists, namely that assumptions of this type are made exclusively for methodological reasons. They are "'heroic abstractions'" and they refer to an "imaginary society," but they are also essential since they make it possible to solve certain problems (Knight 1971: 76). The nine assumptions that Knight discusses in *Risk, Uncertainty and Profit* are the following (ibid.: 76–81). Economic actors are "normal human beings" of the type found in "a modern Western nation" (#1). They act "with 'complete rationality'," something which means that they know what they want, how to get what they want, and also what consequences their acts will have (#2). Economic actors decide themselves what they want, and there are no constraints on their actions in this regard (#3). Nothing can stop the economic actors from carrying out their economic plans, be it physical obstacles or anything else (#4). Perfect competition means, among other things, "perfect ... intercommunication" and that all goods can be divided indefinitely (#5). Economic actors have no social relations with other economic actors, except in the act of exchange (#6). Economic actors only acquire goods through the market and production (#7). Division of labor presupposes a diversification of wants and a specialization of the productive capacity of the individual; resources are unevenly distributed in the world and there is a limit to human mobility (#8). And finally, conditions have to be static, something which means that economic actors can now understand everything about their conduct that they have not already understood (#9).

While Knight does not explicitly state that *homo economicus* can be found only in the market, this is nonetheless the case. The household, in other words, has totally disappeared; and so has all materiality. As to the human body of his economic actors, Knight only makes a few cryptic remarks. People, he says, have "inherited and acquired dispositions"; they also have "wants" (e.g. Knight 1971: 76, 79). But no references beyond this are made to emotions and sexuality, and we do not know if the actors are male or female, old or young, have legs and arms, and so on. Since communication between the actors is assumed to be perfect, language is also eliminated from the analysis.

Knight has even less to say about technology than about the body in his attempt to present the main assumptions of economic analysis. At one point he notes that "material implements of production may be used provided they are either superabundant, and consequently free goods, or else are absolutely joined to their owners (not subject to lease or sale)" (Knight 1971: 80). Knight also refers to the assumption that material implements have to be stable (cf. #9). A very abstract type of technology, in brief, is assumed to be present and it cannot change.

Also the physical and geographic environment leads at best a ghostlike existence in Knight's account of *homo economicus*. The only explicit reference to the environment is to be found in his discussion of assumption #8, where Knight speaks of "the space distribution of the different resources of the earth and the limitations on human mobility" (Knight 1971: 79). The reason for mentioning these two factors is probably related to the state of foreign trade theory at the time when Knight wrote his book.

Finally, material objects are strangely missing from Knight's analysis, despite occasional references to "goods." To some extent this may be related to the disappearance of the term use-value from the vocabulary of modern economics, and the related attempt to replace it with a more subjective terminology, such as "utility" and "preferences." What also makes objects disappear from the analysis is Knight's steady focus on the market at the expense of everything else in economic life, something which means that even though he is aware of the importance of production, whatever happens when goods come into being is ultimately less important than how they are priced—and this takes place in the market. A related example of Knight's attitude to objects is his statement that "we also must assume complete absence of physical obstacles" when it comes to the actor making a decision, and that we have to assume "'perfect mobility'" and "no cost involved in movements or changes" (Knight 1971: 77). When Marx famously said that "all that is solid melts into air," he was thinking of the corrosive impact of bourgeois conditions on feudal values, but his statement also fits the transition from political economy to modern economic theory when it comes to materiality (Marx 1978: 476).

After this discussion of *homo economicus* in its Knightian formulation, I shall proceed to a type of economics that was very strong at Cornell University at the time when Knight presented his dissertation. This is *home economics*, which can be described as the exact opposite in many respects of the theory of *homo economicus*. Where one is abstract and non-materialistic, the other is concrete and materialistic. And while the theory of *homo economicus* has usually been taken very seriously and discussed at great length, home economics has typically been ridiculed and ignored. When male students at Cornell and other universities were gently steered in the direction of economics, the female students were just as gently steered—in the direction of home economics.

The origin of home economics is usually traced to the early 1800s, with Catherine Beecher's *Treatise on Domestic Economy* (1842) as an important landmark. The focus in this type of economics was primarily on the home and skills such as cooking, cleaning and sewing. A few decades later efforts were made to turn home economics into a university subject, something that succeeded in the United States but in no other country, to the best of my knowledge. The reason for home economics being so successful in the United States probably has to do with the fact that this country had a relatively young and flexible university system. Not all the universities, however, accepted the new subject. While the prestigious private universities for women, for example, rejected it, many land-grant universities accepted it. Universities such as Cornell and Wisconsin, where home economics prospered, also had close ties to the government and to the local community. They were also more open to political reforms.

In 1909 the American Home Economics Association was created and a decision made to settle on the term "home economics," as opposed to "household arts," "domestic science," "oecology," and similar terms (e.g. Stage and Vincenti 1997). Home economics received considerable support from the federal government, especially through the Smith-Lever Act of 1914, which allotted funds to improve US agriculture. After a shaky and difficult beginning, the home economics movement reached its peak in the 1910s and the 1920s. Decline began to set in after World War II, and it died a slow death a few decades later. It is often noted that an important reason for its disappearance was that home economics was out of touch with the times by the 1960s and the 1970s, in the sense that gender roles were now very different from what they had been around 1900. Women had in particular begun to move into the labor market, something that made many of the basic assumptions and concerns of home economics less relevant. Home economics was also severely criticized during the 1960s and the 1970s by major figures in the feminist movement.

The development at Cornell, where home economics was a great success, can be used to illustrate the rise and decline of home economics (e.g. Rose 1969; Berlage 1998). In 1907 a Department of Home Economics was created at Cornell that was part of the College of Agriculture, and a very successful degree program for female students was instituted. The Department was popular even if its faculty members were ridiculed and called "cooks" by their male colleagues. In 1925 the Department of Home Economics became its own College with a number of departments, such as the Department of Food and Nutrition, the Department of Economics of the Household and Household Management, the Department of Family Life, the Department of Textiles and Clothing, the Department of Household Art, the Department of Hotel Administration, and the Department of Institution Management. The next few decades were very successful, and more than a hundred doctorates in home economics awarded. By the 1960s, however, times were different, and in 1969 the College of Home Economics was reorganized and had its name changed to the College of Human Ecology.

Home economics has primarily been studied from the perspectives of gender and profession, and it is clear that quite a bit can be said about it from these perspectives. Home economics failed as a profession, and this had much to do with gender relations at the time. But home economics can also be seen as a part of the history of economics and as part of a tradition that goes all the way back to the science of household management in ancient Greece (*oeconomic*). It can be argued that it is precisely from this perspective that home economics makes most sense: as a return of a part of the science of economics that had been rejected and ignored at least since the days of Adam Smith. That home economics also fits nicely into the history of economics as a science about people and their materiality is similarly clear.

That this view is not so dissimilar from the way that the leaders of the home economics movement themselves saw what they were doing, can be illustrated by the following quote from an article in the *Journal of Home Economics* from 1911 by Ellen Richards:

Home Economics means . . . economics in its original significance, household administration, domestic management. Political economists have usurped the word to
mean *production* of wealth. In early times this was largely done within the domain
of the household, but with the taking away of the producing interest through the
rise of factory products, a gap was left in the carrying out of this theory, only now
beginning to be filled by the new science, the economics of consumption. More
than this, the civilization of the past has been developed, we believe, through the
family home, the bond of mutual interest between parent and child, grandparent
and grandchild, brother and sister, which makes cooperation under one roof possible. (Richards 1911: 117)

The type of home economics that existed in the 1800s can be described
as practical knowledge about various tasks in and around the home, especially skills in cooking and sewing. Around the turn of the twentieth century, when women in the United States were for the first time being
admitted into the university, an attempt was made to set home economics
on a scientific footing and thereby make it stronger as well as legitimate.
This was especially the case with the topics of food and cleaning. Chemistry was used to bolster the former, and the science of sanitation the latter.
The results were impressive: advancement toward a scientific understanding of nutrition as well as a considerably better understanding of some
issues related to hygiene and public health. In their work on public health
and sanitation the home economists sometimes took the model of the
household and extended it to the level of the community, going well beyond a concern with the individual home.

The 1920s and the 1930s added the family as a central topic, including
the development of children and parental education. An effort was now
made, for example, to show that boys and girls were not inherently different and made for different tasks in society. The interest among home
economists for technology was also intensified during the course of the
twentieth century, both at the level of the home and at the level of the
community. The introduction of science and technology into the kitchen
constituted an important item on the agenda of home economics, and the
bringing of electricity to the countryside and the family farm another. In
all of these efforts, the practitioners of home economics displayed a deep
and sincere interest for the practical and material aspects of people's everyday life that is unique in economics.

When one looks at the role that the theories of *homo economicus* and
home economics played in university education in the early 1900s, it becomes clear that the young male students were trained in an abstract, im-
practical and very prestigious type of knowledge, while the young female

students were trained in a concrete, practical and often ridiculed type of knowledge. The end of the story adds a further twist: home economics with its interest for the household was by the 1960s seen as outmoded and made to disappear, while mainstream economics, as a sign of its continued vitality, now took on the study of the household. "The new household economics," as it became called, developed precisely around this time, and it extended abstraction, impracticality, and non-materiality to the last stronghold of what had once been called *oeconomic*.

Concluding Remarks

By now it should be clear from what has been said in this chapter that there is more to the texts from ancient Greece on economics and the literature that makes up home economics than has generally been realized. Both of these sources allow us to follow an important theme in the history of economics, namely the evolution of the analysis of the household. What originally constituted the core of economic analysis, later disappeared from mainstream economics—and today lives on under the name of new household economics.

Xenophon, Aristotle, and the authors of home economics also allow us to get a glimpse of what a truly material analysis of economics might look like. The estate of ancient Greece encompassed many activities that cannot be found in the individual household in the United States in the late 1800s and the early 1900s and vice versa, so together these two sources span a broad range of economic activities. A comparison of this type of analysis with that of new household economics could also be interesting in that it could show us the advantages and disadvantages with a material analysis of economic phenomena.

Many economic activities in the modern world clearly take place outside the household, and for a full theory of material economics one would like to see also these analyzed from the perspective of Latour et al. (See figure 2.3.) Advocates of the new materiality have developed a series of concepts and ideas that may come in handy in an enterprise of this kind; some of their analyses also touch on economic phenomena. One can, for example, reread Bijkjer and Pinch's 1987 article on the evolution of the bicycle and Callon's 1987 essay on the attempt to develop an electrically driven automobile in France in the 1970s from this perspective. Concepts such as "actant" and "collective" and the idea that networks involve not only human actors but also objects have as well to be applied to economic phenomena in order to test their analytical strength. And so should Trevor

	Market activities	**Non-market activities**
Non-material approach	Mainstream economic theory	New household economics
Material approach	?	*Oeconomic,* home economics

Figure 2.3
Market and non-market activities.

Pinch's notion that institutions have a material dimension (Pinch 2008). Bruno Latour's forthcoming exploration of Gabriel Tarde's work on the economy will probably provide some answers that are relevant in this context as well (Latour 2008).

Some attempts to use the materialistic approach to analyze economic phenomena, looking especially at the role of technology, already exist. This is especially true for analyses that focus on the market mechanism, which by economic sociologists is usually seen exclusively in social terms ("markets as social structures") and by economists in terms of demand and supply ("the market as a price mechanism"). The way a market operates, however, is also dependent on the role of technology. The ticker, as Alex Preda shows in this volume, will deliver information that is crucial for the formation of market prices that differs from, say, the modern computer. The same is true for the use of telephones in a market, as we know from the work of Fabian Muniesa (this volume). Social technologies, say in the form of the pricing of options, may operate in a similar way (MacKenzie and Milo 2003; MacKenzie 2006).

While a material approach has been developed to analyze the market mechanism, very little has been done in the areas of production and consumption. Both of these are organized somewhat differently in economies where much of the production goes through the market (capitalism), and in economies that primarily draw on the state or some other social agency for distribution (socialism, early economies). Competition and private property clearly lead to certain forms of organization and interactions that differ from, say, socialist, mercantilist, or pre-industrial economies. One would therefore expect that capitalism also has its very own form of materiality. But this may be true only to a certain extent, and it is important to point out that today's emphasis on the role of the market tends to overplay the differences between capitalist and non-capitalist type of economies. To this should also be added the fact that any type of capitalist economy will have pockets of household economies and that these are imperative to its general functioning. The more general point that I am leading up to, and on which I would like to end, takes us back to the main theme of this chapter, namely that a better understanding of materiality is essential in order to properly understand economic life.

Acknowledgments

I thank Patrik Aspers, Mabel Berezin, Fabian Muniesa, and Trevor Pinch for helpful comments.

Notes

1. While the tendency of producing a highly abstract type of economics has its roots in the second half of the nineteenth century, abstract definitions of economic theory are usually dated the 1930s and especially to Lord Robbins's *Essay on the Nature and Significance of Economic Science* (1932; see e.g. Kirzner 1976: 108–45).

2. Other writers on the economy include Protagoras, Antistehenes, and the real authors of *Oeconomia*, which is conventionally ascribed to Aristotle (see e.g. Pomeroy 1994: 7–8, Aristotle 1935). For an introduction to Greek economic thought, see Trever 1916.

3. According to M. I. Finley (1973: 17), "the word 'economics', Greek in origin, is compounded from *oikos*, a household, and the semantically complex root, *nem-*, here in its sense of 'regulate, administer, organize.'" There is no single word in English that can adequately capture *oikos*, according to Pomeroy (1994: 41), who also notes that the *oikos* constituted the foundation of the Greek economy.

References

Aristotle. 1935. *Metaphysics, Oeconomica, Magna Moralia*. Harvard University Press.

Aristotle. 1948. *The Politics of Aristotle*. Clarendon.

Arrow, K. 1987. Rationality of self and others in an economic system. In *Rational Choice*, ed. R. Hogarth and M. Richter. University of Chicago Press.

Bahnisch, M. 2000. Embodied work, divided labour: Subjectivity and the scientific management of the body in Frederick W. Taylor's 1907 "Lecture on Management." *Body & Society* 6: 51–68.

Becker, G. 1976. *The Economic Approach to Human Behavior*. University of Chicago Press.

Berlage, N. 1998. The establishment of an applied Social Science: Home economics, science, and reform at Cornell University, 1870–1930. In *Gender and American Social Science*, ed. H. Silverberg. Princeton University Press.

Bijker, W., and T. Pinch. 1998. The social construction of facts and artifacts. In *The Social Construction of Technological Systems*, ed. W. Bijker et al. MIT Press.

Brunner, O. 1956. Das ganze Haus und die alteuropäische Ökonomik. In Brunner, *Neue Wege der Sozialgeschichte*. Vandenhoeck & Ruprecht.

Callon, M. 1987. Society in the making: The study of technology as a tool for sociological analysis. In *The Social Construction of Technological Systems*, ed. W. Bijker et al. MIT Press.

Featherstone, M. 1982. The body in consumer culture. *Theory, Culture & Society* 1, no. 2: 18–33.

Finley, M. 1970. Aristotle and economic analysis. *Past and Present* 47: 3–25.

Finley, M. 1973. *The Ancient Economy*. University of California Press.

Folbre, N. 1991. The unproductive housewife: Her evolution in nineteenth-century economic thought. *Signs* 16: 463–84.

Folbre, N. 1998. The "sphere of women" in early-twentieth-century economics. In *Gender and American Social Science*, ed. H. Silberberg. Princeton University Press.

Foucault, M. 1985. *The Use of Pleasure*. Volume 2 of *The History of Sexuality*. Random House.

Foucault, M. 1986. *The Care of the Self*. Volume 3 of *The History of Sexuality*. Random House.

Garcia-Parpet, M.-F. [1986] 2007. The social construction of a perfect market: The strawberry auction at Fontaines-en-Sologne. In *Do Economists Make Markets?* ed. D. MacKenzie et al. Princeton University Press.

Kirzner, I. 1976. *The Economic Point of View: An Essay in the History of Economic Thought*. Sheed and Ward.

Knight, F. 1916. Theory of Business Profit. Doctoral dissertation, Cornell University. Kresge Collections of Rare Books and Manuscript Collections.

Knight, F. 1971. *Risk, Uncertainty and Profit*. University of Chicago Press.

Knorr-Cetina, K., and A. Preda, eds. 2005. *The Sociology of Financial Markets*. Oxford University Press.

Koebner, R. 1959. Adam Smith and the Industrial Revolution. *Economic History Review* 11: 381–391.

Latour, B. 1993. *We Have Never Been Modern*. Harvard University Press.

Latour, B. 1999. *Pandora's Hope: Essays on the Reality of Science Studies*. Harvard University Press.

Latour, B. 2008. Never Too Late to Read Tarde. www.ensmp/~latour/

MacKenzie, D. 1996. Marx and the machine. In MacKenzie, *Knowing Machines*. MIT Press.

MacKenzie, D. 2006. *An Engine, Not a Camera: How Financial Models Shape Markets*. MIT Press.

MacKenzie, D., and Y. Milo. 2003. Constructing a market, performing a theory: The historical sociology of a financial derivatives market. *American Journal of Sociology* 109: 107–45.

MacKenzie, D., F. Muniesa, and L. Sin, eds. 2007. *Do Economists Make Markets? On the Performativity of Economics*. Princeton University Press.

Marx, K. 1976. *Capital*, volume 1. Pelican Books.

Marx, K. 1978. Manifesto of the Communist Party. In *The Marx-Engels Reader*, ed. R. Tucker. Norton.

Marx, K. 1978. Theses on Feuerbach. In *The Marx-Engels Reader*, ed. R. Tucker. Norton.

Persky, J. 1995. The ethology of *homo economicus*. *Journal of Economic Perspectives* 9, no. 2: 22–31.

Pinch, T. 2008. Technology and institutions: Living in a material world. *Theory and Society* 37 (in press).

Polanyi, K. 1971. Aristotle discovers the economy. In *Primitive, Archaic and Modern Economies*. Beacon.

Pomeroy, S. 1994. *Xenophon Oeconomicus: A Social and Historical Commentary*. Clarendon.

Richards, E. 1911. The social significance of the home economics movement. *Journal of Home Economics* 3, no. 2: 117–26.

Robbins, Lord. 1984. *An Essay on the Nature and Significance of Economic Science*, third edition. Macmillan.

Rose, F., et al. 1969. *A Growing College: Home Economics at Cornell*. Cornell University Press.

Smelser, N., and R. Swedberg, eds. 2005. Introducing economic sociology. In *The Handbook of Economic Sociology*, ed. N. Smelser and R. Swedberg. Princeton University Press and Russell Sage Foundation.

Smith, Adam. 2000. *The Wealth of Nations*. Modern Library.

Stage, S., and V. Vincenti, eds. 1997. *Rethinking Home Economics: Women and the History of a Profession*. Cornell University Press.

Tilly, L., and J. Scott. 1978. *Women, Work and Family*. Holt, Rinehart and Winston.

Trever, A. 1916. *A History of Greek Economic Thought*. University of Chicago Press.

Weber, Max. 1978. *Economy and Society*. University of California Press.

Xenophon. 1923. Memorabilia *and* Oeconomicus. Harvard University Press.

3 Command Performance: Exploring What STS Thinks It Takes to Build a Market

Philip Mirowski and Edward Nik-Khah

There are two positions we have to abandon. The first is the idea of critique of hard economists, which is intended to show them that they are wrong. And the second position is to describe markets just to say that they are more complicated than economists or political decision makers believe.... Let us stop criticizing the economists. We recognize the right of economists to contribute to performing markets, but at the same time we claim our own right to do the same but from a different perspective.

—Michel Callon, in Barry and Slater 2003 (p. 301)

Once upon a time, many people believed that "The Market" was something that has always existed in a quasi-Natural state, much like gravity or language. It seemed to enjoy a material presence, sharing many of the characteristics of the forces of nature, and as such was deemed a coherent unified phenomenon which warranted a science of its own; and that science was first called "political economy" and then, after roughly 1870, "economics." The modern orthodoxy of that science, the neoclassical tradition, has always taken the nature of this Market as the central province of economics, has it not? Curiously enough, that notion would be premature, as some high-profile modern economists have noted: "It is a peculiar fact that the literature on economics ... contains so little discussion of the central institution that underlies neoclassical economics—the market." (North 1977: 710) "Although economists claim to study the market, in modern economic theory the market itself has even a more shadowy role than the firm." (Coase 1988: 7) Arrow and Hahn's *General Competitive Analysis* asserts in passing that it takes the "existence of markets ... for granted" (1971: 348). In fact, a judicious and unbiased overview of the history of the first century of neoclassical economics would confirm that its adherents had been much more fascinated with the status and nature of *agents* than with the structure and composition of markets. Most of the time, the concept of the market was treated as a general synonym for the phenomenon of exchange itself, and hence rendered effectively redundant (Rosenbaum 2000). Even in the few instances when influential thinkers in the tradition felt they should discuss the actual sequence of bids and asks in their models of trade—say, for instance, Leon Walras with his *tâtonnement* and his *bons*, or Francis Ysidro Edgeworth with his recontracting process—what jumps out at the economic historian is the extent to which the sequence of activities posited therein had little or no relationship to the operation of any actual contemporary market.[1] Mid-twentieth-century attempts to develop accounts of price dynamics were, if anything, even further removed from the increasingly sophisticated diversity of market formats and structures and the actual sequence of what markets accomplish. While there would be many ways to account for this incongruous turn of events, the condition we shall opt to stress here was the strong dependence of the neo-classical tradition on *physics* to provide the respected paradigm of scientific explanation. Not only had energy physics provided the original agent formalism of optimization over a utility field in commodity space (Mirowski 1989); it also supplied the background orientation to which law-governed explanations were presumed to conform. The strong reductionism inherent in modern physics suggested that all agents would of necessity exhibit some fundamental shared characteristics (viz., "rationality") and therefore, for modeling purposes, should be treated as all alike. Furthermore, any differences in market structures where the agents congregated would be treated as second-order complications (viz., perfect competition vs. monopoly) or else collapsible to commodity definitions ("the" labor market; "the" fish market), and therefore "The Market" came to be modeled as a relatively homogeneous and undifferentiated entity. Whether justified as mere pragmatic modeling tactic (for reasons of mathematical tractability) or a deeper symmetry bound up with the very notion of the possibility of existence of "laws of economics," market diversity was effectively suppressed, as one can still observe from modern microeconomics textbooks.

However, since roughly 1980 things have been changing within economics (Mirowski 2007) and also within economic sociology. Various developments in neoclassical economics have fostered a more "constructivist" approach to markets, in the sense that it has become possible for the first time to acknowledge that market formats do indeed differ in significant ways, and furthermore, that it might be possible for economists to intervene in the setup and maintenance of these diverse structures. In part, this was related to the displacement of physics by biology and computer science as the premier exemplary sciences of our postmodern era.[2] But it also has quite a bit to do with the post-1980 progressive commercialization of the sciences in their academic and industrial contexts, as we shall explain below. Where economists once placidly contemplated markets from without, situated in a space detached from their subject matter, so to speak, instead now they are much less disciplined about their doctrines concerning the nature of economic agency, and much more inclined to be found down in the trenches with other participants, engaged in making markets.

It should not be surprising that at least one subset of economic sociologists have noticed that economists have adopted a more hands-on approach, and sought to interpret this as leaving an opening for their own theoretical predilections and potential interventions. The version of economic sociology we shall examine here will be variously referred to as the Paris School, "actor-network theory" (ANT), and ANT 6.5 (to be defined later). It seems that these sociologists believe that their discovery of the active intervention of economists into (some) markets, a phenomenon which they wish to characterize as economists "performing" the economy, is a major validation of the ontological theses for which they have become famous.

We shall seek to argue that what initially has debuted as a local controversy over what has come to be called the "performativity" of the economy by economists, first broached by Michel Callon (1998), endorsed by Latour in his most recent books (2004a: 272, note 11; 2005) and of late given qualified endorsement by MacKenzie (MacKenzie and Millo 2003; MacKenzie et al. 2007), in fact signals the outlines of one practical future envisioned by these authors for the role of science studies in the ecology of the post-Cold War regime of scientific research. In short, after a long period of silence, this version of STS finds it now wants to say something about economics. We believe that the implications of this nascent tendency have not been sufficiently understood. The stumbling block derives from a systemic intellectual problem STS has had with the social sciences almost since its inception, exacerbated by modern structural changes occurring in the social organization of scientific research in the direction conventionally called "commercialization." These two phenomena have become juxtaposed in a rather curious way to produce Callon's version of "performativity," which turns out to be an overture to a prospective alliance to be struck up with neoclassical economists, as directly illustrated by the quotation we used as this chapter's epigraph.

We fear that this proposed pact would be a prescription for disaster for economic sociology. However, in this chapter we elect not to argue against the pact on pragmatic grounds, especially since we have been told we are rather unlikely defenders of the virtues of economic sociology[3];

rather, we propose to first explore what sorts of considerations might have led such prominent spokesmen for the version of STS (sometimes) known as "actor-network theory" to such a precipitous pass, and then to subject what seems to be one of their exemplary empirical instances of "performativity"—which just happens to be the American construction of a certain specific type of auction to allocate communications spectrum under the auspices of the Federal Communications Commission (Callon 2007; Muniesa and Callon 2007; MacKenzie 2002)—to a skeptical audit. The net result of this exercise is to begin to reveal just how little solid in the way of usable analysis for economic sociology can be expected to be derived from the performativity thesis, and therefore what a slender thread it is upon which to hang the argument for a rapprochement with neoclassical economics.

On Feeling "Out of It" in "Social" Situations

Science studies as an academic formation has long harbored a number of reasons to be uncomfortable with the social sciences, and around economics in particular.[4] Right off the bat, there are its largely unacknowledged roots in the Marxist "Social Relations of Science" movement of the 1930s (McGuckin 1984). And then, there is the under-appreciated fact that the British branch of the movement tended to be constituted in opposition to most of what passed for the "sociology of knowledge" in the immediate postwar period, be it Mannheim or Merton or Zilsel. But also significant is the fact that many of its earliest protagonists were recruits from the natural sciences, with little or no formal background in the social sciences. This had the salutary effect of warding off the attacks of the most virulent of initial opponents to science studies, who were adamant that absent formal training in the natural sciences, outsiders had no business saying anything whatsoever about content, much less the operation, of the modern *Naturwissenschaften*. But it also had the unintended consequence that it left the leaders of the nascent field of research with more than a little ambivalence about the intellectual and professional commitments of the social science disciplines, even those to which they sometimes became formally attached within the hierarchical postwar university, since it was rare that STS achieved the status of a free-standing academic department.

There has persisted a Groucho-Marxist quality that has pervaded the postwar history of STS: it was never quite content to join any academic club that would have it as a member. This neurosis goes some distance to explain some of the more curious episodes in the history of STS, such as the intense but short-lived fascination with the problem of reflexivity (Woolgar 1988; Ashmore 1989): Why should we believe in the "Social Construction of X" when you won't apply it to STS? Yet, more to the point, we are convinced that the reception of Bruno Latour's *Science in Action* (1987) by science studies and the subsequent fame of the Paris School have been very much predicated upon the hostility expressed within their precincts to the very idea of a "social explanation" of science, and indeed, to social theory *tout court*. Latour, in his own jocular style, now pleads guilty to removing the word 'social' from the title of the second edition of *Laboratory Life*, "like faces of Trotsky deleted from pictures of Red Square parades" (in Ihde and Selinger 2003: 27); yet the insistence upon the essential illegitimacy of social science explanation has made its appearance in various ANT manifestos for something approaching two decades now. Some of his texts are more disparaging than others: our own personal favorite source is "A Prologue in the Form of a Dialogue between a Student and His (Somewhat) Socratic Professor."[5] A *menu degustation*: "I have no patience for context"; "I have no patience for interpretative sociologies"; "[W]e are in the business of descriptions. Everyone else is trading on clichés"; "Organization Studies, Science and Technology Studies, Business Studies, Information studies, Sociology, Geography, Anthropology whatever the field, they cannot rely, by definition [of ANT], on any structuralist explanation"; "So an actor for you is some fully determined agent, plus a place-holder for a function, plus a bit of perturbation, plus some consciousness provided to them by enlightened social scientists? Horrible, simply horrible...." The following quotation reiterates a position that has now become hardened into boilerplate:

The word 'social' ... does not designate a 'kind of stuff' by comparison with other types of materials.... Are the facts discovered by sociologists and economists so much stronger than the ones constructed by chemists, physicists and geologists? How unlikely. The *explanandum* does not match the *explananda*. More importantly, how could the homogeneous stuff of almighty 'society' account for the bewildering variety of science and technology? Constructivism, at least in our little field of science and technology, led to a completely different program than the one repeated *ad nauseam* by critical sociology. Far from trying to explain the hard facts of science with the soft facts of social science, the goal became to understand how science and technology were providing some of the ingredients necessary to account for the very making and the very stability of society. (in Ihde and Selinger 2003: 28–30)

It may prove helpful to understand the ambitions of the research program formerly known as "actor-network theory"[6] in order to grasp the significance of the recent initiative by Michel Callon concerning "performativity."

Most would agree that ANT, since its inception in the Callon-Latour paper of 1981, has sought to transcend what it has regarded as a raft of problematic dichotomies: Nature/Society, agency/structure, normative/descriptive, doing/knowing, and so forth. We tend to agree with Zammito (2004: 184) that Latour aspires to a "first philosophy" which will resolve some basic problems in science through the promotion of a novel metaphysics; ANT was his attempt to insist "the social possesses the bizarre property of not being made of agency and structure at all, but rather of being a circulating entity" (Latour 1999: 17). It is of course incongruous in the extreme that someone would even attempt such a quest from a position outside of philosophy proper, but further comprehension of that is a "social" question which we leave for another time and place.[7] What is relevant to our current argument is that ANT has been promoted as a Theory of Everything (in the way that physicists commonly use the term) that would permit a view from nowhere, validated, it would seem, entirely on ontological grounds. Somehow explanation would proceed from neither Nature nor Society, and would originate neither with agents nor with structure; instead it would emanate from that vast blank no-man's land situated between those portentous dichotomies. "What makes ANT difficult to grasp is that it fills in precisely the space that is emptied by critical sociologists with the damning words of 'objectification' and 'reification'." (Latour 2005: 77) The ANT analyst would therefore be doing social theory without being a social scientist; she would discuss a generic "Science" without becoming committed to a generic "scientific method"; she would "follow scientists around" without ever becoming subject to the disciplinary codes (and pecuniary accounts) regimenting the scientists. And most paradoxically, although there appear to be no "protagonists" in ANT in the conventional sense (here we nod towards the notorious attribution of symmetry between agents and things, given its strongest statement by Callon), we find ourselves enmeshed in a situation of unmitigated and incessant aggression and war. "The similarity between the proof race and the arms race is not a metaphor. It is literally the mutual problem of *winning*. . . . It is only now that the reader can understand why I have been using so many expressions that have military connotations. . . . I have used these terms because, by and large, technoscience is part of a war machine and should be studied as such." (Latour 1987: 172)[8]

It would take us too far afield to document that there is very little new under the sun, particularly when it comes to social sciences that seek to deny their own status as social sciences in the post-World War II era.[9] In particular, we should like to suggest that many of the trademark philosophical moves of ANT were in fact pioneered in an entirely unexpected location, part academic discipline and part professional intervention, forged in the battles of World War II, which later became the source and inspiration of many of the academic postwar social sciences, from decision theory to artificial intelligence, from management science to computational theory, from logical positivism to American neoclassical economics. That urdiscipline was dubbed "Operations Research" or OR.[10] One of us has written extensively on the history of OR elsewhere, but all we wish to suggest here is that many of the ambitions and attributes of ANT can be found in relatively developed form in OR, which preceded ANT by four decades. This turns out to be an important input for a better understanding the modern appeal of doctrines of "performativity."

The hallmark of OR is that it rapidly became promoted as a Theory of Everything which evinced a distinct interest in blurring most conventional ontological boundaries between the Natural and the Social, between agency and structure. It accomplished this in the first instance by projecting physical models onto agglomerations of men and machines, or as proponents of ANT prefer to call them, technoscience, in order to develop a science of war. It parlayed an expertise in the manipulation of material things like radar sets and gunsights into a supposedly equivalent expertise in the manipulation of men. Crucially, the first operations researchers bore their own contempt for the social sciences, feeling that their training in a natural science endowed them with a portable competence in the "scientific method," which would sanction their pronouncements on any and all mobilizations of men and materiel to achieve specific ends. Furthermore, OR officers managed to "consult" on the conduct of war without having to be responsible for the commands given or even having to become subordinate to the military command structure themselves. They were notoriously given special dispensations to "follow the colonels around." Some of the earliest use of formal network theory was conducted under the rubric of OR; but more to the point, OR modeled all interactions as trials of strength in the face of duplicitous, propagandistic, and unscrupulous opponents. OR served as the main incubator for game theory, which has become the mathematical model of choice in many of the contemporary social and biological sciences—especially within American neoclassical economics.

It does not take much imagination to detect the family resemblances of OR and ANT. "Actor network theory, and for that matter almost every other approach in ST&S, portrays science as rational in a means-ends sense." (Sismondo 2004: 70) In this, it merely conforms to the median format of discourse in much of postwar Western social theory. But one of Latour's interlocutors has insisted: "Your theory defines the types of actants who define their own worlds in specific ways. You focus on antagonisms and goal-oriented rationality metaphorics." (in Ihde and Selinger 2003: 23) Another has complained: "ANT has a tendency to reproduce, in different words, the standard essentialist understanding of what science is." (Erickson 2005: 85) This description fits OR at least as well as it captures ANT. As John Law, one of the major ANT authors, has admitted, the experience of working on military aircraft research and development jolted him into realizing that "the terms used by those working on, in and around the project, were more or less the same that I was using to analyze it.... [It tended] to make similar analytical and lived assumptions about the proper and perhaps the necessary ways of practicing technology." (2003: 6) It was the operations researchers who pioneered the practice of agnosticism about "defining the actors of the world in advance," as well as intervening to bring about the realities their theories describe, not Monsieur Latour & Cie. Indeed, as one modern game theorist has testified,

Cyborgs use an individualistic methodology, because we can thereby construct coherent models that are reasonably tractable. We don't care at what level of organization an individual is defined, provided that its actions are sufficiently consistent that they can be described in terms of maximizing the expected value of a utility function. We know that individual human beings are sometimes irrational, and so don't always behave with the consistency that our theories require of a player. But experiments in the field and in the laboratory confirm that human beings are sufficiently consistent in some contexts that our theories work like clockwork. How else would it be possible for us to use game theory to design the big telecom auctions that recently amazed the world by generating billions of dollars in revenue apparently from nowhere? (Binmore 2004: 481)

So perhaps ANT and its proponents Latour and Callon have not been so wickedly radical or as "amodern" as they first appear to those innocent of the proliferation of science/society hybrids incubated within the military in our recent past. Moreover, this brief glimpse of history suggests that there might be closer consanguinal relationships with certain social sciences— and here, we deliberately point the finger at economics—than might have been suspected, given the self-denying ordinance that ANT has promulgated with regard to the social sciences. Indeed, we think ANT has tended to walk and talk more and more like the very model of a modern neoclassical economist for quite some time now. (We are nowhere near the first to broach this suggestion: see McClellan 1996 and Hands 2001.) Latour has more recently admitted that his replacement program for ANT, which has been promoted under the banner of "political ecology," might appear to outsiders to resemble economics (although not in all respects).[11] Latour has also echoed Callon's plea, quoted in our epigraph, to just stop sniping at the economists:

There has surely been enough complaining about the economizers' hardness of heart.... Dangerous as infrastructure, economics becomes indispensable as documentation and calculation, as secretion of a paper trail, as modelization. (Latour 2004a: 152–153)

Thus the exhortation to "stop worrying and learn to love the Nash equilibrium" turns out to have been percolating deep within the ANThill for some time now. The "performativity thesis" merely brings it more explicitly out into the open. But, most convenient for our present argument, we note that both the advocates of performativity in economic sociology and the modern economists point to the very same set of events—the FCC spectrum auctions in the United States and their European imitators—as providing what they consider to be some of the best evidence supporting their ontological claims. Game theory, writes MacKenzie, "was no longer an external description of the auction, but had become—as Callon would have predicted—a constitutive, performative part of the process" (2002: 22).

Could this shared fascination provide important clues to the real significance of the doctrine of "performativity"?

Callon on Performativity and Economics

Callon's performativity thesis has been called "the most challenging recent theoretical contribution to economic sociology" (MacKenzie and Millo 2003: 107), yet there persists a fair amount of dissension and confusion as to its provenance and significance.[12] In *The Laws of the Markets*, Callon asserts that "the economy is embedded not in society but in economics" (1998: 30), rejecting the assertion that "the market is socially constructed." Later in the same work, Callon proposes the following:

By ridding ourselves of the cumbersome distinction between economics (as a discipline) and the economy (as a thing) and showing the role of the former in the formatting of markets, we find ourselves free from a positivist, or worse still, a constructivist conception of law. Market laws are neither in the nature of humans and societies ... nor are they the constructions or artifacts invented by social sciences. (ibid.: 46)

From our previous section, we can appreciate that any doctrine which so insistently eschews the very existence of a category called 'society' can readily emit such denials. Our critique begins with what Callon and the most current version of ANT understand as constituting "economics." We believe the reader of *The Laws of the Markets* cannot come away from the experience without the conviction that the authors persistently confuse and conflate "economics" with the activities of accounting and marketing. Indeed, one of our complaints below will be that Callon does not attend closely enough to what does and what does not count as legitimate "economics" among the agents.

We would hope that the current version of ANT (henceforth, to avoid repeating this phrase, and to focus on a precisely defined subset of economic sociology, we replace it with ANT 6.5[13]) would not want to be caught trafficking in bootless tautologies, and therefore it becomes imperative to try and understand just what is being asserted about "economics" by ANT 6.5. We think we can discern four related though distinct propositions in the writings of Callon:

(A) Markets are a set of diverse imperfectly linked calculative entities, sometimes conceived of using computer metaphors, and sometimes Darwinian metaphors. Callon (1998: 32) insists upon "the prime importance of the existence and hence the formatting of calculative agencies.... Several types of organized market exist, depending in particular on the nature of the calculations of the calculative agencies ... the market is a process in which the calculative agencies compete and/or co-operate with one another." And Callon et al. (2002: 194) write: "Markets evolve and, like species, become differentiated and diversified. But this evolution is grounded in no pre-established logic."

(B) Once proposition A is acknowledged, there is nothing standing in the way of treating actual existing markets as technoscientific phenomena, much as ANT has been treating speed bumps, scallops, and microbes for years. "Instead of considering 'laboratory' markets, like those studied in experimental economics, as caricatures of real markets, we can explore how a particular calculative element is simulated in a particular way, and how the relationship between a market simulation in a laboratory and the actual 'scale one' market is constructed." (Callon and Muniesa 2003: 197–198) "The natural and life sciences, along with the social sciences, contribute toward enacting the realities that they describe. The concept of performativity affords a way out of the apparent paradox of this statement." (Callon 2007)

(C) Once proposition B is acknowledged, the "scientists" whom science studies should be "following around" are the certified economists, who in turn have been known to claim that they pursue their prognostications in a space outside the "economy" but who in fact by their activities help to produce it. (Professionals who do openly profess to construct economic life, including accountants, lawyers, marketers, government regulators, and corporate managers, would appear not to be suitable targets for this activity: they are treated as secondary.) This, it seems, is the effective content of the "performativity" thesis. "*Homo economicus* really does exist.... He is formatted, framed and equipped with the prostheses which help him in his calculations and which are, for the most part, produced by economics." (Callon 1998: 51) "Without economics the market would not exist.... Economics in the wild is not pure economics; it is mixed with engineering, life sciences and management science." (Callon 2007)

(D) Once proposition C is acknowledged, it follows that ANT 6.5 can't go around challenging the legitimacy and efficacy of the economists, any more than they should challenge the legitimacy and efficacy of the natural scientists they formerly shadowed. This means that economic models are to be approached as "true," although with the caveat that it is the economists and their allied actants who make it so. Conveniently, this implies that adherents of ANT 6.5 can go wherever the economists go, forge many of the same alliances, and be engaged by the same client groups that support the economists. "Professional economists no longer have the direct or indirect monopoly (assuming they did ever have it) on authorized and legitimate discourse." (Callon et al. 2002: 195) "Economists have succeeded in creating alliances with technocrats.... We can imagine economic sociologists co-operating with actors who are interested in thinking about ways of organizing markets in order to counter the role of the mainstream economists. What is very important is to abandon the critical position, and to stop denouncing economists and capitalists and so on." (Callon in Barry and Slater 2003: 301) "I would be reluctant to use this programme to cooperate with governments for the purposes of public administration." (ibid.: 306)

Rather than discussing performativity in a vague way, we believe our restatement sharpens the issue and renders the production of case studies themselves more pointed and apt. In the rest of the chapter, we shall settle on the case identified by Callon and MacKenzie as one illustration of the program of performativity: the FCC spectrum auctions. Contrary to their intuitions, we shall interpret the case as supporting proposition A but calling propositions B, C, and D into question. Foreshadowing our conclusion, study of the events reveals that the evidence does not support the widespread impression, apparently shared by the economics community and the science studies community (Guala 2001; MacKenzie 2002; Parkin 1998), that economists' game-theoretic accounts of auction theory dictated the format of the auctions adopted and therefore rendered the economists' theories "true" by construction. The confrontation of the material world and the social theorists was a much more complicated phenomenon. The auctions as they finally materialized were a curious amalgam of technical achievement and crude politics, but that does not imply that a flat ontology of "actants" and networks would help us understand how they came about. Indeed, in our opinion, so far taking this approach has only served to obscure the actual causes of events—indeed, in precisely the same manner as the economists themselves have misrepresented the causes.

The FCC Auctions

In 1994 the Federal Communications Commission commenced the practice of auctioning communications spectrum licenses to the highest bidder.[14] The process of determining the best method of auctioning rights to control certain frequencies of the electromagnetic spectrum was marked by another innovation: the heavy involvement of academic game theorists—practitioners of one of the most abstract mathematical fields of economics, and a field often thought to exist at a remove from practical problems. Once the first set of auctions was complete, and the dollar tally came in, those economists gleefully took credit for what was initially perceived as a highly successful performance.[15] Within economics the episode has become the textbook example of the practical relevance of game theory, and it was directly responsible for at least one Nobel Prize. One of the most interesting uses of the FCC auction results has been to bolster claims concerning successes ensured by the participation of economists in producing the outcome.

In depicting the FCC auctions as the outcome of an instance of performativity, Callon and MacKenzie follow the work of Francesco Guala, who developed an account of the FCC auctions as "a tour de force from [the] preliminary identification of the target to the final product" (Guala 2001: 455). Congress established as a "target" an auction that would meet several organizational, distributional, and macroeconomic goals. The "final product" was, in Guala's terminology, an "economic machine" representative of "our best science and technology," and it was ultimately judged by

Guala to have been a "success" (ibid.: 473–474). This "economic machine" account works by focusing on a stylized notion of techniques used in product research and development[16] and derives its evidence almost exclusively from a few published accounts of the major game-theory participants. From this vantage point, an R&D process takes place not only in the "abstract realm of theory" but also in the "university lab" (ibid.: 475), the different locations corresponding to different stages in the systematic process of developing a fully functioning "machine." Similarly, the ANT 6.5 narrative is concerned with the construction of "calculative collective devices," and with the methods economists use to construct a "relationship between a market simulation in a laboratory and the actual 'scale one' market" (Callon and Muniesa 2003: 198). The R&D narrative regards itself as following the economists around as they overcome difficulties and obstacles in the development process, some involving the "multiple constraints imposed by the FCC" and others arising from the "nature of the goods" (Callon 2007). We do not think it out of place to point out that neither Guala nor Callon actually followed any economists around in this instance; what they followed was a subset of the economists' own self-serving accounts, published after the fact or related in interviews.[17]

Whereas Callon (2007) emphasizes the inability of economic theorists to provide a "turnkey solution," "the increasing role of experimentation in market engineering," and the need for participants to "adopt a logic of compromise," the economists who participated in the FCC auctions were quite prepared to admit that "the theory does not specify an unambiguous best form for the spectrum auction" (McMillan 1994: 151), that experimental economists participated in the construction of an operational auction (Kwerel 2004; McMillan 1994), and that the final outcome represented successful collaboration by several participant groups (Kwerel and Rosston 2000: 261). Furthermore, those economists—like Callon—identified the characteristics of the spectrum commodity and the establishment by the government of "multiple aims" for the auctions as the primary reasons for abandoning their use of formal methods (McMillan 1994; also see McMillan et al. 1997). Therefore, the most striking aspect of the ANT 6.5 account is how little it adds to the firsthand accounts given by participating economists. The precept of uncritically accepting the account of the firsthand participants is actually granted the status of a virtue by Latour (2005: 36).

This ANT 6.5 account tends to obscure the process of determination of the goals, the methods by which the economists were recruited by interested parties, and the social maneuvers used to deal with the presence of incompatible aims. As Callon (2007) puts it: "It is not the environment that decides and selects the statements that will survive; it is the statements that determine the environments required for survival." In good ANT fashion, the economic setting deliquesces into a gauzy web of networks, hazy and indistinct. ANT 6.5 accounts tend to foster the impression that they situate conflict over goals, trials of strength over the creation of concepts, and struggles over the recalcitrance of phenomena at the very center of the analysis; as Latour has asserted, these accounts open up the "black box" shut by the victors. We think that the track record of ANT 6.5 does not come anywhere near supporting this belief: if everything is an "actant," it is hard to fill in the dance card with identifiable protagonists. In this particular instance, an awareness of the different objectives pursued by the distinct participants is indispensable to understanding the FCC auctions. In our suggested counter-narrative, we identify four relatively distinct and salient participant groups: the government (represented by the FCC), a handful of large "telecoms" (telecommunications firms), and two groups of economists (game theorists and experimentalists), each possessed of a distinct set of objectives. A blend of theoretical, pecuniary, and political motivations resulted in an auction that did not meet any of the originally stipulated objectives yet eventually managed to create the impression that it was, nevertheless, a "success." While we might agree with Latour that "power and domination have to be produced, made up, composed" (2005: 64), we would add that his ontological strictures only serve to obscure and suppress that process, in exactly the same way that the economists themselves have done.

It is commonplace for accounts of the FCC auctions to begin with a discussion of the stipulation of several goals for the auctions by Congress. This is a particularly important feature of the ANT 6.5 narrative, because it gives the impression that the goals for the auctions were propounded independent of the process, before it began. In fact, Congress charged the FCC with the following (U.S. Congress 1993, Title VI):

[T]he development and rapid deployment of new technologies, products, and services for the benefit of the public, including those residing in rural areas, without administrative or judicial delays;

Promoting economic opportunity and competition and ensuring that new and innovative technologies are readily accessible to the American people by avoiding excessive concentration of licenses and by disseminating licenses among a wide variety of applicants, including small businesses, rural telephone companies, and businesses owned by members of minority groups and women;

Recovery for the public of a portion of the value of the public spectrum made available for commercial use and avoidance of unjust enrichment through the methods employed to award uses of that resource; and

Efficient and intensive use of the electromagnetic spectrum.

This list was the outcome of a prolonged debate over the aims of telecommunications policy and the role of the government in promoting access, innovation, and competition. The FCC, however, would eventually take the position that all these complicated considerations involving industrial organization, macroeconomics, and distributional equity should ultimately be reduced to the narrower "economic efficiency," and that the most appropriate goal to pursue should be to award licenses to the highest valued users (FCC 1993: ¶34; 1994: ¶70). One participating economist noted that, while there was some controversy over the drastic collapse of multiform intentions to drab uniformity, the decision represented the adoption of an economist's criterion (Milgrom 2004: 4). Our first observation is that the criterion adopted was certainly not universally respected by economists across the board but was broadly consistent with the preferred understanding of game theorists.

By replacing the goals of Congress with their preferred "efficiency" criterion, the FCC's staff economists were able to ground their policy analysis in game theory. What was significant about this was not, as has been commonly asserted, the substitution of political with "scientific" considerations (McMillan 1994; Milgrom 2004), but rather the enrollment of a specific group of academic game theorists into the FCC's policymaking process. Academic game theorists were first invited to participate after the FCC released a Notice of Proposed Rulemaking (NPRM) for the licensing of Personal Communications Services. In every rulemaking process, the FCC is required to ask for comments from "interested parties"—broadcasters, telephone companies, equipment manufacturers, industry groups, government agencies, and to a far less extent consumer groups—that would be affected by changes in administrative rules. This particular set of rule changes met with heated debate, as Congress punted the most contentious political issues to the FCC (Galambos and Abrahamson 2002: 163–164). In response, FCC Chairman Reed Hundt hit upon the idea of calling for the involvement of game theorists. The appearance in the NPRM of a call for game-theoretic analysis of auction policy was unprecedented, and it gave certain interested parties the idea of hiring academic game theorists to further their objectives.

Those hoping to ground controversial public policy in uncontentious science would soon be disappointed, as the enlistment of an increasing number of game theorists would result in a remarkably diverse array of inconsistent recommendations, and ultimately a failure to produce any clear-cut recommendation. One plan for the auction of licenses called for a sequence of English auctions (Weber 1993a,b); a second called for a sequence of Japanese auctions (Nalebuff and Bulow 1993a,b); a third called for simultaneous sales of all licenses (McAfee 1993a,b; Milgrom and Wilson 1993a,b).[18] Some proposals insisted on admitting bids for bundles of geographically linked licenses (also known as "package bidding" or a "combinatorial auction"); others favored restricting bids to individual licenses only.

The sticking point was that game theory supplied no global discipline with regard to the type of recommendations tendered: a game theorist could legitimately support any of an array of auction forms by stressing one set of information properties over others. Game theory is not and has never been a unified theoretical tradition (Mirowski 2002). Game theorists recruited by the FCC did display a penchant to conceptualize an auction as a Bayesian learning game; this tended to focus attention on the release of information during the auction that would better promote knowledge of the licenses' true value, hence promoting efficiency. Generally, the version of game theory favored by the economics orthodoxy dealt with a single good, and assumed knowledge of the "true value" of this good to be distributed stochastically among participants; the state of play was conceptualized as information being released during the conduct of an auction, which will promote the participants learning the true value of the good. There was, however, no conventionally accepted standard for determining the precise value of the information provided by a given auction, much less the "true" value of any good, and this constituted a problem for attempts to generalize existing results to an environment with multiple heterogeneous goods. Game theorists therefore supported their recommendations not with their own conventionally accepted standards of mathematical proof, but with loose analogy and piecemeal analysis, mooted in seemingly clear but frequently contradictory catchphrases as "the more open, the better," "make sure participants get quality information," and "avoid free rider problems."

Some who participated in the runup to the spectrum auctions have acknowledged that game theory was unable to provide a knock-down argument for the optimality of a specific form of auction (McAfee and McMillan 1996: 171; McMillan et al. 1997: 429). An ANT 6.5 account might attribute the lack of a determinate recommendation to the essential inadequacy of "abstract theoretical reflection" for the development of a working product, but faulting arid abstraction does not begin to get to the heart of the matter. The lack of a determinate recommendation was less a disagreement over the significance of various learning effects than a disagreement over the *aims* for the auction. Although there was ample room for disagreement over the efficiency properties of the auction proposals, firms' narrowly constituted interests clearly played a major role in the policymaking process:

... the business world was fully aware of [the strategic significance of] the rulemaking process and had engaged many groups of consultants to position themselves. Businesses understood that the rules and form of the auction could influence who acquired what and how much was paid. The rules of the auction could be used to provide advantages to themselves or to their competitors. Thus, a mixture of self-interest and fear motivated many different and competing architectures for the auctions as different businesses promoted different rules. (Plott 1997: 606)

The most prominent "consultants" used by businesses to "position themselves" were academic game theorists. Several firms responded to the FCC's NPRM by lobbying for preferred sets of auction rules, and some (mostly Baby Bells[19] and their progeny) enlisted academic economists to draft supporting comments. The telecoms went on a hiring spree. Nynex hired Robert Harris and Michael Katz of the University of California at Berkeley; Telephone and Data Systems (TDS)[20] hired Robert Weber of Northwestern University; Bell Atlantic hired Barry Nalebuff of Yale and Jeremy Bulow of Stanford; Airtouch[21] hired R. Preston McAfee of the University of Texas; Pacific Bell hired Paul Milgrom and Robert Wilson of Stanford.[22] In accepting their role as consultants, they participated at the pleasure of their clients:

[Pacific Bell Attorney James] Tuthill, who organized Pac[ific] Bell's lobbying before the FCC, knew it would be crucial to hire an expert who could figure out where, amid the highly technical details of the auction proposal, PacBell's interests lay.... He wanted someone who could speak plain English and come across to the FCC as more than just an opinion-for-hire. "If it's just another party coming up and telling our line, that isn't going to be effective." ... During the summer before the FCC released its auction plan, Tuthill's staff drew up a list of games [*sic*] theorists.... By the time the FCC's plan was in the hands of Pac Bell's competitors, the company had signed a contract with Milgrom and Wilson. Although Wilson was a more senior professor, Milgrom was assigned the lead role because he was willing to lobby. (Thelen 1995)

The requirements that economists figure out where their clients' "interests lay" and be "willing to lobby" deepened the controversy over auction form while decoupling proposals from the pursuit of anything resembling the public interest. The absence of a global theory of auctions and the internal difficulties of the game theorists' approach provided opportunities for disagreement, but the high-stakes setting of the design process—along with the establishment of consultant relationships with most of the theorists—virtually ensured it.

The clearest example of businesses' using economists to promote different auction architectures is provided by the assortment of comments pertaining to the use of a combinatorial auction. While all participants were in agreement that a combinatorial auction would ease the aggregation of licenses, detractors characterized this easing as "biased," whereas supporters characterized it as "efficient." One economist—a consultant for Pacific Bell—was remarkably candid about the relationship between corporate strategies and the proposals made:

In the US telecommunications spectrum auctions, sophisticated bidders anticipated the effects of packaging on the auction and lobbied the spectrum regulator [the FCC] for packages that served their individual interests. For example, the long distance company MCI lobbied for a nationwide license which, it claimed, would enable cell phone companies to offer seamless coverage across the entire country. MCI knew that if such a nationwide license plan were adopted, it would exclude existing mobile telephone service providers from bidding, because those providers were ineligible to acquire new licenses covering areas that they already served. In the same proceeding, regional telephone companies such as Pacific Bell lobbied for licenses covering regional areas that fit well with their own business plans but poorly with the plans of MCI. (Ausubel and Milgrom 2005: 2)

Firms seeking nationwide coverage—including MCI,[23] Bell Atlantic, and Nynex (Andrews 1994; Galambos and Abrahamson 2002; Skrzycki 1993)—supported nationwide package bidding. Firms pursuing regional strategies—Pacific Bell and Airtouch (Galambos and Abrahamson 2002; Kwerel and Rosston 2000; Thelen 1995)—supported licenses covering regional areas, and opposed package bidding. In between these two groups stood TDS, which favored package bidding, but only for regional groupings across license bands and not for a nationwide license. TDS was pursuing a regional strategy and had no intention to seek a nationwide collection of licenses (Murray 2002: 270; Weber 1997: 534). TDS believed that a "hub and spoke" strategy of securing licenses surrounding major metropolitan areas would best serve its interests, and that sequencing the auctions from highest to lowest population would best facilitate its strategy. The TDS/Weber proposal sparked a debate among the economists over the appropriate method of sequencing auctions:

The primary advantages of this order of sequencing are that it facilitates regional "hubbing," and that it brings substantial valuable information (concerning both pricing and licensee identity) into the public domain quickly. The information will help applicants bidding for [smaller] licenses ... to refine their acquisition strategies, and hence will enhance the efficiency of the final allocation of licenses. (Weber 1993b: 6)

... suppose that the Commission chose to auction spectrum for the New York City area first due to its population size, with other areas following. As auctions progress, participants will learn more about what is going on. Hence, participation in early rounds may be riskier. But a firm like NYNEX might have no choice but to bid in its home region. Therefore, if the Commission does adopt sequential auctions for different geographic areas, it should proceed in random order across trading areas within each block. (Harris and Katz 1993: 17)

The debate over sequencing auctions renders the considerations informing the proposals obvious. Weber argued his sequencing proposal would facilitate the strategy of "hubbing," which was unique to his client (TDS), while Harris and Katz argued it would disadvantage their client (Nynex). Such arguments abandoned any distinction between social welfare and the welfare of their clients.

In an ironic twist, the task of determining the public version of what academic game theory ultimately dictated fell to the FCC. The multiplicity of aims and proposals forced the FCC to display some creativity in conjuring a "consensus" recommendation for the "simultaneous-multiple round-independent" (SMRI) auction—the form of auction that most of the economists opposed.[24] Experimental economists appeared to demonstrate that the combinatorial auction was more effective. But the SMRI auction did possess the virtue of being broadly consistent with the concerns of a distinct group of large telecoms that were united by their fear of being leapfrogged by MCI, which would assume a commanding position if it acquired a nationwide license.[25]

Working out the details of the never-before-implemented SMRI turned out to require far more elaborate competencies and redoubled efforts beyond those deployed in the initial rounds of the public policymaking process. Consequently, experimental economists were recruited to participate in the design of the auction. Though the ANT 6.5 account fosters the impression that it was the pesky abstractness of theory that prompted the inclusion of experimental economists, it was actually the adoption of a seemingly innocuous proposal of some game theorists to computerize the auction that unwittingly endowed experimentalists with their most important role, and put the process on track to build some real machines.

Attempts to produce a prototype auction failed.[26] The FCC was thereby
induced to seek help from the only economists who had actually produced a computerized auction, and it devolved to the experimentalists
to accept major responsibility for coding the auction. It was therefore
computerization—and not the abstractness of the theory—that prompted
the inclusion of experimental economists. What had begun as a mere side-
show rapidly became the main arena of the contest. The decision to computerize the auction would have several unintended consequences, one of
which was to effect a change in the criteria pursued. This point requires
careful development, because the adherents of ANT 6.5 have failed to take
note of it. Experimentalists did not view themselves primarily as software engineers or troubleshooters or bricoleurs, but rather as a distinct
professional group in possession of their own ideas about how to design
markets.[27] For our present (severely telegraphed) purposes, it is possible
to reduce the differences between game theorists and experimentalists to
three primary areas of disagreement.[28]

First, whereas game theorists tended to represent markets as Bayes-Nash
games, experimentalists represent them as combinatorial optimization procedures. Experimentalist market theory has roots in general-equilibrium
theory, and particularly in efforts searching for determinate price-
adjustment processes. They were therefore concerned with the existence of
a competitive equilibrium (Banks et al. 1989: 2–3). In the absence of a competitive equilibrium, prices no longer suffice to coordinate agents to optimal allocations (Ledyard et al. 1997: 656). The attainment of competitive
equilibrium is generally not a concern for game theorists.[29] What absorbs
their attention, rather, is the putative mendacity of participants, who are
the ultimate sources of information about the economy. For game theorists, all the action happens in the mind of the participant, modeled as
an inductive machine assumed to "learn" through Bayesian inference; for
experimentalists, most of the action happens in the price-adjustment process, conceived as a price-discovery device.

Second, game theorists want to improve the "price system" by increasing
the amount of information it provides, whereas experimentalists seek
improvements in its capacity for information processing. Game theorists
focus on methods for discovering and publicizing the information that
they assume to be already dispersed in the minds of participants. While
experimentalists are undeniably interested in the same information, they
focus their efforts mostly on finding procedures—or "smart markets"—
that will make the best use of this increased access to information. This
focus on construction of a tractable optimization program (problematic

	Game theory	**Experimental economics**
Market	Bayes-Nash auction game	Combinatorial optimization problem
Solution	Increase information	Improve information processing
Welfare criterion	Ex post "Pareto optimality"	Ex ante "Pareto optimality"

Figure 3.1
Rival approaches to market design.

for integer programming problems because they are computationally burdensome) encourages experimentalists to treat the market rules as an algorithm. There is no such equivalent imperative for game theorists, who provide only the most stylized descriptions of markets; they conceive of their machines abiding inside peoples' heads. While these experimentalists "black box" cognitive processes to study features of the exchange process, game theorists "black box" the exchange process to focus on treating the mind as an inference engine.[30] As a consequence, it has been the experimentalists and not the game theorists who have tended to foster appreciation of the importance of the sheer diversity of market forms.[31]

Third, whereas game theorists generally judge the success of a market on how it assists learning, experimentalists tend to judge it by the reliability of the successful execution of trades. This is reflected in the different criteria used by the two groups. (See figure 3.1.) Game theorists pursue the criterion of ex post Pareto optimality (the bidder who would create the most value from owning the license wins it); experimentalists pursue ex ante Pareto optimality (the bidder who values the license highest at the outset acquires it). These differences in criteria are responsible for different styles of arriving at a "solution": The experimentalists' prescription is frequently described as the product of a balancing act between "full central processing" of information, which relies on the processing algorithm to use the information, and "decentralization," which relies more on participants to use information. Because game theorists are concerned only with the "processing" that takes place in the heads of the participants, they are concerned only with producing a form that maximizes the amount of information given to the participants.

The controversy that erupted over the combinatorial auction during the intermediate phase provides a perspective from which to observe the rival approaches at work. Both game theorists and experimentalists were concerned with the presence of interdependent values of different geographic spectrum allocations, but they understood the problem this valuation structure posed in a radically different way. Experimentalists argued that the only sort of market algorithm that could be counted on to produce a dependably "optimal" allocation of licenses (by arriving at a competitive equilibrium) required a method for collecting information on the value of *packages*, or combinations of licenses, in addition to the value of individual licenses. They recommended package bidding, and devoted much of their efforts to finding a smart market that would best process the information. By contrast, the game theorists who opposed the combinatorial auction argued that merely asking for information on package values would *reduce* the amount of information collected.[32] While citing what they believed were the informational advantages of their preferred auctions (in the sense of reducing probabilistic uncertainty), the game theorists did not feel compelled to discuss what would be done with this increased information, preferring instead to leave it up to participants to decide how to benefit from this information.

The experimentalists failed to convince the FCC to resort to the combinatorial auction, but when charged with the computerization of the auction they took over responsibility for determining the criteria the algorithms would meet. Concerns with issues of learning were banished, and the criterion of ex post Pareto optimality came to be trumped by "technical" issues of computation and practical imposition of coordination and the criterion of ex ante Pareto optimality. But although the participation of experimentalists would significantly diminish game theorists' effective participation in the process of "putting flesh on the markets," the experimentalists promoted the success claims of game theorists. In coding and testing the market, experimentalists encountered and resolved nagging inconsistencies and ambiguities of the SMRI. In their capacity as coders and software testers, experimentalists would initiate a methodical search for code inconsistencies by employing the "user bounty" method (Plott 1997: 631–632), which paid "sizable bounties of one hundred dollars or more" to find errors and to crash the software (ibid.: 631). One can get a sense of the results of using these methods from an unpublished report generated for the FCC by the Caltech team of John Ledyard, Charles Plott, and David Porter. When they tested the prototype auction, they encountered problems severe enough to "render the technology unusable unless properly fixed," making it "impossible ... to certify that the auction programs and supporting software will function properly when in use" (Ledyard et al. 1994: 1). Many of the patchwork policy fixes offered by game theorists were so ill-conceived as to be useless from an operational stand-

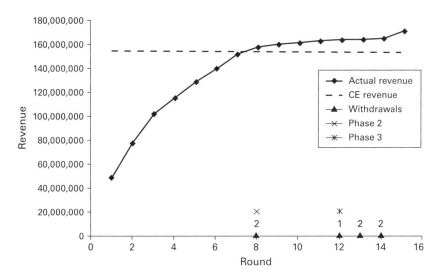

Figure 3.2
FCC Auction Test, September 29, 1994.

point. While the game theorists thought the SMRI to be a simple auction, to the experimentalists the market required the design of "complex software to track, among other things, bidding and bidder eligibility," ultimately rendering it "a very 'complex' simple auction" (Bykowsky et al. 2000: 226). At least one participant credited the experimentalists with implementing the FCC auctions, observing that without their participation "the FCC's first auction might not have succeeded in translating the FCC auction rules into software code" (Kwerel 2004: xxi–xxii).

It is imperative for our narrative that the role of the experimentalists would not be confined to mere software engineers. The inclusion of the experimentalists in the design process confirmed the displacement of the goals attributable to imperatives associated with game theory. Figure 3.2, which is reproduced from the experimentalists' report to the FCC, follows the behavior of revenues (the sum of standing bids) as the auction passes through several rounds, through two "phase changes" (in rounds 8 and 12), and finally to the auction's termination in round 15 near a revenue level marked "CE Revenue." There are three remarkable features of this diagram. The first is its focus on the revenue path. While it is difficult to know what to make of a revenue path from the standpoint of game theory, the trajectory assumes the very precise meaning of an empirical demonstration of the convergence of an optimization program within the idiom of computational mathematics. Second, its designation of the desirable level of revenue as a "competitive equilibrium" (CE) provides an economic interpretation of the convergence of the optimization program. Third, the attention paid to the impact of bid withdrawals and stage changes demonstrates the wish to evaluate the impact of introducing "plug compatible" options to an existing auction platform. The diagram represents, in sum, the competence of the experimentalist program to produce fully functional decision technologies, and displays the range of criteria experimentalists use to evaluate them.

It is easy to become mesmerized by the trees and thus to lose the forest, as often happens to adherents of ANT 6.5 (e.g., Callon 2007). Lest the lessons of this handoff of the auction design to the experimentalists be lost amidst details pertaining to their unique expertise in software engineering, along with their employment of new performance criteria, it is important to make clear at this point that the shift in the criteria cited by the experimentalists does not so much indicate a shift in the overall goals selected for the spectrum auctions as it offers further evidence of the extent to which the dictates of the game-theoretic program in market design ultimately did not matter to the overall auction-design process. *Corporate imperatives demonstrably played the decisive role in determining the auction.* As with the game-theoretic program, the imperatives of the experimentalist program would matter only to the extent that they could be made to seem compatible with corporate strategies. Experimentalists would be allowed to address communication within markets so long as such communication was consistent with strategies common to the Baby Bells; they would not be permitted to implement a smart market. They wanted to deploy their preferred ex ante efficiency welfare criterion to decisively determine the best auction form, but found severely limited opportunities to do so: Plott was permitted to present experimental evidence that was perceived to be supportive of the Pacific Bell/Milgrom-Wilson proposal, while Porter and Ledyard's experimental evidence was ignored.[33] It should be apparent that none of the economists was in any position to make the world fit his preferred model.

A proposal sponsored by the National Telecommunications and Information Administration (the federal agency responsible for managing government spectrum usage) and authored by Mark Bykowsky and Robert Cull with the help of the Caltech experimentalist John Ledyard provides a vantage point from which to view the possible alternatives for market design that would have been attainable if other entities had funded the research. Their smart market offered the possible benefits of increased auction revenues and improved allocative efficiency over the alternatives by increasing bidding competition, and perhaps assigning a nationwide license along the way. As a government agency, the NTIA had no strategic acquisition concerns, and did not care whether the Baby Bells, MCI, or some other entity emerged victorious. Rather, the NTIA was most concerned with gaining credit for helping implement a successful auction, which it tended to understand in terms of maximizing auction revenues (Irving 1995: 44). The smart market was therefore a good fit for the NTIA (proponents had repeatedly cited its revenue-maximizing potential), but not for the telecoms who would bid in the auction.[34]

To review, the ANT 6.5 narrative informs us that the FCC set the goals for the economists to attempt to achieve, subject to congressional constraints. The economists proceeded to imagine a world, then set about to make their words and equations flesh. "To make a formula or auction system work, one has to have tools, equipment, metrological systems, procedures, and so on.... A host of professions competencies and nonhumans are necessary for academic economics to be successful.... They are engaged in the construction of a world described and performed by statements and models we readily agree belong to the world of economics, in the strict sense of the word." (Callon 2007) Well, no: we don't all agree. Our narrative finds fault with such an account for its portrayal of the economists, telecoms, and government officials as a largely undifferentiated swarm of actants united in pursuit of the pragmatic operability of a "machine." Adherents of ANT 6.5 missed most of what went on in the FCC auctions; they misunderstood what got to count as "good" economics, as well as how societies do and do not work. The FCC thought the economists might help them exert some control over the process of the allocation of spectra, but they may have been a bit naive. Game theorists and experimentalists were not necessarily "on the same page," seeking to bridge the inevitable gap between pure science and its applied contexts. Until very late in the game, nobody was really sure about where the machine would even be situated (between the ears? on the silicon chip? in the patented algorithm? at the corporate merger specialist's office?), much less about what it accomplished. Everyone was busily trying to recruit everyone else, although some "actants"—viz., the telecoms—were unequivocally "more equal" than everyone else. Once the diversity of aims and understandings has been accounted for, we are left with a story in which some economists managed to redefine the goals for the government to achieve, subject to the telecoms' veto, while letting a different set of economists bask in the limelight and take the credit. Is this an instance of "performativity," or is

it yet another instance of bigger forces' determining the economic outcomes while masking their activities with a fog of learned disputation and
superfluous mathematics, a hoary old chestnut that Latour professes to despise? More disturbingly, do advocates of ANT 6.5 feel more compelled to
find constituencies than to find causes?

Game theorists have been loudly trumpeting the success of the FCC auctions for more than a decade, and this has led to the explosion of the subfield of "auction theory." And advocates of ANT 6.5 have endorsed these
claims despite considerable contrary evidence built up in the interim. The
original congressional mandates have, of course, been conveniently forgotten.[35] Many businesses buying licenses defaulted on their down payments (Murray 2002: 274–275), which led to considerable "administrative
delay" in re-awarding licenses.[36] The lion's share of licenses won by
"small" and "entrepreneurial" businesses went to entities bankrolled by
large telecoms—a failure to get licenses into the hands of a "wide variety
of applicants."[37] The auctions have not lived up to their promise to promote "rapid deployment [in] rural areas," as both large telecoms and
smaller firms have tended to concentrate their effort on large metropolitan
areas (Copps 2004; Meister 1999: 76–77). Overall, the allocation of licenses
produced by the auctions proved unstable, as the industry has gone
through a spate of merger and acquisitions and telecom failures, ultimately
leading to a high degree of license concentration (Murray 2002: 289–291).
True, the auctions did capture a tidy sum for the government coffers, but
perhaps they did so at the expense of any solid foundations for the economic health of the industry over the medium term. Yet despite the failure
to implement public policy, the FCC auctions were, as one participant
noted, "a huge success for the auction theorists involved" (Cramton 2002:
3). One of the most interesting upshots of the auctions was the development of companies—with many of the key participant game theorists
taken on as partners—devoted to the construction of markets. As Alvin
Roth (2002) has noted, the FCC auctions opened up "a new way for game
theorists to earn their livings, as consulting engineers for the market
economy."

Perhaps by focusing so insistently on the narrative structure of the ANT
6.5 account, we have missed out on its advocates' true aims. Callon points
out "the increasing role of R&D and experiments in the conception of markets or in the regulation of interventions on their modes of functioning,"
and suggests it is part and parcel of economics becoming a "truly experimental science" (Callon and Muniesa 2003: 33). But this message was surely not intended purely as a contribution to the methodology of the social

sciences. Experimental economics has already found a secure niche within the economics profession, not to mention the world of corporate research, and certainly has no need of science studies to provide it with some convenient rationale. (No one harps on "good science" more than Vernon Smith.) Rather, Callon is actually more interested in engaging in "R&D," largely because he wants to argue that "the role of the sociology and anthropology of economies is precisely to design tools and to provide actors with such tools" in order to "influence or structure institutions" (Callon in Barry and Slater 2002: 300). What Callon seems to be doing is arguing for creation of a space for his preferred disciplinary reference groups to participate in "social engineering." But why should we expect that science studies scholars would prove any more nimble than the game theorists, or any less naive than the FCC?

It should be clear from our account that much redirection of goals takes place in the process of social engineering, but only by those who have something to offer in the way of vision and resources to powerful interested parties. Who will sponsor ANT 6.5's performance? And what exactly do its exponents bring to the table? Game theorists have had to engage in a great deal of self-promotion to clinch the deal. Regarding his appearance on the television show *CNN Business Morning*, Paul Milgrom reports:

On the eve of the FCC PCS spectrum auction #4, the author made a television appearance on behalf of Pacific Bell telephone, announcing a commitment to win the Los Angeles telephone license, and successfully discouraging most potential competitors from even trying to bid for that license. (Milgrom 2004: 23)

Because Milgrom argues that "marketing a sale is often the biggest factor in its success," and because he acknowledges that attempts to "discourage others from bidding, hoping to get a better price" (ibid.: 23) undermines efforts to market a sale, it is impossible to understand his television appearance as anything but an acknowledgment of his willingness to place the interests of his client first. The lesson has sunk in: game theorists have gleefully noted *The Economist*'s conclusion that "for the firms that want to get their hands on a sliver of the airwaves, their best bet is to go out first and hire themselves a good game theorist" (1994: 70, quoted in McAfee and McMillan 1996: 159).

Is it possible to salvage anything for understanding markets from the performativity thesis, for example by augmenting the ANT 6.5 account with a better description of the various "actants" involved? If the analysis of ANT 6.5 is deficient, is it possible to imagine a "good" use of performativity? One candidate is provided by the legal scholar Lawrence Lessig, whose work on issues of regulating intellectual property, the Internet, and communications has attracted some recent attention from STS (e.g., Pinch 2007: 274). Lessig's (1999) call for an analytical treatment of how things (for example, "architectures" constructed out of software code) regulate behavior would seem to resonate with "materiality," while his advocacy (1998: 674–675) for reengineering markets to promote certain desirable ends is similar to the position of ANT 6.5. More pertinent to the present case, Lessig's willingness (2001: 233) to credit "an idea about property [with] doing all the work" in supporting the current regime of licensing spectrum resembles the performativity thesis: here economics shapes the world in a process that "enrolls" both living things (such as government agencies, courts, and congress) and nonliving things. Though not uninterested in technical features, this account avoids getting mired in the R&D process, and instead stresses that the various methods of changing human behavior all amount to "regulation" that, at this point in time, is being used by "existing and powerful interests" for the purpose of engineering the economy to suit their own business models (2001: 223). From this vantage point, the decisions to continue licensing spectrum and to auction the licenses cannot be appreciated without accounting for the strategies of the telecoms and making a judgment about the significance of their implementation for society. We do not deny the great potential for generating accounts like Lessig's that use concepts related to materiality to discuss regulation. But in light of the track record, we do not look to ANT 6.5 to provide them. Having already found the black box of technology empty (Winner 1993), adherents of ANT 6.5 have opened up the black box of "the market" and found it empty as well.

Bringing "Society" Back In

The time has come to try and make sense of all the talk about construction and performativity, specifically with regard to the relationship of science studies to economics. It should be obvious by now that we find ourselves unable to agree with Latour that "political ecology [his neologism for ANT] alone is finally bringing the intrinsically political quality of the natural order into the foreground" (2004a: 27–28). That Natural Order is dragooned to political purposes is old news, at least as old as Leviticus (Douglas 1984, 1986). The fact that order (natural or otherwise) is made, not found, is equally unprepossessing. What bothers us is that advocates of ANT 6.5 seem uninterested in the details of how order is actually wrought.

Latour has written: "If the social remains stable and is used to explain a state of affairs, it's not ANT." (2005: 10) We agree, but we draw a different conclusion. Our account of the FCC auctions cannot be "incorporated" or otherwise co-opted to the ANT project. It seems to us that for the bulk of the history of the neoclassical orthodoxy in economics, the comparison of the price system to a natural mechanism existed precisely in order to repress how nature was being used to stabilize one version of the social (Mirowski 1989). In the neoclassical tradition, Markets were Natural, pitched somewhere out beyond the bounds of the social. The sea change, if indeed one can speak in such terms, has come about only recently, when neoclassical economists have conceived of the ambition to *fabricate* markets and not simply treat them as States of Nature. This has created all sorts of tensions and barely acknowledged contradictions in their current projects and self-image (Mirowski 2007). It appears that advocates of ANT 6.5, and Callon in particular, view this as a golden opportunity to bring the economists around to their own research program by getting them to see the attractions of a "constructivist" approach. "The natural and life sciences, along with the social sciences, contribute toward the production of the realities they describe. The concept of performativity affords a way out of the apparent paradox of that statement." (Callon 2007) The alliance is made all the more plausible by the very real family resemblances between ANT 6.5 and modern neoclassicism, which share a consanguinal ancestry with Operations Research and which both nurture a jaundiced opinion of Society, as was noted above. But Operations Researchers knew how to attach their program to the strongest actor in the 1940s and the 1950s: the military. By contrast, Callon proposes that once economists are brought to a more refined level of appreciation for the nature of science, and once they acknowledge that their theories are powerful because they make them so, they will graciously make room for science studies and economic sociology to have their say.

There are two or three things that are wrong with this glorious vision of the radiant future. First, the neoclassical story is so utterly flawed that it cannot be made to "work" for much longer than it takes to come up with another (possibly contradictory) story to take its place. Over the course of the twentieth century the neoclassical orthodoxy with regard to its core price theory has "flipped" at least three times (Marshallian supply and demand / Walrasian general equilibrium / Nash non-cooperative equilibrium), not to mention a host of further slapdash alternate variants (Mirowski and Hands 2006). The spectacle has been far less dignified than the furious activity in Neurath's Boat. Second, the neoclassical school has nonetheless maintained its appearance of monolithic continuity and placid confidence. This is not due to anything particularly conceptual that the economists have said or done, but is rather directly attributable to durable structures disparaged by Latour: the nation-state, the corporation, and the military. As we have seen from our retelling of the saga of the FCC spectrum auctions, only when you leave out the government, the telecoms, and that notorious quasi-material shape-shifter the computer can you make it appear that the auctions were the result of the free play and creative tinkering on the part of the economists, even folding into the account a little help from their friends. It is that despised entity Society and its doppelganger Nature that lends rigidity and structure to what otherwise might seem a fluid and circulating aether. *This is not at all isomorphic to the performativity thesis*, at least as we have attempted here to render it precise.

But if Society may not so easily be banished, then perhaps it follows that the adherents of ANT 6.5 are not quite so free themselves to forge alliances and pursue their constructivist programs as they wish. For instance, the very idea that neoclassical economists would consort openly with them, much less deign to share their sources of support with them, appears to us risible. American-trained economists are notoriously allergic to self-reflection, and they stoop to learn anything about the other social sciences only as a prelude to moving in as an occupying power. Scholars of science studies are kidding themselves if they ever think that the present orthodoxy in economics would ever consent to treat them as equals, much less permit adherents of ANT 6.5 to horn in on their livelihoods.

Therefore, returning to our quadrapart characterization of ANT 6.5, we agree with proposition A that much of economic theory is predicated on computer metaphors, and we tend to approach markets as calculative devices (Mirowski 2007). However, recourse to scientific metaphors does not dictate that (proposition B) economic theory can be approached in the same way that other technoscientific phenomena have been framed within science studies. Too much concentration on machinic metaphors and lively things tends to distract critical attention from some of the most important social processes going on underneath, as we have tried to argue above. Furthermore, isolating the economists as the appropriate protagonists to "follow around" (proposition C) again tends to distract attention from those who may be the major players involved in the construction and shoring up of the "economy." In the case of the FCC auctions, it led both Callon and Guala to ignore the pivotal role of the telecoms in orchestrating the outcome, not to mention slighting the actual intellectual history of game theory and the sad saga of the co-optation of the FCC. Finally, it seems that prescription D, namely that science studies make pact with the neoclassical economists, is at least as potentially disastrous as the alliance that the FCC thought it was forging with the game theorists. Helping promote the fiction of *homo economicus* might have all sorts of blowback for science studies, which should be thought through much more carefully.

Acknowledgment

This is a revised and amended version of a previous paper, "Markets Made Flesh." It is based on Nik-Khah's 2005 thesis, Designs on the Mechanism.

Notes

1. A symptom of the general oblivion to market structures is the urban myth about the early neoclassical theory of Walras being inspired by the Paris Bourse. A good historian such as Walker (2001) makes short work of this fairy tale.

2. This is an important issue we must regrettably bypass in the present paper. See, however, Mirowski 2002, 2004, 2007.

3. Although one of us has done something similar for the modern predicament of the philosophy of science: see Mirowski 2004. To prosecute the argument on a purely philosophical level, as has been the wont of Latour for more than a decade now, would actually clash with one of our reasons for rejecting ANT in the first place, as we argue below in the conclusion.

4. Here we wish to register our gratitude to Steve Fuller, who has been one of the few science studies scholars to insist that this stands as one of the endemic problems within STS. See Fuller 2000a; Fuller 2000b; Barron 2003.

5. The selections are taken from Latour 2002 and from a slightly altered published version (2004b).

6. Latour notoriously repudiated the ANT designation, only to reverse himself once again to embrace it (2005: 9). We suspect this tergiversation is itself a symptom of a deeper indeterminacy of the attempt at a theory.

7. Latour, as usual, is candid on this issue: "... although I teach sociology, I have always considered myself as a philosopher at heart" (in Ihde and Selinger 2003: 15).

8. The current attempt to unilaterally "declare peace" by peremptorily swapping "democratic" for military metaphors (Latour 2004a) deserves its own consideration but would take us too far afield.

9. For some examples, see Mirowski 2002 and Ross 2005.

10. The argument linking the history of OR to the above social sciences can be found in the following: Mirowski 1999, 2002, 2004; Collins 2002; Kirby 2003.

11. "To all appearances, however, [economics] deals with all the topics we have evoked up to now under the name of political ecology. It bears on groupings of humans and nonhumans … it too seeks to take into account the elements that it has to internalize in its calculations; it too wants to establish a hierarchy of solutions, in order to discover the optimum in the allocation of resources; it too speaks of autonomy and freedom. … Apparently, then, the collective that we have deployed does no more than rediscover the good sense of political economics." (Latour 2004a: 132) Latour then goes on to denounce aspects of what he understands as modern economics because of its naturalism, which he believes he has escaped.

12. See, e.g., Slater 2002; Miller 2002; Fine 2003.

13. We take the name from Latour (2005: 207): "I often find that my reader would complain a lot less about my writings if they could download ANT version 6.5 instead of sticking with the beta."

14. Prior to the auctions, the FCC relied on comparative hearings and lotteries to assign spectrum licenses.

15. Many aspects of this sequence of events will be related in only the most cursory manner in this chapter. However, they are covered in the detail one has come to expect from science studies in Nik-Khah 2005.

16. For instance, Guala conflates the way a general equilibrium economist uses the terminology of 'mechanisms' with the way it is used by philosophers of science such as Nancy Cartwright and John Dupré. The terminological conflation is not harmless, we might suggest. For a better history of postwar mechanism design in economics, see Lee 2004 and Nik-Khah 2005.

17. Guala appears to have different aims than the ANT 6.5. Guala believes that "interpretations of a scientific theory (in the natural and the social sciences) should take applied science as their point of departure" (2001: 453), and there uses that method to provide a philosophically motivated intervention to the debate over rational choice theory. His argument is that rational choice theory can be made to work with an understanding of its "real capacities."

18. An English auction is one in which prices increase, the bidder placing the highest bid winning the item. A Japanese auction is similar to an English auction, but all participants are considered active bidders until they drop out. Studies of the formal properties of ascending auctions often substitute the Japanese auction for the English auction.

19. The Baby Bells were (monopolistic) regional wireline telephone service providers created from the 1982 breakup of AT&T. As a condition of the breakup, the Baby

Bells had also received licenses to operate the previous generation of mobile telephones in their own geographical area, and therefore held a commanding position in wireless telephony as well. The Bells tended to view the acquisition of additional licenses as a way of increasing the wireless side of their businesses while at the same time keeping out potential entrants.

20. TDS is a member of the American Personal Telecommunications family of cellular providers that today goes by US Cellular.

21. At the time of the proposal, Airtouch was a wholly owned subsidiary of Pacific Bell with plans to spin off prior to the auctions.

22. Charles Plott was hired by Pacific Bell to run a few experiments to corroborate some theoretical conjectures of Milgrom and Wilson. Plott's experiments, first reported in the conference proceedings contained within NTIA 1994, were later published as Plott 1997.

23. MCI was not a Baby Bell but rather a long-distance service provider, and was therefore a newcomer to mobile telephony. MCI was almost universally regarded as the most formidable of the potential entrants.

24. For a detailed discussion of the FCC's decision-making process, see Nik-Khah 2005.

25. MCI's decision not to participate in the auction was the direct result of the successful persuasion by game theorists of the FCC to reject nationwide combinatorial bidding (Thelen 1995).

26. The extent of this failure is on vivid display in the experimentalists' report to the FCC of their tests of the auction software (Ledyard et al. 1994).

27. There is a relationship between this observation and Galison's (1997) point that experimentalists as a group have conceptual traditions themselves not determined by the beliefs of theorists. The route of the experimentalists to market design through Walrasian mechanism design (and not game theory) is discussed in Lee 2004.

28. On the full contrast, see Nik-Khah 2005.

29. There has been considerable misunderstanding of this point. For example, Guala tends to conflate Nash game theory with Walrasian general equilibrium theory: "Complementarities are one of economists' nightmares, because models of competitive markets with goods of this kind in general do not have a unique equilibrium and are unstable. No theorem in auction theory tells you what kind of institution will achieve an efficient outcome." (2001: 458) The ramifications of complementarity for uniqueness and stability have no place in auction theory, only in general equilibrium theory. However, one should admit that textbooks often elide this distinction to foster the impression of the unity of microeconomics.

30. Game theorists displayed no appreciation of the computational features of the market. On how experimentalists tend to neutralize the vagaries of the minds of their subjects, see Mirowski and Lee 2003.

31. This case is made with greater specificity in Mirowski 2007. This also is signifi-cant for the claims broached at the beginning of the paper, since it was not the game theorists who have promoted many of the constructivist themes found in con-temporary economics.

32. The argument propounded by game theorists is in the form of an analogy with the well known "free rider" problem. There was considerable dispute among econo-mists whether this was a general problem of combinatorial auctions (McMillan 1994: 156), or the artifact of a particular representation (Chakravorti et al. 1995: 364). The reader should bear in mind, however, that some game theorists supported package bidding.

33. Charles Plott's experiments actually compared a Japanese auction with a simulta-neous auction prototype. Plott's results were deemed by Pacific Bell to be supportive of its preferred auction form, and were then presented to the FCC (Milgrom 1995).

34. Experimentalists were cognizant of the conflict of interests, and offered a not-so-veiled accusation that corporate imperatives quashed "package bidding" (Ledyard et al. 1997: 656–660). We have arrived at this interpretation of events as a result of a conversation between John Ledyard and one of the authors at Notre Dame.

35. This case is made in much greater detail in Nik-Khah 2005.

36. The original plan called for allocating licenses in three auctions, to be conducted over a two-year period. The FCC was eventually forced to conduct eleven auctions over a ten-year period. The process of re-auctioning finally concluded in February 2005.

37. The success of the large telecoms in co-opting small and entrepreneurial tele-coms is on vivid display in the sudden emergence of nationwide entrepreneurial telecoms based in, of all places, Alaska. Cook Inlet, Salmon PCS, and Alaska Native Wireless emerged as major bidders in the FCC auctions (Cramton et al. 2002; Laba-ton and Romero 2001; Lee and Martin 2001). The popularity of Alaska as a wireless entrepreneurial hotbed becomes less mysterious once one realizes that firms desig-nated "Native Alaskan corporations" are not liable for any penalties that arise from selling licenses obtained with the FCC's bidding credits for designated entities (FCC 2000: ¶13). These Alaskan telecoms all "partnered" with large telecoms: Cook Inlet with VoiceStream, Salmon PCS with Cingular, and Alaska Native Wireless with AT&T Wireless. In a breathtakingly audacious statement, an executive at Cingular acknowledged "We are going to be doing all our bidding through our designated en-tity, Salmon PCS. That will allow us to bid on all eligible licenses, including a num-ber of those set aside just for small businesses" (Anonymous 2000). Commenting on the success of large companies in displacing and co-opting small and entrepreneurial

firms, one anonymous FCC official candidly observed, "this certainly does make us look like a bunch of idiots" (Labaton and Romero 2001).

References

Andrews, E. 1994. 2 phone concerns seeking to merge wireless services. *New York Times*, June 29.

Anonymous. 1994. Revenge of the nerds. *The Economist*, July 23: 70.

Anonymous. 2000. Cingular, Triton not bidding in auction as standalone companies. *Mobile Communications Report* 14, no. 23. Available at www.factiva.com.

Arrow, K., and F. Hahn. 1971. *General Competitive Analysis*. Holden-Day.

Ashmore, M. 1989. *The Reflexive Thesis*. University of Chicago Press.

Ausubel, L., and P. Milgrom. 2005. Ascending proxy auctions. In *Combinatorial Auctions*, ed. P. Cramton et al. MIT Press.

Banks, J., J. Ledyard, and D. Porter. 1989. Allocating uncertain and unresponsive resources: An experimental approach. *RAND Journal of Economics* 20, no. 1: 1–25.

Barry, A., and D. Slater. 2003. Technology, politics and the market: An interview with Michel Callon. *Economy and Society* 31, no. 2: 285–306.

Barron, C. 2003. A strong distinction between humans and non-humans is no longer required for research purposes: A debate between Bruno Latour and Steve Fuller. *History of the Human Sciences* 16, no. 2: 77–99.

Binmore, K. 2004. Review of *Machine Dreams*. *Journal of Economic Methodology* 11, no. 4: 477–483.

Bykowsky, M., R. Cull, and J. Ledyard. 2000. Mutually destructive bidding: The FCC auction design problem. *Journal of Regulatory Economics* 17, no. 3: 205–228.

Callon, M., ed. 1998. *The Laws of the Markets*. Blackwell.

Callon, M. 2007. What does it mean to say that economics is performative? In *Do Economists Make Markets?* ed. D. MacKenzie et al. Princeton University Press.

Callon, M., and F. Muniesa. 2003. Les marches economiques comme dispositifs collectives de calcul. *Resaux* 21, no. 122: 189–233. Translated version available at http://www.coi.columbia.edu.

Callon, M., and B. Latour. 1981. Unscrewing the big Leviathan. In *Advances in Social Theory and Methodology*, ed. K. Knorr-Cetina and A. Cicourel. Routledge and Kegan Paul.

Callon, M., C. Méadel, and V. Rabeharisoa. 2002. The economy of qualities. *Economy and Society* 31, no. 2: 194–217.

Chakravorti, B., W. Sharkey, Y. Spiegel, and S. Wilkie. 1995. Auctioning the airwaves: The contest for broadband PCS spectrum. *Journal of Economics and Management Strategy* 4, no. 2: 345–373.

Coase, R. 1988. *The Firm, the Market, and the Law*. University of Chicago Press.

Collins, M. 2002. *Cold War Laboratory: RAND, the Air Force, and the American State*. Smithsonian Institution Press.

Copps, M. 2004. Statement of Commissioner Michael J. Copps. In Report and Order and Further Notice of Proposed Rulemaking. FCC docket 04-166.

Cramton, P. 2002. Introduction to Chapter. In *Game Theory in the Tradition of Bob Wilson*, ed. B. Holmstrom et al. Bepress. http://www.bepress.com.

Cramton, P., A. Ingraham, and H. Singer. 2002. The Impact of Incumbent Bidding in Set-Aside Auctions: An Analysis of Prices in the Closed and Open Segments of FCC Auction 35. Criterion Economics Working Paper 02-07.

Douglas, M. 1984. *Purity and Danger*. Ark.

Douglas, M. 1986. *How Institutions Think*. Syracuse University Press.

Erickson, M. 2005. *Science, Culture and Society*. Polity.

FCC (Federal Communications Commission). 1993. Notice of Proposed Rulemaking. FCC docket 93-455.

FCC. 1994. Second Report and Order. FCC docket 94-61.

FCC. 2000. Order in re Applications of Cook Inlet and VoiceStream. DA docket 00-2820.

Fine, B. 2003. Callonistics: A disentanglement. *Economy and Society* 32, no. 3: 478–484.

Fuller, S. 2000a. *Thomas Kuhn: A Philosophical History for Our Time*. University of Chicago Press.

Fuller, S. 2000b. Why science studies has never been critical of science. *Philosophy of the Social Sciences* 30, no. 1: 5–32.

Galambos, L., and E. Abrahamson. 2002. *Anytime, Anywhere: Entrepreneurship and the Creation of a Wireless World*. Cambridge University Press.

Galison, P. 1997. *Image and Logic*. University of Chicago Press.

Guala, F. 2001. Building economic machines: The FCC auctions. *Studies in the History and Philosophy of Science* 32, no. 3: 453–477.

Hands, D. 2001. *Reflection without Rules: Economic Methodology and Contemporary Science Theory*. Cambridge University Press.

Harris, R., and M. Katz. 1993. A Public Interest Assessment of Spectrum Auctions for Wireless Telecommunications Services. Comments of NYNEX. PP docket 93-253, FCC.

Ihde, D., and E. Selinger, eds. 2003. *Chasing Technoscience*. Indiana University Press.

Irving, L. 1995. Spectrum management: A balancing process. *IEEE Communications Magazine* 33, no. 12: 44–46.

Kirby, M. 2003. *Operational Research in War and Peace: The British Experience from the 1930s to 1970*. Imperial College Press.

Kwerel, E. 2004. Forward. In P. Milgrom, *Putting Auction Theory to Work*. Cambridge University Press.

Kwerel, E., and G. Rosston. 2000. An Insiders' view of the FCC spectrum auctions. *Journal of Regulatory Economics* 17, no. 3: 253–289.

Labaton, S., and S. Romero. 2001. Wireless giants won FCC auction unfairly, critics say. *New York Times*, February 12.

Latour, B. 1987. *Science in Action*. Harvard University Press.

Latour, B. 1999. On recalling ANT. In *Actor Network Theory and After*, ed. J. Law. Blackwell.

Latour, B. 2002. A Prologue in the Form of a Dialogue between a Student and His (Somewhat) Socratic Professor. Available at www.ensmp.fr.

Latour, B. 2004a. *The Politics of Nature*. Harvard University Press.

Latour, B. 2004b. On using ANT for studying information systems: A (somewhat) Socratic dialogue. In *The Social Study of Information and Communication Technologies*, ed. C. Avgerou et al. Oxford University Press.

Latour, B. 2005. *Reassembling the Social*. Oxford University Press.

Law, J. 2003. Networks, Relations, Cyborgs: On the Social Study of Technology. Centre for Science Studies, Lancaster University. Available at www.comp.lancs.ac.uk.

Ledyard, J., C. Plott, and D. Porter. 1994. A Report to the Federal Communications Commission and to Tradewinds International, Inc.: Experimental Tests of Auction Software, Supporting Systems and Organization. PP docket 94-12, FCC.

Ledyard, J., D. Porter, and A. Rangel. 1997. Experiments testing multiobject allocation mechanisms. *Journal of Economics and Management Strategy* 6, no. 3: 639–675.

Lee, K. 2004. Laboratory markets as information systems: Vernon Smith designs some mechanisms. Paper presented to HES conference, Toronto.

Lee, L., and R. Martin. 2001. Wireless in Alaska. *The Industry Standard*, March 19: 58–61.

Lessig, L. 1998. The new Chicago School. *Journal of Legal Studies* 27, no. 2: 661–691.

Lessig, L. 1999. *Code and Other Laws of Cyberspace*. Basic.

Lessig, L. 2001. *The Future of Ideas*. Vintage.

MacKenzie, D. 2002. The Imagined Market. *London Review of Books* 24, no. 21: 22–24.

MacKenzie, D., and Y. Millo. 2003. Constructing a market, performing theory: The historical sociology of a financial derivatives exchange. *American Journal of Sociology* 109, no. 1: 107–145.

MacKenzie, D., Muniesa, F., and L. Siu, eds. 2007. *Do Economists Make Markets?* Princeton University Press.

McAfee, R. 1993a. Auction design for personal communications services. Comments of PacTel. PP docket 93-253, FCC.

McAfee, R. 1993b. Auction design for personal communications services: Reply comments. PacTel Reply comments. PP docket 93-253, FCC.

McAfee, R., and J. McMillan. 1996. Analyzing the airwaves auction. *Journal of Economic Perspectives* 10, no. 1: 159–175.

McClellan, C. 1996. The economic consequences of Bruno Latour. *Social Epistemology* 10, no. 2: 193–208.

McGuckin, W. 1984. *Scientists, Society and the State*. Ohio State University Press.

McMillan, J. 1994. Selling spectrum rights. *Journal of Economic Perspectives* 8, no. 3: 145–162.

McMillan, J., M. Rothschild, and R. Wilson. 1997. Introduction. *Journal of Economics and Management Strategy* 6, no. 3: 425–430.

Meister, A. 1999. Evaluating the Performance of the Spectrum Auctions: A Case Study of the PCS Auctions. Ph.D. thesis, University of California, Irvine.

Milgrom, P. 1995. Auctioning the Radio Spectrum. Manuscript.

Milgrom, P. 2004. *Putting Auction Theory to Work*. Cambridge University Press.

Milgrom, P., and R. Wilson. 1993a. Affidavit of Paul R. Milgrom and Robert B. Wilson. Comments of PacBell. PP docket 93-253, FCC.

Milgrom, P., and R. Wilson. 1993b. Replies to comments on PCS auction design. Reply comments of PacBell. PP docket 93-253, FCC.

Miller, D. 2002. Turning Callon the right way up. *Economy and Society* 31, no. 2: 218–233.

Mirowski, P. 1989. *More Heat Than Light*. Cambridge University Press.

Mirowski, P. 1999. Cyborg agonistes. *Social Studies of Science* 29, no. 5: 685–718.

Mirowski, P. 2002. *Machine Dreams: Economics Becomes a Cyborg Science.* Cambridge University Press.

Mirowski, P. 2004. The scientific dimensions of social thought and their distant echoes in 20th century American philosophy of science. *Studies in the History and Philosophy of Science A* 35, no. 2: 283–326.

Mirowski, P. 2007. Markets come to bits. *Journal of Economic Behavior and Organization* 63, no. 2: 209–242.

Mirowski, P., and K. Lee. 2003. The Purest Form of Rationality Is That Which Is Imposed. Working paper, University of Notre Dame.

Mirowski, P., and Hands, D. eds. 2006. *Agreement on Demand.* Duke University Press.

Muniesa, F., and Callon, M. 2007. Economic experiments and the construction of markets. In *Do Economists Make Markets?* ed. D. MacKenzie et al. Princeton University Press.

Murray, J. 2002. *Wireless Nation.* Perseus.

Nalebuff, B., and J. Bulow. 1993a. Designing the PCS auction. Comments of Bell Atlantic. PP docket 93-253, FCC.

Nalebuff, B., and J. Bulow. 1993b. Response to PCS auction design proposals. Reply comments of Bell Atlantic. PP docket 93-253, FCC.

Nik-Khah, E. 2005. Designs on the Mechanism. Ph.D. thesis, University of Notre Dame.

North, Douglas. 1977. Markets and other allocation systems in history. *Journal of European Economic History* 6, no. 3: 703–716.

NTIA (National Telecommunications and Information Administration). 1994. Ex parte submission. PP docket 93-253, FCC.

Parkin, M. 1998. *Economics,* fourth edition. Addison-Wesley.

Pinch, T. 2007. The sociology of science and technology. In *21st Century Sociology,* ed. D. Clifton and D. Peck. Sage.

Plott, C. 1997. Laboratory experimental testbeds: Application to the PCS auction. *Journal of Economics and Management Strategy* 6, no. 3: 605–638.

Rosenbaum, Eckehard. 2000. What is a market? *Review of Social Economy* 58, no. 4: 455–482.

Ross, Don. 2005. *Economic Theory and Cognitive Science.* MIT Press.

Roth, A. 2002. Preface to "The Redesign of the Matching Market for American Physicians." http://www.bepress.com.

Sismondo, S. 2004. *An Introduction to Science and Technology Studies*. Blackwell.

Slater, D. 2002. From calculation to alienation: Disentangling economic abstractions. *Economy and Society* 31, no. 2: 234–249.

Skrzycki, C. 1993. FCC prepares to carve up new portable phone frontier. *Washington Post*, September 16.

Thelen, J. 1995. Milgrom's progress. *The Recorder*, May 8: S5.

U.S. Congress. 1993. Omnibus Budget Reconciliation Act of 1993.

Walker, D. 2001. A Factual account of the functioning of the 19th century Paris Bourse. *European Journal of the History of Economic Thought* 8, no. 2: 186–207.

Weber, R. 1993a. Comments on FCC 93-455. Comments of TDS. PP docket 93-253, FCC.

Weber, R. 1993b. Reply to comments on FCC 93-455. Reply comments of TDS. PP docket 93-253, FCC.

Weber, R. 1997. Making more from less: Strategic demand reduction in the FCC spectrum auctions. *Journal of Economics and Management Strategy* 6, no. 3: 529–548.

Winner, L. 1993. On opening the black box and finding it empty. *Science, Technology and Human Values* 18, no. 3: 362–378.

Woolgar, S., ed. 1988. *Knowledge and Reflexivity*. Sage.

Zammito, J. 2004. *A Nice Derangement of Epistemes*. University of Chicago Press.

II Infrastructure

4   The Finitist Accountant

David Hatherly, David Leung, and Donald MacKenzie

Sociology, suggests Peter Miller (2001), has forgotten accounting. It played a significant role in the classic analyses of the development of capitalism by Weber and Sombart (see Carruthers and Espeland 1991), but in modern sociology departments accounting is a research topic that is encountered only rarely. A sizeable body of accounting scholarship—represented above all in articles such as Burchell et al. 1980—has sought to build a bridge to sociology, but this enthusiasm has not been reciprocated widely.[1] Thus, Vollmer (2003: 353) notes, there is "not a single entry on accounting" in the index of the first edition of *The Handbook of Economic Sociology* (Smelser and Swedberg 1994). Collaboration between accountants and sociologists in sociology departments remains unusual: the joint work best known to a wider sociological audience is probably Miller's with Nikolas Rose (e.g., Rose and Miller 1992).

Among the theoretical resources from sociology drawn on by scholars in accounting is the work of Foucault, and the resource is certainly appropriate: accounting is indeed a "technology of government," a way of constructing the "accountable" and "calculable" person (Miller and O'Leary 1994: 99). In this chapter, however, we argue for the relevance of another resource: finitist sociology of knowledge. ("Finitism" is explained below.) We do so by examining a topic central to accounting, both as practice and as scholarship, that is of great interest from the viewpoint of economic sociology: financial reporting, in particular the "ethnoaccountancy" of profit (MacKenzie 1996, 2003a)—in other words, the processes of the construction of corporate earnings figures.

Among the antecedents to our argument in the literature of accounting are the delightful parable by Hines (1988) and the Baudrillard-inspired analysis by Macintosh, Shearer, Thornton, and Welker (2000).[2] Valuable though these studies are, they have not been followed on any large scale by ethnographic research on how financial reporting is conducted in practice. (There has been important work on, for example, the setting of accounting standards, but much less on how those standards bear upon practice, as the 2006 review by Cooper and Robson reveals.) The work of more "positivist," quantitative accounting researchers is also useful; we will draw on it below, but we will also argue that it can usefully be supplemented by investigations focused on issues suggested by finitism.

Scandals such as the bankruptcies of Enron and WorldCom help to show why financial reporting is of interest. The scandals can, however, be interpreted simply as examples of how companies lie (Elliott and Schroth 2002), an interpretation that leaves the normal, non-scandalous practice of financial reporting unexamined. Is there a clear-cut truth that corporations could tell if they chose, or if legal penalties were severe enough and enforcement strict enough? We argue that there is not. Although "profit" and "loss" are the central categories of a capitalist economy, they are not self-evident facts. "Profits" are, quite literally, constructed by accountants and by others, and discretion in the way they are constructed is ineliminable.

Financial reporting is at the heart of economic governance in what Giddens (1990) called "high modernity." Resources flow toward "profitable" activities and away from "loss-making" ones, with profound consequences for the behavior and the lives of those involved. An appealing analysis of the societies of high modernity, expressed most sharply by Porter (1995) but echoed also in the works of Giddens and others, is that in such societies quantification has replaced absent relations of interpersonal trust. Unable to trust people, we place our trust in numbers. A finitist perspective, however, suggests that beneath the phenomena rightly pointed to by this analysis is a profound difficulty. Quantification displaces, rather than solves, modernity's problem of distrust.

Finitism

Finitism is an account of meaning, which it views above all through the prism of classification. Finitism has philosophical roots—in particular in the work of Wittgenstein (see below) and of Hesse (especially Hesse 1974)—but the version of finitism drawn on here has been developed most fully by the sociologists of science Barry Barnes (1982) and David Bloor (1997).

At the root of finitism is an analysis of how people classify activities, items, and states of all kinds: objects, living beings, processes, circumstances, situations, and so on. Finitists argue that the terms used in such classifications do not have inherent meanings: there is no fixed division of the infinite universe of activities, items, and states of affairs into instances of X and instances of not-X. "Meaning is use," as the Wittgensteinian slogan reminds us,[3] and any term has been used only a finite number of times in the past (by an individual, or even by an entire culture). The finite set of past usages does not determine future usages: "We *decide* how to develop the analogy between the finite number of our existing examples of things and the indefinite number of things we shall encounter in the future." (Barnes, Bloor, and Henry 1996: 54)

Finitism goes beyond the assertion that meanings are conventions, because that assertion (a truism of social science) can be interpreted as compatible with the view that once conventions are "chosen" they "determine our subsequent taxonomic activity" (ibid.: 55). If this view were correct, financial reporting would be an unproblematic matter, but the view is not correct and accounting is not simple. In contrast to the view that conventions determine classifications, finitists argue that "the future applications of terms are open-ended." There is "no specification or template or algorithm fully formed in the present, capable of fixing the future correct use of [a] term, of distinguishing in advance all the things to which it will eventually be correctly applicable" (ibid.: 55).

All acts of classification are thus in principle defeasible. "[A] classification is applied to the next case by analogy with existing ones" (ibid.: 56), but analogies can always be contested, and we can always decide that past classifications have been wrong. Classificatory acts are interconnected. "In a collective, terms are applied by different individuals at different times in different contexts: the exemplary instances of proper applications of a term are collectively established." Acts of applying a term affect applications of other terms: "No system of classification is so many separate, independent pieces" (ibid.: 57–59).

Consider, for example, the term 'murder'. No matter how much effort we might devote to seeking to define 'murder', such effort would not on its own unequivocally "cut" the universe of all past and future killing into disjoint sets of "murders" and "non-murders." The finite number of killings that have so far been classified as "murders" (or as "non-murders") do not suffice to determine future acts of classification. Classifications of a killing as murder are defeasible and revisable, and such classifications intertwine with the applications of other terms: 'dead' (consider the breathing but "brain-dead"), 'person' (consider debates over abortion), 'responsible' (the "insanity" defense, for example), 'self-defense', 'mercy killing', 'manslaughter', and so on.

'Murder' is an instance in which the finitist case is intuitively plausible, but finitists argue that finitism holds for all terms: from everyday observational terms such as 'red' to mathematical terms such as 'polyhedron', 'edge', 'vertex', and 'face'. (For a classic, implicitly finitist analysis of mathematics, see Lakatos 1976.) Finitism applies not only to classification in the sense of the sorting of particulars into discrete categories—important though that always is—but also to measurement. As Lakatos shows, what can be at issue is not just whether a given three-dimensional structure is "really" a polyhedron, but also how many edges, vertices, and faces it has. The quantitative as well as the qualitative aspects of science and technology can be analyzed finitistically (MacKenzie 1981, 1990).

Barnes's and Bloor's finitism was an outgrowth of earlier work on the sociology of scientific knowledge such as Bloor's (1973, 1976) "strong program." Three issues about the sociology of scientific knowledge therefore must be addressed.

The first is a misconception that can easily be amplified by the finitist emphasis on classification: that "strong program" sociology of knowledge involves a view of knowing as a process separate from action, in other words a view that ignores "practice" (Pickering 1992, 1995). The accusation strikes us as false, but here is not the place to debate it at a general level. Let us emphasize simply that when we discuss the classification of economic transactions we do not have in mind transactions first taking place and then, later and separately, being classified. That can happen, but in large corporations accounting classification is normally part of economic action, not separate from it.

A second issue is whether finitist sociology of knowledge focuses too exclusively on human beings, giving insufficient weight to the interweaving of the human and the non-human. (See, for example, Bloor 1999 and Latour 1999.) Certainly, to ignore that interweaving in the case of accounting would be absurd. The inscriptions of accounting need to be durable and often portable too, which requires recording in materials more lasting and more easily transported than human bodies and brains: clay tablets, papyri, ledgers, and their modern electronic equivalents. Like all economic action, accounting is distributed cognition in the sense of Hutchins (1995a,b). Unaided human beings could not possibly do the accounts of a complex modern corporation, and the process would be hopelessly inefficient if conducted only with the pens and ledgers of the nineteenth century. However, that corporate accounting is now a highly automated, computerized process does not eliminate the need for classification. Accounting software can process transactions only if they are coded to indicate what kind of transaction they are. As a factual matter, coding is still a human province (though the use of artificial neural networks or similar systems is conceivable). Furthermore, no corporation simply presents its investors with a pile of print-outs from its accounting software: additional, highly consequential, human-initiated processing takes place.

The third issue concerning the sociology of knowledge is by far the most widespread misconception about it. Pervasive in the "science wars" initiated by critiques such as Gross and Levitt 1994, it is that strong-program and finitist sociology of knowledge views knowledge as a "mere" social construction unaffected by "reality." In fact, causal input from the material world has always played a part in strong-program sociology of knowledge (Bloor 1976: 31 and passim). Knowledge is shaped by the material world and by the biological characteristics of human beings, as well as by psychological and sociological processes: it is in this sense a co-construction, not reducible to social processes alone. As we shall see, finitism is wholly consistent with the view that there is an "economic reality" to corporations that is affected by accounting classifications but not constituted in its entirety by those classifications.

The most relevant connection between finitism and the philosophy of Wittgenstein is the latter's famous discussion of following a rule. If we think of a rule as a set of words—consider for example, the sixth commandment, "Thou shalt not kill"—then following a rule seems to involve an act of interpretation of what the words refer to or "mean." Does 'kill' include the killing of enemy soldiers, of non-combatants, of human fetuses, of terminally ill people who have expressed a wish to be helped to die, of animals (for purposes of experimentation), of animals (for food), and so on? Of course, one can write rules for interpretation, but if finitism is correct these rules themselves must be interpreted, and we are at the start of a potentially endless regress. If rules are simply verbal formulas, then, because the flexibility of interpretation can never be eliminated entirely, "no course of action could be determined by a rule, because every course of action can be made out to accord with the rule" (Wittgenstein 1967: 81e).

We all know, however, that invoking interpretative flexibility might not allow us in practice to "get away with murder." A crucial point of theoretical contestation arises here, separating the finitism of Barnes and Bloor from ethnomethodology, which is also an inheritor of Wittgenstein's finitism but which interprets its bearing on sociological inquiry quite differently. Space constraints prohibit anything approaching a full account,[4] but at issue is what to make of the way in which, as a finitist in the tradition of

Barnes and Bloor would put it, the logical open-endedness of the application of terms to particulars and the logical under-determination of behavior
by rules are foreclosed in practice.

Wittgenstein writes: "To obey a rule, to make a report, to give an order,
to play a game of chess, are *customs* (uses, institutions) ... there is a way of
grasping a rule which is *not* an *interpretation*, but which is exhibited in what
we call 'obeying the rule' and 'going against it' in actual cases.... When I
obey a rule, I do not choose. I obey the rule *blindly*." (1967: 81e, 85e) The
central difference between Barnes's and Bloor's finitism and ethnomethodology is Barnes's and Bloor's preparedness to invoke psychological and sociological processes to explain the foreclosure in practice of interpretive
flexibility. Ethnomethodology, in contrast, does not see it as a foreclosure,
and does not invoke social processes in an explanatory way (Lynch 1992;
Sharrock 2004).[5] In this respect, the analysis below follows not ethnomethodology but Bloor:

> According to meaning finitism, we create meaning as we move from case to case. We
> *could* take our concepts or rules anywhere, in any direction.... We are not prevented
> by "logic" or by "meanings" from doing this.... The real sources of constraint [are]
> our instincts, our biological nature, our sense experience, our interactions with other
> people, our immediate purposes, our training, our anticipation of and response to
> sanctions, and so on through the gamut of causes starting with the psychological
> and ending with the sociological. (Bloor 1997: 19–20)

Financial Reporting in Finitist Perspective

Such issues may seem far removed from the apparently mundane practice
of accounting, but that is not so, as can be seen by considering financial
reporting. Publicly held companies are obliged periodically to prepare financial statements and to have at least some of those statements certified
by auditors: in the United States, for example, public companies have to report quarterly and must have their annual reports audited. Such financial
statements now typically include (a) an "income statement" (in the United
Kingdom, a "profit and loss account"), which records, for the time period
in question, the corporation's revenues and the expenses incurred in generating those revenues; (b) a "statement of financial position" (in the United
Kingdom, a "balance sheet"), which records the corporation's assets and
liabilities at the end of the time period in question; and (c) a "cash flow
statement," which records cash paid and received by the corporation over
the time period in question. From the viewpoint of many analysts and
investors, the key is the first of these, the income statement, because it

defines "profit" or "earnings" (in other words the difference between revenues and expenses). The income statement has, however, to be reconciled with the balance sheet and cash flow statement, so these are important too. Furthermore, companies also have to report earnings to tax authorities, and in some jurisdictions (such as the United States and the United Kingdom) earnings as reported for tax purposes can, perfectly legally, differ considerably from earnings as reported to investors.

Producing income statements and other corporate accounts involves, above all, the classification of transactions, and this is the viewpoint from which the relevance of finitism is most clearly seen. Suppose a corporation buys something (an object, a building, a service, an employee's time, and so on). The resultant payment needs to be classified. Is it an expense (which must enter into the corporation's income statement and thus directly reduces its earnings or profits), or is it the purchase of an asset (to be recorded on its balance sheet and to affect income statements only in the form of changes in value or of depreciation: see below)? Buying a building, for example, might seem clearly the purchase of an asset, but what about renovating the building? What about painting it? What about the interest paid on money borrowed to buy it? Paying staff salaries may seem clearly to be an expense, but what if the staff involved are researchers? Is expenditure on research and development an expense or purchase of an asset, albeit perhaps an intangible one? To take a classificatory issue of another kind, suppose a corporation grants its managers options entitling them to purchase its stock at a set price in a given future time period. It is giving them a right that may be valuable, yet it has spent nothing. There has been fierce controversy in the United States over whether the value of the options should be calculated (for example using an option-pricing model) and classed as an expense, thus reducing corporate earnings.

To identify "assets" and "liabilities" involves classifying entities according to whether they are inside or outside an organization's boundaries. If a corporation's subsidiaries and "related entities" can be seen as separate from it, their liabilities are not its own (as Enron's accountants were aware). More generally, as Hines (1988: 258) points out, "financial accounting controversies are controversies about how to define the organization. For example what should 'assets' and 'liabilities' include/exclude; at what point does an asset/liability become so intangible/uncertain/unenforceable/ unidentifiable/non-severable, etc. that it ceases to be considered to be a 'part' of an organization? The answers to questions such as these, define the 'size,' 'health,' 'structure' and 'performance,' in other words the reality of an organization."

Current financial reporting is "accruals-based," which means that there is no necessary correspondence between a corporation's cash flow in a given time period and its revenues and expenses in that period. As one textbook puts it, a "naive, non-accountant" would imagine that Profit = Receipts of cash − Payments of cash (Perks 2004: 173–174). Instead, in current financial reporting, Profit = Revenues earned − Costs incurred in earning those revenues. But when is revenue earned? As Hines points out (1988: 253),

> Sometimes we recognize revenue when the goods are completed; sometimes when they are partly completed; sometimes when the customer is invoiced; or even when he telephones and places an order; or sometimes when he is billed; or when he pays. And even these are not clear-cut. When is a building "finished," for example? What percentage of a building … is "completed"? When does a customer "pay": when his cheque is received; when it is honoured?

There may be a risk that goods will be returned after the corresponding revenue has been "recognized," or that payment for them may not be received; in complex transactions (for example, involving financial derivatives[6]) the stream of payments may be contingent on the movement of asset prices or interest rates. So questions arise as to the provisions to be made for bad debts and other contingencies, and perhaps also how large a revenue to "book" (to enter into the income statement) in respect to a sale.

The need in modern financial reporting to match costs to the corresponding revenues makes the classification of costs by time period also an important matter. Suppose a corporation sells an item from its inventory of similar items. What is the corresponding cost? Is it the cost of the most recently produced such item ("last in, first out") or of the oldest such item ("first in, first out"), or a weighted average of costs? If prices are changing, the difference between the answers may be consequential. Advertising, to take another example, is often a major expenditure. Modern financial reporting requires the judgment whether advertising is generating sales in the current period (in which case it should be recorded as an expense in that period), or whether it will lead to sales in future time periods (in which case its recognition as an expense should be deferred to those time periods).

A further crucial issue is the valuation of a company's assets: should they be included at cost or at current "fair value"?[7] This can affect not only a company's balance sheet but also its income statement (and thus its earnings or profits). If "fair value" is chosen, the question obviously arises as to how value is to be determined. If, alternatively, assets are to be included at their cost (the traditional treatment), then that cost needs to be allocated across time periods in the form of a depreciation charge: it would not be regarded as reasonable to allocate the entire cost of an asset to the first year of its lifetime if it will remain useful for many years. How should depreciation be calculated if, for example, an airline buys a new aircraft? Will the aircraft have a useful life of 20, 30, or 40 years? Should it be regarded as depreciating by the same amount in each year, or by larger amounts in its early years?[8]

Accounting for Economic Reality

Many further contingencies affecting accounting classifications could be listed, but let us move on to what might structure such classifications. The purpose of financial reporting, it would widely be agreed, is to represent the economic situation of a corporation, for example so that its existing investors, its creditors, and other stakeholders can assess matters such as whether their money is being well used or whether they will be paid, and potential investors can decide whether or not to entrust the corporation with their capital. In other words, just as the physicist reports on physical reality, so the accountant reports on economic reality.

As emphasized above, the variety of finitism we advocate accepts both that "material reality" exists and that it affects our beliefs about it. Can the same be said of "economic reality"? Certainly, we must bear in mind the strong feedback from "report" to "reality." Financial reporting directly affects the economic health of corporations (a corporation that appears sound and profitable is attractive to investors and to lenders, but a bank that appears unsound is vulnerable to a bank run),[9] while astrophysicists' models of the nuclear reactions in stars do not seem to affect those reactions. Indeed, the basic category of all financial reporting—money—is the prototypical example of a "social-kind" term (Barnes 1983). A metal disk, a piece of paper, or an electronic record counts as money because we treat it as money and believe that it will continue to be treated as money.

Nevertheless, for all that economic reality is constitutively social, it is still reality—for example, a powerful constraint—from the viewpoint of the individual person or individual corporation. All readers will be aware that there are some purchases they cannot make, and some patterns of expenditure they could not sustain without increasing their income; and something similar holds for corporations. A corporation can become unable to meet its financial obligations just as an individual can, and financial reporting—however optimistic—may not prevent this happening. (Accounting scandals often take the form of the sudden bankruptcy of an apparently profitable corporation. The possibility of "apparent" profitability indicates the flexibility of financial reporting, but the possibility of bankruptcy suggests that economic reality is not shaped by accountants' reports alone.) It is not our task to attempt to define what "economic reality" means for a corporation—to do that is to do accounting, not to analyze it sociologically—but we entirely accept that there is a reality to the financial situation of corporations that is constituted only partially by accountants' reports on it.

That physicists report on a reality that is constituted only partially by their activities does not render a sociology of physics impossible (Collins 2004). One way of developing the sociology of scientific or technological knowledge is to identify local cultures and local traditions of the practice of science and technology (Barnes, Bloor, and Henry 1996: 26–31). Participants in each local culture may well believe—sometimes fervently—that it captures "reality" (or, in technology, the best way of doing something), but the sociologist often finds that local cultures differ in their practices, sometimes radically.

Accounting, too, has its local cultures and local traditions: the shared goal of capturing economic reality is insufficient on its own to determine the practice of accounting. The most obvious such traditions are national. Formal rules and standards are increasingly being harmonized across countries, but the process is still incomplete, and practice can still vary considerably even when rules are similar or identical. Thus the rules governing the depreciation of fixed assets were similar in the United Kingdom and France, but the typical implementation of the rules was quite different (Walton, Haller, and Raffournier 2003: 23). The adoption in 2005 by the European Union, and increasingly by other countries as well, of common rules and principles—the "International Financial Reporting Standards" laid down by the International Accounting Standards Board (2004)—has not eliminated variation. One survey found that "a company's country of domicile, and its previous national accounting standards appear to have the greatest influence on the choices it makes" (KPMG 2006: foreword).

As a result of national differences, when the assets and profits of a corporation are calculated according to the practices of more than one country, the resultant figures can differ considerably. In 1993, for example, Daimler-Benz AG was listed on the New York Stock Exchange, and until 1996 (when it began to use exclusively US rules) it prepared two sets of accounts, US and German. The value of its shareholders' equity (the difference between the valuations of Daimler-Benz's assets and its liabilities) was 40–45 percent higher in its US accounts. Its earnings also differed, and the most dramatic difference (in 1993, Daimler-Benz's German accounts showed profits of 615 million DM, while its US accounts recorded a loss equivalent to 1,839 million DM) seems to have been caused mainly by revaluations designed to reduce discrepancies in asset values (Bay and Bruns 2003: 397–399).

Sometimes national differences in accounting practices arise for relatively clear reasons. In Germany, Austria, and Italy, for example, the earnings figures that determine corporate taxes must closely follow earnings as reported to investors. In the United Kingdom and the United States, in contrast, tax accounting and financial reporting are, as noted above, largely divorced. This is one reason for financial reporting typically having been more conservative in jurisdictions such as Germany and Austria than in the United States and the United Kingdom: an optimistic presentation would attract higher taxes. Assets may, for example, be assumed to depreciate more rapidly, and provisions for bad debts and other contingencies may be larger (Walton, Haller, and Raffournier 2003).

Members of different local cultures of financial reporting may indeed believe strongly that their practices best capture reality. For example, a Continental European corporation's accountants may have felt they were taking proper account of a rapidly changing and uncertain world, when to their British and American counterparts they seemed to be salting away large undeclared profits. That such convictions can be passionate means that the recent harmonization of international accounting standards across the European Union and the ongoing harmonization between Europe and the United States have been fraught. Such harmonization is intended to make it easier for global investors to compare corporations that report in different jurisdictions, but the person most central to these efforts forecast "blood all over the streets" as they came to fruition (Sir David Tweedie, quoted by Tricks and Hargreaves 2004).

Particularly controversial was International Accounting Standard 39 (IAS 39), which governs the valuation of financial instruments such as derivatives.[10] The most salient issue was the bearing of the standard on situations in which derivatives are used to hedge a risk, for example when a bank offers its customers fixed-rate mortgages or guaranteed interest rates on their deposits, and uses derivatives to offset its consequent exposure to changes in interest rates. Banks typically take the view that in such situations the economically realistic accounting treatment is "hedge accounting," in which fluctuations in the market value of hedging instruments are not recognized in their balance sheets and income statements, on the grounds that those gains and losses are offset by fluctuations in the value of the items being hedged.[11] Opponents of IAS 39 argue that its rules governing the permissibility of hedge accounting are too restrictive, for example in failing adequately to take into account the way in which banks hedge risks such as interest-rate exposure in aggregate, not item by item. The danger, they argue, is that what in economic reality are risk-reducing hedging transactions will be made to appear risky by injecting spurious, artificial volatility into their earnings. In 2004, concerted lobbying by banks led the European Commission to endorse the standard only in part, a decision condemned sharply by the UK Accounting Standards Board, which was reported as suggesting that UK companies "should ignore it" (Tricks and Buck 2004).

It might be imagined that disputes over whether accounting rules reflect "economic reality" could be settled by turning to the acknowledged experts on the latter, economists. In fact, the small minority of economists who have taken research in accounting seriously have rarely been able decisively to settle the issues at stake. Perhaps the single most important question in financial reporting is the definition and measurement of "income" (or "earnings" or "profits"). The great British economist John Hicks provided what has become perhaps the canonical definition of "income,"[12] but he admitted it was not precise. Making it precise—in particular, separating income unequivocally from capital—might be "chasing a will-o'-the wisp," said Hicks. Economists, he wrote, "shall be well advised to eschew *income*." The concept was a "bad tool ... which break[s] in our hands" (Hicks 1946: 176–177). Accountants, however, were in no position to duck one of their central classificatory responsibilities. As Dennis Robertson put it, "the jails and workhouses of the world are filled with those who gave up as a bad job the admittedly difficult task of distinguishing between capital and income" (quoted in Kay 2004).

Rule-Governed Accounting?

If "economic reality," important though it is, does not suffice to determine how accountants should classify it, perhaps convention does? In other words, perhaps the rules of accounting can be made tight enough to eliminate local variation and discretion. It might not be possible to prove that those rules capture economic reality optimally, but consistency might be achieved. For example, investors would know that all corporations in a given country—and perhaps eventually all corporations in the world—were doing their accounting in an identical, comparable way.

The extent of formal, written rules of accounting varies with time and place, and there has sometimes been strong opposition among accountants to such rules. In Britain (and perhaps especially in Scotland, the original home of an organized profession of accounting) there has often been a conviction that the requirement to capture economic reality—to give "a true and fair view" of the financial situation of companies, as the United Kingdom's Companies Acts require[13]—necessitates "a custom-built document" requiring "the exercise of an informed judgment" with which others, even accountants' own organizations, should not "interfere" (Slimmings 1981: 14). Such a perspective emphasizes professional status: one of the hallmarks of a "professional"—as distinct, say, from a "clerk"—has been taken to be the exercise of "judgment" (Porter 1995: 91).

Nevertheless, until very recently the direction of change, driven above all by accounting scandals, has been toward rules. After the Great Crash of 1929, the economic reality of many US corporations was seen to have been at variance with their financial reporting. In part to ward off government intervention (possibly even compulsory government auditing of corporate accounts), the American Institute of Accountants made at least a symbolic sacrifice of some of the accountant's individual discretion, and began to promulgate formal accounting standards (Zeff 1984). The effort did indeed help to keep accountants in charge of formulating standards— in 1938, the Securities and Exchange Commission delegated its standard-setting powers to the Institute's Committee on Accounting Procedure— but it marked the beginning of a proliferation of formal standards. The episode serves as one of Porter's prime examples of the rise of the "ideal of mechanical objectivity, knowledge based completely on explicit rules" (Porter 1995: 7), and rightly so. The six brief "rules or principles" formulated by the American Institute of Accountants in 1934 have by 2007 become the Financial Accounting Standards Board's 159 standards, some of which exceed 100 pages.[14]

To what extent, though, do these extensive, detailed formal rules govern the practice of financial reporting in the United States? It would be pleasing at this point to be able to cite an extensive corpus of ethnographic literature on the processes of financial reporting—in the United States or elsewhere—but no such corpus exists. An extensive literature search by the second author revealed only a very limited number of such studies.[15] Fortunately, however, there is a considerable body of quantitative research that enables the question to be addressed indirectly. This research concerns the practice of "earnings management," influentially defined by Schipper (1989: 92) as "purposeful intervention in the external financial reporting process, with the intent of obtaining some private gain (as opposed to, say, merely facilitating the neutral operation of the process)." Schipper's definition does not say so explicitly, but earnings management is normally taken to be permissible, legal forms of this intervention. "Fabricating invoices to create fictitious sales revenue" (ibid.: 93) is seen as fraud, not as earnings management.

The relevance of earnings management is thus that it is behavior "within the rules": its prevalence is an indicator of the extent to which discretion can still be exercised even when, as in the United States, the financial reporting process is governed by extensive, formal rules. What "private gain" might induce managements to engage in earnings management? Probably most important is what appears to be a widespread belief among corporate managers that stock analysts and investors prefer corporations whose earnings rise predictably to corporations whose earnings fluctuate substantially (even if around the same underlying trend). If the rewards enjoyed by corporate senior managers reflect stock prices, as in recent decades they increasingly have, there is an incentive for "income smoothing," in other words for exploiting permissible discretion to reduce the volatility of earnings. Clearly, too, there is usually, though not always (see below re "Big baths") an incentive to avoid reporting losses, and it is often very important to meet or to surpass stock analysts' predictions of corporate earnings.

It is not productive to try to detect income smoothing or other forms of earnings management by comparing reported earnings with "true," unsmoothed income or unmanaged earnings: even if the necessary detailed data were available, which they are not, the measurement of "true" income by an accounting researcher would be no less contestable than management's original figures. Instead, research on earnings management employs a variety of less direct methodologies. None are beyond criticism, but their results suggest that, despite attempts to make US financial reporting rulebound, significant discretion remains.

One approach is to identify situations in which there is a clear but temporary incentive to manage earnings; to scrutinize corporate accounts for patterns consistent with earnings management; and to examine whether those patterns correlate with situations in which the incentive is present. A pioneering study of this kind was Jennifer Jones's (1991) examination of the financial reporting of firms in industries that were petitioning the US International Trade Commission to recommend tariffs and import restrictions. Such petitions stood a chance of being granted only if there was evidence that domestic industry was being "hurt" by overseas competition. Jones focused on accruals: balance-sheet changes for which there is no immediate cash-flow counterpart such as depreciation, changes in the valuation of property, plant, and inventory, and estimates of accounts payable and receivable. She estimated the discretionary component of such accruals by subtracting from total accruals a regression-based estimate of "normal," "non-discretionary" accruals. Aggregating results for five industries (automobiles, carbon steel, stainless and alloy tool steel, copper, footwear), she showed statistically significant negative discretionary accruals in the years of International Trade Commission investigations.

Initial public offerings (IPOs) of stock are another case in which there are temporarily strong incentives to "window-dress" accounts (in this case to portray financial strength). A comparison of the "unexpected" accruals of companies engaged in IPOs with a matched control group of similar companies found that 62 percent of the IPO firms had higher accruals than the corresponding control (Teoh, Wong, and Rao 1998: 187, table 3). Since chance processes would suggest 50 percent, "this implies that roughly 12 percent of the issuing firms manage earnings" (Healy and Wahlen 1999: 373).

A different approach to the detection of earnings management is to examine the statistical distribution of earnings, looking for discontinuities or "kinks" at earnings levels that correspond to particularly strong incentives to earnings management: zero earnings (and thus the divide between making a profit and registering a loss), earnings in the previous year or corresponding previous quarter, and corporations' or analysts' earnings predictions. Such kinks turn out to be substantial (figure 4.1). For instance, analysis of US corporate earnings for 1976–1994 suggests that "30% to 44% of the firms with slightly negative pre-managed earnings exercise discretion to report positive earnings" (Burgstahler and Dichev 1997: 124).

The detection of earnings management abounds with conceptual and methodological difficulties (McNichols 2000). Analyses based on "discretionary" or "unexpected" accruals are extremely sensitive to the model of "normal," non-discretionary accruals that is employed (if, for example, earnings management is widespread, "normal" levels of accruals may already reflect such management) and they cannot detect techniques of earnings management that do not involve accruals. Nor are distributional analyses unequivocal. A distributional "kink" is not in itself evidence of earnings management. It may be, for example, that anticipated small losses are turned into small profits not by changes in accounting classifications but by "real" interventions (sales drives, cuts in expenditure on maintenance or on research and development, and so on). Burgstahler and Dichev (1997) attempt to overcome the problem by investigating cash flow from operations and levels of accruals around reference points such as zero

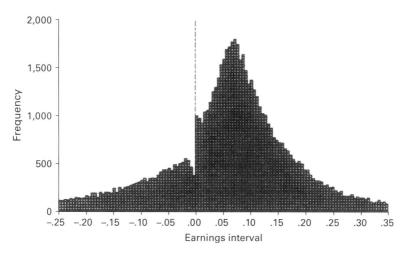

Figure 4.1

Frequency distribution of reports of net annual income by U.S. corporations, 1976–1994. Reprinted from David Burgstahler and Ilia Dichev, "Earnings management to avoid earnings decreases and losses," *Journal of Accounting and Economics* 24: 99–126, copyright 1997, with permission from Elsevier. Burgstahler and Dichev draw their data from Computstat, and the population of corporations excludes the financial sector and regulated industries. Income is scaled by division by the corporation's market value at the start of the year; interval width is 0.005. The dashed line is the zero-earnings point.

earnings, but this kind of analysis may not be entirely robust (Dechow, Richardson, and Tuna 2003). There are even potential issues of reflexivity. Some sophisticated investors are already employing academic earnings-management detection models (Henry 2004), and it would be surprising if regulators were not doing so too, so there is now an incentive to manage earnings in ways that the models cannot detect.

Nevertheless, the overall thrust of the literature on earnings management is consistent with pervasive anecdotal evidence (most authoritative is Levitt 1998) of widespread earnings management by US corporations, at least in the 1980s and the 1990s. (Empirical study of periods prior to the 1980s is too sparse to allow any definitive conclusions. Because there is a lag in the availability of the necessary data—especially the "Compustat" corporate financial reporting databases—and in analysis, one also cannot be sure of the situation in very recent years. Henry (2004) suggests that earnings management is still widespread, but his conclusion has yet to be tested by extensive published quantitative work.) The scandals of the early 2000s appear not to be isolated instances of rule-breaking, but (extreme) manifestations of the widespread exercise of deliberate discretion. As a finitist analysis would predict, the construction of the world's most rule-intensive system of financial reporting did not eliminate discretion from corporate accounting.

Conclusion

Corporate financial reporting is a major gap in sociological understandings of contemporary economic processes. It is a crucial aspect of those processes, but is almost *terra incognita* from the viewpoint of research by sociologists in sociology departments. Even the sociologically oriented literature in accounting, which is rich on topics such as professionalization, management accounting, and the regulation of accounting, is sparser in regard to the practices underpinning financial reporting: "One of the disappointing characteristics of field studies in organizations is how few have examined how accounting and audit decisions are made." (Cooper and Robson 2006: 435)

How might the gap be filled? As Cooper and Robson imply, the most attractive form of research is (for reasons we outline below) ethnographies of corporate financial reporting. By this we mean observational studies of the material processes (human and technological) of classification and measurement by which economic transactions are constructed and processed into audited corporate accounts, and of the ways these procedures and their outcomes are shaped by the contingencies that finitism points to: "our interactions with other people, our immediate purposes, our training, our anticipation of and response to sanctions," and so on (Bloor, as quoted above).

Among the reasons ethnography is likely to be necessary is the probable central role of "interactions with other people": for example, interactions among and between bookkeepers and accountants, between finance-department staff and other employees (including senior managers), and between a firm's accountants and its auditors. It is hard to envisage gaining access to such interactions in any detail other than ethnographically. Almost certainly, though, the reason ethnographic studies are sparse is difficulty of access and the sensitivity of the data that would be collected: the ethnographer could, for instance, easily become privy to information relevant to a corporation's stock price that is not publicly known. Nevertheless, under some circumstances (for example, with a legally binding agreement governing disclosure) access can be negotiated. Thus the third author has completed a small-scale pilot study (MacKenzie 2008), although it was not fully ethnographic: it was conducted *in situ*, but was interview-based. Leung (forthcoming) has been able to conduct a full-scale, fully ethnographic study, involving seven months of participant observation. Of course, it is likely that the findings of any particular ethnographic study will be to a degree specific to the site being studied, so many more such studies are needed before any more general picture can be painted with confidence.

We conjecture that in at least one respect the findings of such work will have an emphasis that differs from those of the classic formulations of finitism, highlighting a factor that is not explicit in Bloor's list of "sources of constraint," quoted above: technology. The technical systems of accounting—which can be stand-alone systems, but which increasingly, at least for larger firms, are aspects of continuously evolving Enterprise Resource Planning systems such as Oracle or SAP (Quattrone and Hopper 2006)—are neither merely neutral media nor simply means of increasing the efficiency of what unaided human beings might do. Multiple people are needed to do the accounts of any large entity, and technical systems link their work in *structured* ways. In part to reduce opportunities for fraud, accounting systems are, for example, normally designed to constrain the ways in which any given person can alter the results of the work of another (or indeed to prevent any other than a limited set of people making such alterations), and to leave an ineliminable trace when an alteration is made.

Such deliberate "technical" barriers then become "social" constraints. The vast bulk of accounting classifications are made not by senior managers or professional accountants, but by staff lower in corporate hierarchies such as bookkeepers. Technical systems then "solidify" those classifications by restricting what more senior staff can then do to alter them. This matters, we hypothesize, because it is normally senior staff (not those lower in the hierarchy) whose remuneration is most affected by stock prices and whose interests might thus be served by earnings management. Is a senior manager going to attempt to influence in advance thousands or millions of "primary" classifications made by dozens or hundreds of bookkeepers, or subsequently to attempt to alter those classifications (especially if he or she does not have the access permissions to do so and if the alterations leave a visible audit trail)? The record of cases such as Enron or WorldCom suggests that these are not the most attractive paths. Earnings management seems more often to be accomplished by leaving the primary classifications mostly or entirely intact, and performing higher-level reclassifications.

Of course, technical constraint is never absolute. Any system's controls can be subverted if the technical staff in charge of a system's access-control matrix can be persuaded to alter it, or if others can be persuaded to disclose their passwords. Furthermore, the creation of an audit trail is a deterrent only if an auditor is likely to scrutinize it and to contest the reclassifications it reveals, which returns us to the sphere of human constraints, of Bloor's "interactions with other people" and "anticipation of and response to sanctions."

Nevertheless, such considerations do suggest that ethnographies of financial reporting will need to pay close attention to its technological bases. They also point to the importance of studying the work of bookkeepers as well as of professional accountants. If our conjectures are correct, it is bookkeepers—not accountants—who produce much of accounting's equivalent of science's observational base. Bookkeepers' classifications are just as open to finitist analysis as those of accountants (and there is no "theory-independent observation language" in accounting or in science), but their classificatory work has almost never been examined in observational detail.[16]

Another issue worth attention *is* present explicitly on Bloor's list: training. Finitism suggests that classification and concept-application are based on relations of similarity and difference that, ultimately, are learned ostensively—that is, by exposure to authoritative examples of "correct" classification and "appropriate" concept application. For instance, a scientific paradigm is at root a set of concrete, exemplary solutions to scientific problems. Scientific training consists in good part of learning of how to perform these exemplary solutions and how to extend them to similar cases (Kuhn 1970; Barnes 1982). Socialization into the "paradigm" in the broader sense of an overarching disciplinary framework is not achieved solely by the framework being learned explicitly (if finitism is right, it could not be learned in its entirety in this way), but by repeated, authoritative ostension.

Accounting and bookkeeping, we conjecture, are also learned in good part ostensively.[17] Accountants do learn many explicit rules, but they also have to learn how to apply these rules to particulars. Some of this training takes place in formal educational settings; much of it takes place "on the job." If training in accounting is like scientific training, we would expect it to consist largely in repeated experience of solving problems for which there are authoritative "right" and "wrong" answers. If the analogy with science holds, the result of such training will go beyond technical competence in any narrow sense. It will be found to be socialization into a way of viewing the world that is not wholly explicit, but is not for that reason any less powerful. That it is unlikely to be entirely explicit is another reason why ethnographic (rather than, for example, interview-based) research is needed. Those who practice the technical cultures of bookkeeping and accounting may simply be unable to give a full verbal account of what they do and why they do it.

In particular, prolonged ostensively based socialization can "achieve realism." To neophytes, we conjecture, classifying items (in accounting, science, or elsewhere) will frequently "feel" like classification: here is an item; here are possible classifications $(X_1, X_2, X_3, \ldots, X_n)$; which shall I choose? The experienced practitioner, in contrast, will often feel "this item *is* an X_3," just as the experienced bird-watcher glances at a bird and thinks "that is an oystercatcher," not "I am classifying that bird as an oystercatcher." The classification can still be analyzed as a choice (or so finitism insists), but to those involved it no longer feels like a choice, or indeed even as a classification. Again, this points to a likely limitation of studies based solely on methods such as interviewing that rely on participants' own formulations. It is, for example, much more straightforward to interview accountants about situations in which they know they exercise "judgment" than about classifications they make without conscious reflection.

The extent to which accounting and bookkeeping are in practice conducted in nominalist "choice among classifications" mode or in "realist" mode is an empirical question. Our conjecture is that both modes will normally be present. Routine, familiar items, for example, may evoke "realist" mode; unfamiliar items provoke explicit choices. An accountant engaged in earnings management can be expected to operate in "choice" mode, but will also need to take into account the classificatory impulses of those in "realist" mode. Thus the crisis at WorldCom was triggered by an internal auditor refusing to accept that the costs of the unused proportions of network capacity leased from other firms could properly be classified as purchase of an asset. There is no evidence of strategic intent on her part: she seems simply to have felt strongly that the classification was wrong, and that these costs *were* expenses, not assets (see MacKenzie forthcoming).

Another issue worthy of more research is the audiences for corporate accounts and the way in which their anticipated reactions feed back into the production of accounts. Auditors are of course an audience whose likely response is of crucial, immediate importance, but others include stock analysts, investors, and tax authorities. Do such stakeholders treat reported earnings as "facts," or do they—as efficient-market theory would suggest in regard to investors—anticipate and discount earnings management? The quantitative evidence on the point is ambiguous. For example, the earnings management literature has produced "compelling evidence" (Healy and Wahlen 1999: 372) that it is common for banks to manage their earnings via adjustments to provisions for losses on loans, but the record of banks' stock returns is consistent with investors discounting "abnormally low loan loss provisions." On the other hand, earnings management prior to IPOs and other equity offerings appears to be successful in achieving high stock valuations that are reversed only later (Healy and Wahlen 1999: 374).

Again, though, this quantitative research is often not definitive: there are, for example, other explanations of poor post-IPO stock returns. We would conjecture that qualitative work (ethnographic or interview-based) would find a deep tension in the reception of earnings figures by stock analysts and professional investors. On the one hand, such actors seem to orient their activities to a large extent around accounting data. Analysts devote considerable attention to "forecasting" earnings, and the ratio of a corporation's stock price to its earnings is the single most widely used investment metric. On the other hand, it seems to us inconceivable that analysts and sophisticated investors in the United States were not aware of the substantial elements of discretion in the construction of accounting data well before it was highlighted by the collapse of Enron. Earnings management was front-page news in the *Wall Street Journal* as early as 1994 (Smith et al. 1994). A March 1997 *Fortune* article discussed specific earnings management practices it claimed were used by named corporations (Fox 1997). By 1998, "accounting hocus-pocus" and the colloquial terms for its techniques—heavy "big bath" losses, attributed to corporate restructuring, which create "cookie jar reserves" to boost future earnings, and so on— were prominent in a widely reported speech by the chair of the Securities and Exchange Commission (Levitt 1998).

A productive focus of research would therefore be on how actors conceive of and treat figures that are both enormously consequential and also, we anticipate, known to be subject to substantial discretion. One of the very few relevant existing sociological analyses is a fascinating study by Zorn (n.d.). He shows that while in the 1980s it was typical for around half of US corporations to meet or beat stock analysts' consensus forecasts of their earnings,[18] that figure climbed sharply in the 1990s (figure 4.2). Zorn finds a correlation with firms' internal structure: corporations in which the second-in-command was designated "Chief Operating Officer" were less likely to exceed analysts' forecasts than firms in which the second-in-command was a "Chief Financial Officer."

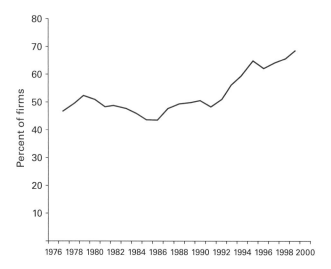

Figure 4.2
Three-year moving averages. Source: Zorn (n.d.). Reprinted courtesy of Dirk Zorn.

One probable reason for Zorn's findings is that earnings management became more active in the 1990s, particularly among firms whose growing "financialization" (Fligstein 2001) was indicated by a prominent role for the Chief Financial Officer. Testing that interpretation would require the quantitative techniques of the earnings management literature. Another, compatible, explanation would be that corporations (again, especially those with strong financial-market orientations) became increasingly successful at managing analysts' forecasts downwards. Perhaps, we conjecture, the price of continuing to receive the "nudges and winks" (Collingwood 2001: 73; Zorn n.d.: 38) from corporate executives that help an analyst to a career-enhancing, more-or-less precise earnings estimate was an implicit undertaking not to make one's estimate more accurate by second-guessing the—at times quite predictable, as Collingwood (2001: 70) suggests—extent to which the corporation would exceed the estimate. With sufficiently strong guarantees of anonymity, stock analysts might conceivably be prepared to talk about processes such as this.

Research of this kind will bear directly on broader questions of economic sociology. We anticipate that it will be found that the processes of financial reporting and of the reception of such reports are interwoven intimately with more general forms of economic life. Conservative reporting procedures in "stakeholder capitalism" in Germany, for example, seem to have reflected the protection of the interests of corporations' creditors (notably banks) and the exigencies of taxation, rather than prioritizing the kind of information that mobile international investors might seek. (Conservative accounting procedures facilitate the management of earnings, because generous provisions for contingencies can be unwound at appropriate moments. Zimmermann and Gontcharov (n.d.: i) suggest "substantial earnings management occurs in the whole sample" of German firms they study, although it must be admitted that their methodology is less convincing than those of the earnings-management studies discussed above.[19])

Another potential example of the interweaving—one that has been noted in the literature of economic sociology—is the possibility that a good part of the "mergers and acquisitions" movement in the United States and the United Kingdom in the 1980s can be explained by the way in which takeovers expand accounting discretion (Espeland and Hirsch 1990). Although there have been energetic regulatory efforts to curtail the advantages of "merger accountancy," a pessimistic accounting treatment of the situation and prospects of the acquired firm can be a substantial source of subsequent reported earnings by the acquiring corporation. This can be a powerful incentive for acquisitions, and thus an important factor fueling the "market for corporate control" that has helped transform Anglo-American economic life in the direction of "financialization."

As the "audit society" (Power 1999) and high modernity's "mathesis" (Townley 1995) spread, we expect questions analogous to those raised by corporate earnings management to become more salient in other spheres. For example, national accounts covering matters such as total public debt and fiscal deficits are of increasing salience (Suzuki n.d.). In Greece, for instance, the reported year 2000 deficit of 2 percent of gross domestic product was doubled to 4.1 percent in a revision following the election victory of the New Democracy—a level that would have denied Greece participation in European Monetary Union. At stake were issues familiar in corporate accounting: the valuation of pension fund surpluses, and whether to classify spending on defense equipment as capital investment (thus contributing to the deficit only as it depreciates) or as current expenditure (Munchau 2004). To take another example, the United Kingdom's Public Finance Initiatives, and their analogues overseas, are the analogue of many of the private sector's numerous "special purpose entities": they shift liabilities off balance sheets.

In short, sociology—not just economic sociology, but other areas too—cannot continue to forget accounting, and sociologists working in sociology departments need to build far stronger links to their colleagues in accounting departments. We hope that this chapter will contribute to those links by demonstrating the potential contribution of one particular sociological perspective, finitism. "Governing by numbers" (Miller 2001) requires that those who generate the numbers are themselves governable, but if finitism is right there is a sense in which their discretion is in principle ineliminable. This makes the resultant empirical questions—how discretion manifests itself, with what consequences, and the roles in constraining it of interactions with other people, of technological systems, of classificatory impulses, and so on—crucial topics for sociologists as well as for accountants. The finitist accountant is at the heart of modern economic life, and the material processes (human and technological) that enable and constrain her or his discretion deserve the most intense empirical investigation.

Acknowledgments

The research reported here was supported by a Professorial Fellowship (RES-051–27-0062) awarded to MacKenzie and associated PhD studentship awarded to Leung by the UK Economic and Social Research Council. For helpful comments on a previous version of this paper we are grateful to Alex Arthur, David Bloor, Anthony Hopwood, Colwyn Jones, Mike Power, Richard Swedberg, and Hendrik Vollmer.

Notes

1. For reviews of the sociologically oriented literature on accounting, see Vollmer 2003, Cooper and Robson 2006, and Napier 2006. Hopwood and Miller 1994 remains the most useful single collection of articles.

2. See also McGoun 1997, although it is less focused on accounting.

3. Effectively the only sociological article to draw on the earnings management literature is Zorn n.d.

4. Wittgenstein 1967: 14e and passim.

5. For a helpful introduction, see Kusch 2004 and subsequent articles in the August 2004 issue of *Social Studies of Science*.

6. One potential advantage of financial reporting as a topic for exploring these issues is the extent (with few parallels even in high modernity) its rules are explicit and can be pointed to—for example, by directing one's browser at www.fasb.org or by purchasing their international analogues, such as International Accounting Standards Board 2004. There is thus a proximate separation between the "rule" and the "practice" (q.v. Sharrock 2004: 604).

7. A derivative is a contract or security the value of which depends upon the price of an underlying asset or on the level of an index, exchange rate, interest rate, or other measure.

8. "Fair value" has been defined as "the amount for which an asset could be exchanged, or a liability settled, between knowledgeable, willing parties in an arm's length transaction" (International Accounting Standards Board 2004: 2169).

9. Similar issues arise in regard to the valuation of liabilities (e.g. should they be included at their original amount, or at current "fair value"?), which for reasons of brevity we ignore.

10. The latter is, of course, Merton's (1948) famous example of self-fulfilling prophecy. As Hines (1988: 256) indicates, it makes a huge difference whether a corporation's assets are valued on a "going concern" basis or on estimates of what they would fetch if the corporation were liquidated. The former is the standard practice; adoption of the latter would make many corporations seem no longer viable.

11. IAS 39 can be found in International Accounting Standards Board 2004: 1635–2003.

12. A typical objection is thus to an outcome in which the items being hedged (for example, a portfolio of fixed-rate mortgages) are not "marked-to-market," but the instrument used to hedge them has to be.

13. "[A] man's income [is] the maximum value which he can consume during a week, and still expect to be as well off at the end of the week as he was at the beginning" (Hicks 1946: 172). The difficulty of this definition lies in making precise what is meant by "as well off." As Hicks pointed out, that leads into issues such as future interest rates and prices, and the depreciation of durable goods. On the episode, see Hopwood and Bromwich 1984.

14. This phrasing, in force since 1947, replaced the earlier requirement (in the 1879 Companies Act) for "a true and correct view" (Myddelton 1995: 9).

15. www.fasb.org. For example, Standard 133 ("Accounting for Derivative Instruments and Hedging Activities") stretches over 212 pages.

16. See Leung 2004. Perhaps the most relevant such work from the viewpoint of this chapter is a pioneering study of auditing in the UK that combines survey data with six case studies of companies based on "matched interviews" with the company's finance director and the auditor's corresponding "engagement partner" (Beattie, Fearnley, and Brandt 2001: xvii).

17. The ethnomethodology-influenced corpus of workplace ethnographies offers the closest approach to the study of bookkeepers' classifications. See Suchman 1983; Button and Harper 1993.

18. Unfortunately the literature on the education of accountants (such as Power 1991 and Anderson-Gough, Grey, and Robson 1998) does not offer a clear-cut answer to the question of the extent of ostensive learning.

19. The original source of such "consensus estimates," and the database employed by Zorn, was I/B/E/S (Institutional Brokers Estimate System), which has provided such estimates since the early 1970s. By the late 1990s others such as First Call were also providing them (Fox 1997: 49).

20. Zimmermann and Gontcharov construct a "smoothing ratio," defined, following Pincus and Rajgupal (2002), as "the ratio of the standard deviation of non-discretionary earnings to the standard deviation of earnings" (Zimmermann and Gontcharov n.d.: 7), and they test for values that exceed 1.0 by statistically significant amounts. "Non-discretionary earnings" are "the sum of cash flow from operations and normal (non-discretionary) accruals from the Jones model [Jones 1991]" (Zimmermann and Gontcharov n.d.: 7).

References

Anderson-Gough, F., C. Grey, and K. Robson. 1998. *Making Up Accountants: The Organizational and Professional Socialization of Trainee Chartered Accountants*. Ashgate.

Barnes, B. 1982. *T. S. Kuhn and Social Science*. Macmillan.

Barnes, B. 1983. Social life as bootstrapped induction. *Sociology* 17: 524–545.

Barnes, B., D. Bloor, and J. Henry. 1996. *Scientific Knowledge: A Sociological Analysis*. Athlone and University of Chicago Press.

Bay, W., and H.-G. Bruns. 2003. Multinational companies and international capital markets. In *International Accounting*, ed. P. Walton et al. Thomson.

Beattie, V., S. Fearnley, and R. Brandt. 2001. *Behind Closed Doors: What Company Audit Is Really About*. Palgrave.

Bloor, D. 1973. Wittgenstein and Mannheim on the sociology of mathematics. *Studies in the History and Philosophy of Science* 4: 173–191.

Bloor, D. 1976. *Knowledge and Social Imagery*. Routledge.

Bloor, D. 1997. *Wittgenstein, Rules and Institutions*. Routledge.

Bloor, D. 1999. Anti-Latour. *Studies in the History and Philosophy of Science* 30: 81–112.

Burchell, S., C. Clubb, A. Hopwood, J. Hughes, and J. Nahapiet. 1980. The roles of accounting in organization and society. *Accounting, Organizations and Society* 5: 5–27.

Burgstahler, D., and I. Dichev. 1997. Earnings management to avoid earnings decreases and losses. *Journal of Accounting and Economics* 24: 99–126.

Carruthers, B., and W. Espeland. 1991. Accounting for rationality: Double-entry bookkeeping and the rhetoric of economic rationality. *American Journal of Sociology* 97: 31–69.

Collingwood, H. 2001. The earnings game: Everyone plays, nobody wins. *Harvard Business Review* 79, no. 6: 65–74.

Collins, H. 2004. *Gravity's Shadow: The Search for Gravitational Waves*. University of Chicago Press.

Dechow, P., S. Richardson, and I. Tuna. 2003. Why are earnings kinky? An examination of the earnings management explanation. *Review of Accounting Studies* 8: 355–384.

Elliot, A., and R. Schroth. 2002. *How Companies Lie: Why Enron Is Just the Tip of the Iceberg*. Brealey.

Espeland, W., and P. Hirsch. 1990. Ownership changes, accounting practice and the redefinition of the corporation. *Accounting, Organizations and Society* 15: 77–96.

Fligstein, N. 2001. *The Architecture of Markets*. Princeton University Press.

Fox, J. 1997. Learn to play the earnings game (and Wall Street will *love* you). *Fortune*, March: 48–52.

Giddens, A. 1990. *The Consequences of Modernity*. Polity.

Gross, P., and N. Levitt. 1994. *Higher Superstition: The Academic Left and Its Quarrels with Science*. Johns Hopkins University Press.

Healy, P., and J. Wahlen. 1999. A review of the earnings management literature and its implications for standard setting. *Accounting Horizons* 13: 365–383.

Henry, D. 2004. Fuzzy numbers. *Business Week* October 4: 50–54.

Hesse, M. 1974. *The Structure of Scientific Inference*. Macmillan.

Hicks, J. 1946. *Value and Capital: An Inquiry into Some Fundamental Principles of Economic Theory*. Clarendon.

Hines, R. 1988. Financial accounting: In communicating reality, we construct reality. *Accounting, Organizations and Society* 13: 251–261.

Hopwood, A., and M. Bromwich. 1984. Accounting research in the United Kingdom. In *European Contributions to Accounting Research*, ed. A. Hopwood and H. Schreuder. Amsterdam: Free University Press.

Hopwood, A., and P. Miller, eds. 1994. *Accounting as Social and Institutional Practice*. Cambridge University Press.

Hutchins, E. 1995a. *Cognition in the Wild*. MIT Press.

Hutchins, E. 1995b. How a cockpit remembers its speeds. *Cognitive Science* 19: 265–288.

International Accounting Standards Board. 2004. International Financial Reporting Standards (IFRSs).

Jones, J. 1991. Earnings management during import relief investigations. *Journal of Accounting Research* 29: 193–228.

Kay, J. 2004. Ignore the wisdom of accounting at your own risk. *Financial Times* September 7: 21.

KPMG. 2006. The Application of IFRS: Choices in Practice. www.kpmgifrg.com.

Kuhn, T. 1970. *The Structure of Scientific Revolutions*, second edition. University of Chicago Press.

Kusch, M. 2004. Rule-scepticism and the sociology of scientific knowledge: The Bloor-Lynch debate revisited. *Social Studies of Science* 34: 571–591.

Lakatos, I. 1976. *Proofs and Refutations: The Logic of Mathematical Discovery*. Cambridge University Press.

Latour, B. 1999. For David Bloor …, and beyond: A reply to David Bloor's "Anti-Latour." *Studies in the History and Philosophy of Science* 30: 113–129.

Leung, D. 2004. The Social Construction of Corporate Earnings: A Union of Isomorphism and Finitism. M.Sc. thesis, University of Edinburgh.

Leung, D. Forthcoming. Accounting in the Wild: Ethnoaccountancy of a Life Science Company. Ph.D. thesis, University of Edinburgh.

Levitt, A. 1998. The "Numbers Game." http://www.sec.gov.

Lynch, M. 1992. Extending Wittgenstein: The pivotal move from epistemology to the sociology of science. In *Science as Practice and Culture*, ed. A. Pickering. University of Chicago Press.

Macintosh, N., T. Shearer, D. Thornton, and M. Welker. 2000. Accounting as simulacrum and hyperreality: Perspectives on income and capital. *Accounting, Organizations and Society* 25: 13–50.

MacKenzie, D. 1981. *Statistics in Britain, 1865–1930: The Social Construction of Scientific Knowledge*. Edinburgh University Press.

MacKenzie, D. 1990. *Inventing Accuracy: A Historical Sociology of Nuclear Missile Guidance*. MIT Press.

MacKenzie, D. 1996. *Knowing Machines: Essays on Technical Change*. MIT Press.

MacKenzie, D. 2003a. Empty cookie jars. *London Review of Books* 25, no. 10: 6–9.

MacKenzie, D. 2008. Producing accounts: Finitism, technology and rule-following. In *Knowledge as Social Order: Rethinking the Sociology of Barry Barnes*, ed. M. Mazzoti. Ashgate.

MacKenzie, D. Forthcoming. *Material Markets: Facts, Economic Actors, and Politics*. Oxford University Press.

McGoun, E. 1997. Hyperreal finance. *Critical Perspectives on Accounting* 8: 97–122.

McNichols, M. 2000. Research design issues in earnings management studies. *Journal of Accounting and Public Policy* 19: 313–345.

Merton, R. 1948. The self-fulfilling prophecy. *Antioch Review* 8: 193–210.

Miller, P. 2001. Governing by numbers: Why calculative practices matter. *Social Research* 68: 379–396.

Miller, P., and T. O'Leary. 1994. Governing the calculable person. In *Accounting as Social and Institutional Practice*, ed. A. Hopwood and P. Miller. Cambridge University Press.

Munchau, W. 2004. National accounts disguise a blacker reality. *Financial Times*, October 18: 17.

Myddelton, D. 1995. *Accountants without Standards? Compulsion or Evolution in Company Accounting*. Institute of Economic Affairs.

Napier, C. 2006. Accounts of change: 30 years of historical accounting research. *Accounting, Organizations and Society* 31: 445–507.

Perks, R. 2004. *Financial Accounting for Non-Specialists*. McGraw-Hill.

Pickering, A., ed. 1992. *Science as Practice and Culture*. University of Chicago Press.

Pickering, A. 1995. *The Mangle of Practice: Time, Agency, and Science*. University of Chicago Press.

Pincus, M., and S. Rajgopal. 2002. The interaction between accrual management and hedging: evidence from oil and gas firms. *Accounting Review* 77: 127–160.

Porter, T. 1995. *Trust in Numbers: The Pursuit of Objectivity in Science and Public Life*. Princeton University Press.

Power, M. 1991. Educating accountants: Towards a critical ethnography. *Accounting, Organizations and Society* 16: 333–353.

Power, M. 1999. *The Audit Society: Rituals of Verification*. Oxford University Press.

Quattrone, P., and T. Hopper. 2006. What is IT? SAP, accounting, and visibility in a multinational organisation. *Information and Organization* 16: 212–250.

Rose, N., and P. Miller. 1992. Political power beyond the state: Problematics of government. *British Journal of Sociology* 43: 173–205.

Schipper, K. 1989. Commentary on earnings management. *Accounting Horizons* 3: 91–102.

Sharrock, W. 2004. No case to answer: A response to Martin Kusch's "Rule-Scepticism and the Sociology of Scientific Knowledge." *Social Studies of Science* 34: 603–614.

Slimmings, W. 1981. The Scottish contribution. In *British Accounting Standards: The First 10 Years*, ed. R. Leach and E. Stamp. Woodhead-Faulkner.

Smelser, N., and R. Swedberg, eds. 1994. *The Handbook of Economic Sociology*. Princeton University Press.

Smith, R., S. Lipin, and A. Kumar Naj. 1994. Managing profits. *Wall Street Journal Europe*, November 4–5.

Suzuki, T. n.d. Accounting for the Euro: Operationalisation of the Political Economy. Typescript, Saïd Business School, Oxford.

Teoh, S., T. Wong, and G. Rao. 1998. Are accruals during initial public offerings opportunistic? *Review of Accounting Studies* 3: 175–208.

Townley, B. 1995. Managing by numbers: Accounting, personnel management and the creation of a mathesis. *Critical Perspectives on Accounting* 6: 555–575.

Tricks, H., and T. Buck. 2004. ASB tells UK companies to ignore EU ruling on accounting standards. *Financial Times*, October 12.

Tricks, H., and D. Hargreaves. 2004. Accounting watchdog sees trouble. *Financial Times*, November 10.

Vollmer, H. 2003. Bookkeeping, accounting, calculative practice: The sociological suspense of calculation. *Critical Perspectives on Accounting* 3: 353–381.

Walton, P., A. Haller, and B. Raffournier, eds. 2003. *International Accounting*. Thomson.

Wittgenstein, L. 1967. *Philosophical Investigations*. Blackwell.

Zeff, S. 1984. Some junctures in the evolution of the process of establishing accounting principles in the U.S.A.: 1917–1972. *Accounting Review* 59: 447–468.

Zimmermann, J., and I. Gontcharov. n.d. Do Accounting Standards Influence the Level of Earnings Management? Evidence from Germany. Typescript, University of Bremen.

Zorn, D. n.d. No Surprise Anymore: Securities analysts' Forecasts and corporate Profit Reporting, 1981–2000. Typescript, Department of Sociology, Harvard University.

5 Global Financial Technologies: Scoping Systems That Raise the World

Karin Knorr Cetina and Barbara Grimpe

Imagine the trading floor of a large investment bank in one of the world's global financial cities. You may see between 200 (Zurich) and 800 (New York) traders engaged in stock, bond, and currency trading involving various trading techniques and instruments. Up to 20 percent of the traders will deal in foreign exchange at desks grouped together on the floors. Assume you are interested in this market.[1] With an average daily turnover of approximately US$3.2 trillion (Bank for International Settlements 2007) it is the world's largest market—and, insofar as foreign exchange trades inherently are cross-border transactions, it is also the most global market. The traders at these desks in inter-bank currency markets are not brokers who mediate deals but rather market makers. They take their own "positions" in the market in trying to gain from price differences while also offering trades to other market participants, thereby bringing liquidity to the market and sustaining it—if necessary, by trading against their own position. Foreign exchange deals made through these channels start at several hundred thousand dollars per transaction, going up to $100 million and more. The deals are made by investors, speculators, financial managers, central bankers, and others who want to profit from expected currency moves, or who need currencies to help them enter or exit transnational investments (e.g. in mergers and acquisitions). In doing deals, all traders on the floors have a range of technology at their disposal (see also Beunza and Stark, this volume; Zaloom 2003)—most conspicuous, the computer screens (as many as five) that display the market and serve to conduct trading. Traders' eyes are glued to these screens even when they talk or shout to each other. The market constitutes itself in these produced-and-analyzed displays.[2]

What do the screens show? The central feature of the screens and the centerpiece of the market for traders are the dealing prices displayed on the "electronic broker system" (EBS), a special screen and automated dealing service that sorts orders according to best bids and offers. It displays prices for currency pairs (mainly dollars against other currencies such as the Swiss franc or the euro), and deals being possible at these prices. Traders often deal through the electronic broker, which has largely replaced the "voice" (real-life) broker. The prices on the electronic broker influence the dealing prices traders offer to callers approaching them on another special screen, the "Reuters conversational dealing," through which they also trade. In the Reuters dealing system deals are concluded in and through bilateral "conversations" conducted on screen. These resemble email message exchanges for which Reuters dealing is also used in and between dealing conversations. On another screen, traders watch prices contributed by different banks worldwide; these prices are merely indicative, as they express interest rather than being dealing prices as such. Traders may also watch their own current position in the market (e.g. their being long or short on particular currencies), with the history of deals made over recent periods and their overall account balances (profits and losses over relevant periods) at their disposal on this or another workstation. Finally, the screens provide headline news, economic commentaries, and interpretations which traders watch. An important source of information which also appears on these screens, but is closer to traders' actual dealing in terms of the specificity, speed, and currentness of the information, are internal bulletin boards on which participants enter information (see also Bruegger 1999).

How can we conceptualize the assemblage of hardware, software, and information feeds that traders work with? The answer we focus on in this chapter challenges concepts of contemporary information technologies as implying network structures and network society notions (Castells 1996, 2003). The systems involved are scoping systems (Knorr Cetina 2003, 2005), a concept that takes seriously the reflexively projected reality the systems generate and the scopic coordinating mechanism this implies. We illustrate these systems by two examples: that of the trading and information systems used in the foreign exchange market (FOREXS[3]) and that of a debt management and financial analysis system (DMFAS) currently used by 66 countries and developed by a program of the United Nations Conference on Trade and Development. Both systems are global in reach and character. Finance appears to be a particularly fertile breeding ground for the development of global technologies, by which we mean technologies that assimilate and "scope up" national or local differences (DMFAS) or that simply bypass the political and social geographies of the world (FOREXS). The comparison of two technological systems will also allow us to say something about the architectures of global forms that correspond to these systems. Global, one-world, exclusive technologies, when they are live (see below), lead to the flow architecture of the financial market involved. Global assimilating technologies, on the other hand, struggle with similar difficulties as programs of national and cultural assimilation. While they must also be seen as the discursive and technological scaffolds on which global forms can "stand" and a global world can be raised, the global financial culture that emerges from them remains partial, contested, and fragmented.

What Are Scoping Systems? Distinguishing Scopes from Networks

The word 'scope', derived from the Greek 'scopein' (meaning to see), when combined with a qualifying notion, means an instrument for seeing or observing, such as a periscope. To explain what a scope-based mechanism implies, we distinguish it from a network. Networks suggest a very different mechanism of coordination. A network is an arrangement of nodes tied together by relationships which serve as conduits of communication, resources, and other coordinating instances that hold the arrangement together by passing between the nodes. Cooperation, strategic alliances, exchange, emotional bonds, kinship ties, "personal relations," and forms of grouping and translation can all be seen to work through ties and to instantiate sociality—or, in actor-network theory, "alliances"—in networks of relationships. But we should also think in terms of reflexive mechanisms of observation and projection, which the relational vocabulary does not capture.[4] Like an array of crystals acting as lenses that collect light and focus it on one point, such mechanisms collect and focus activities, interests, and events on one surface, from whence the result may then be projected again in different directions. When such a mechanism is in place, coordination and activities respond to the projected reality to which participants become oriented. The system acts as a centering and mediating device through which things pass and from which they flow forward. An ordinary observer who monitors events is an instrument for seeing. When such an ordinary observer constructs a textual or visual rendering of the observed and televises it to an audience, the audience may begin to react to the features of the reflected, represented reality rather than to the embodied, pre-reflexive occurrences.

In the foreign exchange markets we investigate in this paper, the reflexive mechanism and "projection plane" is the computer screen. Along with the screen come software and hardware systems that provide a vast range of observation, presentation, and interaction capabilities sustained by information and service provider firms. Given these affordances, the pre-reflexive reality is cut off and replaced; some of the mechanisms that we take for granted in a lifeworld, for example its performative transaction possibilities, have been integrated into the systems, while others have been replaced by specialized processes that feed the screen. The technical systems visually compile a lifeworld simultaneously projecting it. In the case of the foreign exchange market, they also "apresent" (meaning bring near; the term is adapted from Schutz and Luckmann 1973) and project layers of context and horizons that are out of reach in ordinary lifeworlds—they deliver not only transnational situations, but a global world spanning all major time zones. They do this from trading floors located in global cities (Sassen 2001), which serve as the support structures of the architecture of financial markets. Raised to a level of analytic abstraction, the configuration of screens, capabilities, and contents that traders in financial markets confront corresponds to a global scoping system (GSS). A GSS denotes a reflexive form of coordination that is non-hierarchical[5] in character and based on a comprehensive, aggregate view of things—the reflected and projected context and transaction system. This form of coordination contrasts with network forms of coordination which, according to the present terminology, are pre-reflexive in character—networks are embedded in territorial space, and they do not suggest the existence of reflexive mechanisms of projection that aggregate, contextualize, and augment the relational activities within new frameworks that are analytically relevant to understanding the continuation of activities. With the notion of a GSS, we offer a simplifying term for the constellation of technical, visual, and behavioral components packaged together on financial screens that deliver to participants a global world in which they can participate on a common platform, that of their shared computer screens. On a technological level, the GSS mechanism postulated requires that we understand as analytically relevant for a conception of financial markets not only electronic connections, but computer terminals and screens—the sorts of teletechnologies (Clough 2000: 3) that are conspicuously present on trading floors and the focus of participants' attention—as well as the trading floors themselves, where these screens cluster and through which markets pass.

Consider now the infrastructural side of these trading floors. All financial markets today are heavily dependent on electronic information and communication technologies. Some markets, for example the foreign exchange market that we investigate here, are entirely electronic markets. As over-the-counter markets of inter-bank trading, currency markets rely on electronic technologies that enable the dealer-to-dealer contacts and trading services across borders and continents. Reuters, Bloomberg, and Telerate connections wire together these markets, as do intranets that internally connect the trading room terminals and other facilities of particular banks and groups of banks in global cities. Reuters, Bloomberg, and Telerate are news providers and service providers. In the year 2001, Reuters had more than 300,000 terminals installed worldwide in all markets and facilities, and Bloomberg more than 150,000. Revenue from leases of their systems amounted to approximately $2.5 billion each at the end of 2001.[6] With the terminals come a sophisticated software, dealing and information systems, worksheet, email and customization capabilities, electronic brokerage and accounting services (see also Muniesa 2000), some of which—like EBS—have been developed by the banks themselves. The connections, the intricate and expensive hardware and software delivered by providers, and the banking institutions themselves constitute the material architecture of these financial markets.

How does this bear on the difference between a network form of coordination and the reflexive, scoping form of coordination discussed in this chapter? First, it will be obvious from the description thus far that the material infrastructure of financial markets includes much more than electronic networks, the cable and satellite connections between banks and continents. Above all, it includes the technological systems present on the trading floors in global cities that are the financial centers in the three major time zones: London, New York, Tokyo, Zurich, Singapore, and a few others (see Sassen 2001, chapter 7; Leyshon and Thrift 1997). The trading floors are the central locations of a global market that moves from time zone to time zone with the sun. The centerpieces of the interconnected floors are their federations of terminals that feature the sophisticated hardware and software capabilities discussed. When talking about the electronic infrastructure of financial markets, we should not lose sight of the hardware and software of the trading floors themselves and the terminal structures that "ready" these floors for trading (see also Beunza and Stark 2004; Zaloom 2003). Second, the electronic interconnections which are part of this federation and link the participating institutions are not simply coextensive with social networks through which transactions flow. As electronic networks, they correspond to different construction criteria, they involve electronic nodes and linkages irrelevant to social relationships, and much of what flows through them does not derive from social and financial relationships; examples are EBS deals, which are traders' responses to anonymous buying or selling offers provided by an automated electronic broker system. Third and most important, the terminals deliver much more than just windows to physically distant counterparties. In fact, they deliver the reality of financial markets—the referential whole to which "being in the market" refers, the ground on which traders step as they make their moves, the world which they literally share through their shared technologies and systems. The thickly layered screens laid out in front of traders provide the core of the market and most of the context. They come as close as one can get to delivering a stand-alone world that includes "everything" (see below) for its existence and continuation: at the center the actual dealing prices and incoming trading conversations, in a second circle the indicative prices, account information and some news (depending on the current market story), and further headlines and commentaries providing a third layer of information. It is this delivery of a world assembled and drawn together in ways that make sense and allow navigation and accounting which suggests the globally reflexive character of this form of coordination—and the scopic nature of traders' screens. The dealing and information systems on screen visually "collect" and present the market to all participants.

Scoping Debts: The DMFAS System

Let us now consider DMFAS, the debt management and financial analysis system developed and maintained by the "DMFAS Programme" of the United Nations Conference on Trade and Development.[7] The historical mandate of the Conference reflected in the DMFAS program is to give developing countries a voice in the global economy. Other goals also structured into the Conference's Program can be found in its private sector principles, manifest, for example, in the attempt to partly recover the cost of the IT development process for DMFAS from the countries that use the system,[8] and in "transparency" principles demanded by the International Monetary Fund and the World Bank as institutions "watching" and overseeing the global economic order. For example, in the wake of the "Asian Crisis" of 1997 the IMF demanded more financial and economic transparency—more specified data and statistics that countries needed to collect within shorter time periods and make available to themselves and global financial institutions to safeguard against the destructive impact of sudden financial breakdowns that spilled over into other countries and threatened the world economy. DMFAS can be seen as the attempt to implement such a system focused on national debt and the data needs of those debtor nations that experience the greatest payment difficulties—mostly developing countries. Though the system is older than the recent

IMF effort—the origins of DMFAS date back roughly 25 years[9]—it is part of earlier attempts to create an international financial architecture that takes into account the risks emanating from national financial situations by providing "technical assistance" to countries that need it. When the Conference on Trade and Development began to focus on this technical assistance in 1979, the first difficulty it found was a lack of information: "How much did the country owe? To which creditors? In what currencies? When were the payments falling due, and in which currencies? Who were the national debtors besides the central government? The idea of creating a computer based Debt Management System (originally called CBDMS) emerged very naturally from this experience." (DMFAS 1999/2000: 1)

DMFAS, then, is debt recording and financial analysis software that allows users to assemble and aggregate these and many other pieces of information.[10] It provides the possibility to record all data that arise in the "typical life cycle of a debt agreement."[11] The main financial instruments which DMFAS deals with are loans, bonds, and grants. For example, DMFAS records contract information and loan terms, real and estimated drawings, payment of principal, interest and commissions, etc. Based on tranche information for loans[12] or series information for bonds, the system automatically produces "amortization tables," the schedule for the repayment of principal and payment of interest to pay off a loan or a bond by maturity. Reference files also record information about the participants of a debt agreement, the concrete projects connected to it, their budget lines, exchange rate information, etc.

DMFAS registers debt obligations and details on the level of individual data and on an aggregate, analytic and strategic level. It allows higher-ranked debt management officers to carry out sophisticated debt management procedures at regular intervals. This part of DMFAS includes functions that aggregate data and information. It includes query modes that allow questions about records to be created and answered, forecasting (i.e., debt projection) functions, statistical bulletin functions, and forms that output the information in ways recognized by the World Bank, to which countries report their debt situation. The newest version of DMFAS (version 5.3) gives increased emphasis to capital markets instruments such as bonds. With the respective module, typical bond concepts such as yields, discounts, average prices, capitalized interests as well as bimonthly and weekly repayments can be handled. DMFAS 5.3 also includes functions that automatically download exchange rates from the Internet for reports in a currency different from the national currency or calculation methods for the Asian Development Fund needed by countries that are debtors of this fund, and other "improvements." By providing the respective menus, categories and calculation means, DMFAS serves as a tool for assembling together and formulating on one platform all the relevant parameters of a country's debt and loan situation.

Like FOREXS and its components, DMFAS is a scoping system, but one that operates on a national and global level. On a national level, it reflexively aggregates, contextualizes, and augments what loan officers and debt managers in debtor countries know and what individual files in national agencies contain. The very goal of the respective systems is to provide a constant mirror appropriately focused on a country's debt and financial situation. The mirror is also designed to provide a comprehensive picture according to criteria continually defined by global institutions. Thus the debt representations the system generates are oriented toward what world institutions want to know and can integrate with their own reports and country-specific information. In this sense, scoping national debt means "scoping it up" to the standards and requirements of world financial institutions.

Examples of such "scopic" features in DMFAS version 5.3 are the classification schemes for debt statistics being revised in 2004. As the software developers in Geneva state, these new classifications are intended to "significantly" simplify "compliance with international standards for the production of debt statistics" (DMFAS/UNCTAD 2004: 2). Closely related to this software redesign was the programming of a separate module for debt statistics that would help to build "clear and relevant tables" that would be "consistent with the latest international standards" (DMFAS/UNCTAD 2004: 4). Ethnographic data and document analysis suggest that these technical measures are linked to a global institutions' debt discourse much shaped by the World Bank, the IMF, and a few others, including UNCTAD. One important recurring theme in this discourse is the concern to monitor countries' debt situations in view of potential international financial crisis. The foreword of the "Debt Guide" (IMF 2003), a document various actors referred to during the fieldwork period in controversies about debt classifications, as if it was an arbiter of last resort, is illustrative in this regard. Prepared by eight world institutions and containing 300 pages of debt definitions and categories, this compilation is said to be based on the experiences of the "international financial crises in the late 1990s" and the consequent need for "the early detection of countries' external vulnerability" (IMF 2003: ix). The DMFAS handbook for the just mentioned statistics module takes up this discourse fragment by "bridging the gap between its [the Debt Guide's] recommendations and the actual production of the statistics," arguing that the "recent financial crises … exposed the lack of timely and reliable data" (DMFAS 2004: 4, 5). Similarly, during a statistics workshop of DMFAS software users in Buenos Aires in March 2005, a representative of the IMF statistics department coupled the concerns of IMF and World Bank and of Software providers like the DMFAS program in the following way (translated from Spanish): When the Debt Guide was prepared in 2003, for the first time the two "classic institutions" (IMF and World Bank) who "like discussing concepts or create normativities" would have "sat around a worktable" together with institutions like UNCTAD which would have been assuring that national debt managers could actually "implement the recommendations." Thus, discursively, financial crisis, surveillance needs, questions of classification and concrete DMFAS software functions are linked together, connecting the levels of global and national debt monitoring. To quote a final example, in software developer's practice, part of this discursive pattern reappeared during the testing phase of the new statistics module at the end of 2004, when one of the developers (DMFAS developer A) emphasized the need to change the "[system] specifications" in favor of a Debt Guide "definition," fitting the material reality to the ideal classification (the "system specifications" is a document written in half-technical language defining the programming needs):

DMFAS developer A: You look at all the documents (DMFAS developer B: mhm), when you say public—the definition in the Debt Guide (B: mhm), … is—"General Government" (B: mhm)—"Public Financial Companies," and "Public Non-financial Companies." So when you have "Public" and "Other Public" fo—it's a confusing concept (B: mhm). … I attempt to change this to "General Government." (B: mhm) … You see? … [I]t has been accepted in the, in our [system] specifications … but I— (B: mhm) know that the Debt Guide (B: yes) does not, have this, same definition. … You see that is what we don't want to propagate! What we want to propagate is an international standard! (B: mhm) And "Local Government" is part of "General Government."

DMFAS is thus scopic in the sense that it is a material component in an ongoing effort of global institutions to provide for debt surveillance. On closer examination, the discourse is not unanimous, and the actual self-monitoring by countries varies. Thus, in user practice the scoping mechanism is fragile and does not run smoothly. However, we maintain that, in a design sense, global scoping activities clearly exist in discursive patterns and in software developers' programming and reprogramming activities. They exist less clearly on the level of national user practice, though, as users do not always follow global institutions' monitoring intentions.

There are of course also important differences between FOREXS and DMFAS. First, the FOREXS technologies are live, performative systems in several senses. One is that the FOREXS systems are not recording technologies but provide platforms for trading. Prices on the electronic broker move according to algorithms that sequentially structure and match supply and demand. The systems thus enable the performance of markets and register the relevant moves by market participants. For example, no one needs to retrieve from elsewhere or "update" by hand the relevant price information. The temporal effect of this is that FOREXS provides automatic instantaneous updates of trading volumes and prices upon which participants can act in a global online market. Since the FOREXS market exists only in these systems, every transaction performed online will be instantaneously reflected in the online system. DMFAS, on the other hand, is not a live, performative technology within which debt-making occurs; rather, it is a system that makes possible the representation and assemblage of externally occurring activities and their outcomes on a national level. Accordingly, DMFAS requires debt officers in individual countries to collect and input the data the system requires. The time schedules built into DMFAS reflect external principal repayment and interest payment requirements involving periods of months and years rather than seconds. The smallest interval is the day, the period for which exchange rates are updated. However, the system does include functions that make it possible to obtain a picture of national debt and its associated features on demand: changes entered at the recording end of the data base will be semi-automatically processed at the reporting end, such that users can generate on demand accurate reports that contain the latest available information.[13] Second, DMFAS is a complex software in the eyes of its developers and users that often behaves "opaquely" and routinely needs debugging (there are now 54 patches for version 5.3—some are enhancements and some error corrections). "Helpdesks" and coaching are available to users, and personal missions to user countries by Geneva experts routinely complement the implementation of systems and updates. In contrast, FOREXS, though complex, runs smoothly from a user perspective. Updates, system improvements, and the occasional debugging of terminals are of course also necessary. But these are in the care of specialists that operate behind the backs of agents, preparing things such that implementations and changes can occur practically without user participation and interruption of market activities. Finally, the FOREXS systems package together hardware, software, exclusive intranet connections that link together a banking network of institutions, and news and information feeds. DMFAS users, on the other hand, are not structurally linked together by intranets. Though DMFAS may be integrated with other systems (e.g. the budget system) in domestic networks, no such integration exists on a transnational level. With FOREXS, firms such as Reuters, Bloomberg, and Telerate provide the content that traders watch on screen: these corporations "apresent" (bring near, the notion is adapted from Schutz 1955) the global financial world that provides the context for traders' transactions, and they enable, through their software and hardware development, online transactions that register automatically. In other words, the world within which traders act and which they watch on screen is outsourced to information providers that collect the information and feed the screen. DMFAS users, on the other hand, must not only provide the input for their systems themselves but also supplement this input with their own sources of information.

None of the aforementioned differences contradict the scopic character of both systems. But we can now be more specific about how scoping is achieved in both cases. With FOREXS, the scoping of the market is a continuous reflexive accomplishment that occurs through a mixture of apresentational, performative and temporal means: external events relevant to the markets are delivered into the markets and registered on their common platform nearly instantaneously, trading is performed within the systems in real time, and traders may communicate through the systems and reflexively register their own market prices and observations. The DMFAS software, on the other hand, provides a tool for the globally standardized representation of national debt-making and repayment activities that occur through dispersed, external paths; national debt officers must participate in the representational activities to create the mirror of debt that is desired on a global level. The temporalities of scoping are clearly different: scoping occurs in a reflex-like, instantaneous way in the case of FOREXS; with DMFAS, scoping is geared to measurable intervals—those at whose end points reports are due and data inputs must be updated. The length of these intervals is at least partially determined by IMF and World Bank schedules. For example, the data included in the "debtor reporting system" of the World Bank (published in *Global Development Finance*[14]) and the IMF's General Data Dissemination System must be updated annually, whereas countries reporting to the IMF's Special Data Dissemination System must provide quarterly updates. Yet both systems, FOREXS and DMFAS, create a response reality—they assemble and project a complete "state" of affairs to which participants and users react, taking the assembled reality as a starting point for further activities and considerations. Once a segment of reality becomes effectively scoped, that is assembled, augmented and projected within a technological medium for all participants, a shared reality—and to a degree shared understandings—are created that act as a mechanism of coordination. The audience becomes oriented to the features of the reflected reality, may learn to agree on some of its meanings, and starts to observe it and rely on it in the continuation of activities.

The aforementioned differences also suggest how globality is achieved in currency markets and debt reporting. FOREXS is a paradigm case of a homogeneous, "one world" system upon which a global social form can rest. DMFAS's globality, on the other hand, is more mediated and fragmentary. The global world, in this case, is continually problematic and must continually be negotiated and accomplished through strategic decisions and practices of assimilation and adaptation.

Two Global Histories

We have now linked the theme of this chapter, global financial technologies, to scoping systems, trying to tear them loose from the restrictive understanding of information technologies as simply implying social network structures. If we trace out the actual use of technologies in important areas of finance we come to an expanded conception of global coordination that provides a choice: "If you can scope it, you don't need to network it," one might say. Networks, we said, are embedded in territorial space, and they do not rely on reflexive mechanisms of aggregation and projection that have the potential of "upgrading" the system to a new level of organization. Networks can easily be concatenated into larger structures without fundamental change of its elements. Metaphors that cast globalization in terms of increased connectivity suggest such concatenations. But the global financial systems investigated here appear to rest on a different project: with scopes, it becomes possible to transpose activities embedded in dispersed contexts to shared global spheres. In the last section of this chapter we will emphasize that upgrading systems to a global level by means of scoping technologies need not eliminate the use of social networks for specific means. For example, it is plain and will be emphasized below that global scoping projects often "piggyback" on cable connections and other infrastructural linkages.

The point we want to make in this section is that both DMFAS and FOREXS were, from the beginning, global projects. "Globalization," in this case, has not been the effect of unrelated historical and evolutionary processes or of a world that grows together automatically based on independent technological developments. Instead, the global forms now in evidence in this area evolved from intentional business plans of a globally operating company (Reuters) in the case of FOREXS, and from economic policy concerns of global institutions and groups of countries in the case of DMFAS.

The recent history of foreign exchange markets starts at about the same time as the history of DMFAS. First the United States (1971), then major European countries, including Britain (by 1979), and finally Japan (in the early 1980s) abolished exchange controls, effectively eliminating the 1944 Bretton Woods Agreement of fixed exchange rates and allowing foreign exchange trading for purposes of speculation. Before the breakdown, foreign exchange markets also existed: foreign exchange deals are cross-border exchanges of currencies. Such exchanges were born with the dawn of international trade, and persisted through all ages. But in the 30 years of the Bretton Woods Agreement, foreign exchange deals reflected by and large the real requirements of companies and others that needed foreign exchange to settle bills and pay for goods. When exchange controls were removed, currency trading itself became possible as a market where exchange reflected price movement anticipation. In 1986 the dealing rooms of the world had taken off, with an average of US$150 billion and as much as $250 billion being traded around the globe, double the volume of five years before (Hamilton and Biggart 1993). As indicated before, in April 2004, according to the Bank of International Settlement's Triennial Survey, the average daily turnover in traditional global foreign exchange instruments had risen from $36.4 billion in 1974 to approximately $3.2 trillion (Bank for International Settlements 2007). Two-thirds of this volume derives from "over-the-counter transactions"—i.e., inter-dealer transactions in a global banking network of institutions. Banks had responded quickly to the business opportunities that arose with the freedom of capital that the breakdown of the Bretton Woods system initiated. They also responded to an increasing demand stimulated by volatile exchange and interest rates reflecting various crises (e.g. the energy crisis of 1974) and to the tremendous growth in pension fund and other institutional holdings that needed to be invested.

When exchange controls were removed in 1971, the current foreign exchange market was born. Traders, however, had no computers, and trading was a question of finding and negotiating this market, which lay hidden within geographical space. In a very real sense, dealing was a matter of establishing and using network connections. The most important technology on the trading floor at the time was the telephone. Besides the phone, there was the "ticker," a device which churned out "50 meters a day"[15] of news headlines and price pointers, as a former participant put it (for its specific history, see Preda, this volume), and calculating machines. Activities on the floor centered around "finding the market," that is finding out what the price of a currency was and who wanted to deal. In the following quote, a former chief of trading recalls how he continually chased after the market:

P: … so you had to constantly find out what the rates were in countries.
KK: And you did this by calling up banks?
P: By, yes. And there were also calls on the telex by other banks who either wanted to trade or wanted to know, simply wanted to know where dollar-Swiss was.

A partial attempt at making markets present in a scopic fashion occurred before the introduction of screens: the prices written down by hand on the "big sheets" to which P refers in the above quote were displayed on wall boards. When screens appeared, they were at first no more than substitutes for the "big sheets": displays on which the handwritten price sheets put together by female clerks were projected on the basis of pictures taken of the sheets on the floor. This form of presentation rested upon a chain of activities that was in important respects indistinguishable from the one that fetched prices in pre-screen times: it involved narrowing down where the market was by calling up or telexing banks, writing down the responses by hand (and perhaps recalculating prices in national currencies), and making this information available for internal purposes through a form of central, scopic presentation. All this changed when the British news provider Reuters developed a computerized foreign exchange system that it aptly called Monitor[16] and that became the basis for this electronic market (Read 1992, chapter 12). We cannot present here the history of Reuters's transformation into a financial service provider firm that Monitor embodied; but we can say that Reuters had perceived the uptake in volumes of trading as an opportunity to expand more strongly into the financial service area and take advantage of the large international infrastructure it had put together for the purpose of transmitting news between continents and countries.[17] In doing this Reuters built on an earlier technology that came between the big sheets and Monitor, but was used in the stock and commodity market. Reuters had provided commercial news and economic services to these markets for quite some time and was in fact making excellent profits from these services. In 1964, it teamed up with what we would now call an electronic start-up company, the Ultronic Systems Corporation of New Jersey, in a joint venture. Ultronic had developed a computerized stock quoting system called Stockmaster. The idea behind Stockmaster illustrates what the first electronic scoping technologies after the "big sheets" implied and how it replaced the ticker that had dominated before: Once it became widespread, the Ultronic system "absorbed" tickertape signals from all major stock exchanges and other markets by feeding them into master computers that processed the material and relayed it to subsidiary computers and finally to the offices of brokers. The system made more than 10 000 stock and commodity prices available to brokers on small desk units (originally teleprinters, later displays on screen) at the push of a button (see Muniesa 2000 for the development of a different electronic quotation system).[18] Instead of the local scoping that the big sheets provided, this was a form of global scoping.[19] Monitor's design was based on the experience with this technology, but it also included a crucial new concept. It was based on the notion of installing computer terminals in the trading units of banks on which traders were able to insert their foreign exchange rates into the system directly, thus making them available at the push of a button instantaneously to everyone connected to the system. In other words, Monitor was a reflexive system that relied on the contribution of participants in constructing the market as one central system and in allowing it to operate at instantaneous speed.

When Monitor was launched, in 1973, it presented the market only partially, since it only provided "indicative" prices upon which one could not trade directly. It did, however, from the beginning, augment and contextualize market prices by including news. Actual dealing remained extraneous to screen activities and was conducted over the phone and telex until 1981, when a new system also developed by Reuters that included dealing services went live to 145 institutional customers in nine countries. The system was extended within a year to Hong Kong, Singapore, and the Middle East, resulting in a market with a worldwide presence (Read 1992: 283ff., 310–311). Reuters knew of course that the foreign exchange market was by nature a transnational inter-bank market; for any system to be effective in this market it had to be available transnationally to all major players and, if it was to make dealers' contributed data available and be used for actual trading, it had to link these players. Its project of becoming a financial service provider for this market was based on the awareness that a well established news agency that had access to exclusive (not publicly available) information transmission lines might be well positioned to stitch together this market and sell to it the political and other information it routinely collected.

What we call FOREXS in this chapter is now based on a much further developed version of the original Monitor, called Reuters 3000. It includes other systems—for example, an electronic broker system developed in the 1990s by banks themselves. Foreign exchange market scoping is itself a process proceeding at times by slow accumulation and at times in qualitative leaps forward. Its beginnings predate Monitor, as indicated, and an end to the process is not in sight. Nonetheless, with the inclusion of dealing services and the more recent inclusion of electronic broker systems (Reuters provides its own), these markets appear fully scoped, that is presented and contextualized on screen to a degree to which dealing can now proceed without extraneous means. We can say that the foreign exchange world has migrated into these systems and is now constituted within them. With the help of the deliberate deregulation of these markets by relevant nation-states, it also has become disembedded and decoupled from national political and economic variables and concerns.

No such decoupling and disembedding happened in DMFAS's case. We take the degree to which a system is exclusive of national circumstances and environments or remains inclusive of them to be the major differentiating variable when it comes to global technologies. The nature of this difference can be gleaned, in a first step, from looking further into the history of DMFAS than we have introduced before. Like FOREXS, DMFAS was, from the beginning, a global project. But it was not, like FOREXS, an exclusive project, that is one whose momentum was based more on bypassing and sidestepping national environments and their regulatory and cultural make up than on penetrating these environments (see next section). Rather, DMFAS emerged from a transnational effort involving groups of developing countries, the United Nations, and a growing concern since the early 1960s about the place of developing countries in international trade. The concern led to calls for a full-fledged conference "specifically devoted to tackling these problems and identifying appropriate international actions." It resulted in the first United Nations Conference on Trade and Development (UNCTAD), held in Geneva in 1964; at the same time, developing countries established the Group of 77 (which today has 131 members) to voice their concerns. UNCTAD has since remained a framework for assisting developing countries in a wide range of areas that include trade-related issues, trade negotiator training, investment policy reviews, the promotion of entrepreneurship, competition law and policy, trade and the environment, and debt management. The first technical assistance project on external debt was initiated in 1979 during the preparations for a meeting of the "Paris Club," an informal group of official creditors whose goal is stated to be "to find coordinated and sustainable solutions to the payment difficulties experienced by debtor nations."[20] The first DMFAS software was installed in three countries in 1982. According to official descriptions, the climate at the time was characterized by transformations of economic thinking in a direction where development strategies became more market-oriented and focused on trade liberalization and privatization of state enterprises. Debt crises in developing countries accompanied this development, and despite structural adjustments by the IMF, countries were not able to recover quickly from the crises.[21] The creditors' meeting in Paris, the attempt to create and implement a standardized debt management software, and debtor nations' effort at coordinating their voice, must all be seen in this light. DMFAS had not been set up as a program of the IMF but by a forum explicitly dedicated to technically assisting developing countries. Acceptance of the program was not mandatory for loan extensions or procurement and countries were not forced to accept the program. Nonetheless, it appears plain that existing debt crises made cooperation palatable, and macroeconomic management concerns as well as the aim to "empower" developing countries in dealing with their debt got the program off the ground.[22] The Reuters system had been initiated as a global business idea that fitted into the expansion and reconfiguration plan of an already global corporation and that promoted the technological remaking of markets that operated transnationally. DMFAS appears to have been initiated as a reflexive global tool by the United Nations Conference for Trade and Development and the institutions charged with monitoring and maintaining the world's financial order. It appears to have been initiated on a global level within a field of negotiations between debtor and creditor countries and their global institutional representatives. To proceed with these negotiations, and presumably also to help debtor countries, one needed not only a way to synthesize and "produce" debt and credit data so that countries themselves and their creditor institutions could see and confront an individual debt situation as if in a mirror. One also needed this production matrix to be standardized, so that debts and debt-related dangers could be compared and aggregated across countries and regions and used in global risk assessments, international repayment arrangements, and financial policy and aid decisions.

Inclusive and Exclusive Global Technologies

Globality, in the case of DMFAS, rests upon and comes about through the global distribution of a standardized system of debt reporting that permits and encourages aggregation, comparison, and reflexivity on a national and global level. Let us focus more directly on this now, spelling out what we

mean when we say that DMFAS is a globally inclusive system. We use
the distinction between exclusive and inclusive systems in this context to
point to differences in global strategy. With an inclusive strategy, one
attempts to penetrate the geopolitical landscape of the world and integrate
different national, regional or local environments into some common unit
or plan. With an exclusive strategy, one makes no such attempt at integration or penetration and rather sets out to create what one might call a parallel world.

Inclusive strategies have to contend with the differences between and
diversities of national environments and national cultures. They may try
to reach into these environments and transform some of their elements,
they may try to implant in them external modules on which they can rely
for transnational purposes, and they may try to persuade the countries or
their population of the value of cross-national standardization. For example, integrating local markets may require one to adapt to local consumer
preferences or to find ways to change these preferences. It may mean adapting to different regulatory environments, recruiting and training local suppliers, and many more global-local adaptations of the kind described by
the term 'glocalization'. A globally inclusive financial stock market would
be one where individual investors in any country are able to trade shares
freely across national boundaries. Such a system requires computer availability in all investor locations (e.g. households), language capabilities or
unification, Web architectures, payment and clearing arrangements between stock exchanges, regulatory approvals, and national pension systems
that support individual financial planning. Such systems are in the process
of being created in some regions (e.g. Western Europe), but they are far
from being in place on a worldwide basis. DMFAS is an inclusive system in
tendency. Inclusiveness, in this case, translates into creating a compromise
between unique adequacy—the appropriateness of a system with respect to
local idiosyncrasies—and global standardization. It implies the attempt to
integrate all relevant countries (and potentially the whole world) into the
world financial architecture by providing standardized scripts for debt accounting and reporting. These scripts must be commensurable enough to
allow transnational aggregation and comparison while also being detailed
and flexible enough to provide single countries with an adequate mirror of
their often idiosyncratic debt situations.

Globally exclusive systems and strategies are of a different nature. Since
they lack the goal of inclusive penetration or integration, they can make
short shrift of the complexities that arise from geopolitical and cultural
diversities, and simply impose their own style (originating from earlier

practices) on the new systems developed. FOREXS exemplifies this strategy; it is a system that first included only some elements of earlier practices (indicative prices and news) but grew to embody and reorient the whole style of the market. Problems, in the case of the foreign exchange market, have more to do with creating and articulating the market than with assimilating national diversity and local customs. Of course, exclusiveness and inclusiveness are relative categories, and the question how inclusive or exclusive one can be may have to be negotiated in particular cases. For example, European integration is a project that attempts to carve out a space between full inclusion and full exclusion. We use these concepts here not as definitive categories but as analytic tools to highlight crucial differences between FOREXS and DMFAS as global systems.[23]

Let us return to DMFAS and say more about the meaning and consequences of its inclusiveness. As a global technology, DMFAS depends on the standardization of the system. Standard categories universalize the data content DMFAS collects and reports. But standard categories are caught in a tension between the goal of providing countries with a mirror image of their debt situation, and the goal of feeding transnational statistics and risk assessment and accounting goals. This tension cannot be eliminated; it was present from the beginning of DMFAS development and is visible in the projected version 6.0—the first modular version in the sense that user countries will be able to choose the modules that are most important to them from a standard package. The most pronounced examples of what this tension involves may not come from DMFAS but from older, hoarier and well researched cases, for example that of international disease classifications (existing since the nineteenth century). All organizations and actors in all countries, including medical practitioners and statisticians, would have to agree on how to collect and code the relevant information for disease statistics to provide valid data on a transnational and global level. But numerous variables have always intervened in this project, as illustrated in detail by Bowker and Star (e.g. 1999: 141 ff.). Let us list just a few of the relevant dimensions for which they provide examples. First, medical practitioners and institutions (or non-medical diagnosticians) would have to have the same sophisticated knowledge and technology to perform the required diagnoses. Yet it is plain that even in the developed world, rural areas find it difficult to provide this treatment. Second, all national schools of medicine would have to be eliminated in favor of one common school. But again, even today, these schools persist in the most advanced Western countries, and manifest themselves for example in disagreements over the desirability of surgical procedures. Third, culture would have to stop playing a role in disease reporting. Yet as some epidemiologists studying the low incidence of fatal heart attacks in Japan suggest, it may still play a major role: conceivably, what Americans call fatal heart attacks may often be described as strokes in a country in which heart attacks suggest a low status life of physical labor while strokes suggest a high status cause of death such as an overworked brain (Bowker and Star 1999: 142). Fourth, moral values and government regulations would have to support a universal classification. But this is difficult to expect from countries in which single early childhood diseases trump all other diseases, and the medical and epidemiological need of the nation to combat the disease eat up the resources that would be needed to effectively implement an externally imposed classifications. Fifth, the diverse needs of statistical bureaus, the medical establishment, government agencies, and practicing specialists with little time to spare would have to be reconciled. In other words, heterogeneous professional logics with very different demands on the length, stability, and orientation of a classification would have to be made commensurable. For example, what should the classification be based on—causes of the illness, treatment options, environmental factors, the manifestation of the disease, ethical-political considerations, multiple categories, and if so in which of the numerously possible combinations?

We need not elaborate this example further to give a taste of the problems standardization attempts at a global level face. The DMFAS case appears simpler than that of disease classifications in several ways. For example, DMFAS presentations suggest that the relevant countries did not yet have elaborate debt accounting systems in place when DMFAS was created (DMFAS 1999/2000: 1). Thus DMFAS may not have been running up against entrenched national-bureaucratic ways of dealing with debt, local investments in particular accounting procedures, and actors in strong professional positions that wanted to cling to alternative ways.[24] Nonetheless, DMFAS must be understood as a culturally, politically and technically negotiated outcome. Some of these negotiations occur when DMFAS's project officers engage in communications with at least one "interlocutor" in all of its 80 receiving institutions, as they routinely do. As one project manager put it, "my work stands and falls with the existence of at least one reasonably trustworthy person" with whom the project can be discussed. This negotiated character is also manifest in the current effort DMFAS's information technology experts are making to design the next version 6.0 in a modular way, such that user institutions with different interests may acquire only the modules they want. For example, African countries do not need bond modules, since they do not have developed capital markets; only certain countries (e.g. Argentina) need the module that registers "local government debt," i.e., loans or bonds granted to or issued by local governments or other sub-national entities; only countries that are debtors of the Asian Development Fund need its special calculation methods. And so on. On one level, the DMFAS process exemplifies what constructivist sociologists of technology have shown with respect to other objects: negotiated outcomes and multidirectional rather than linear models of innovation (Pinch and Bijker 1987). But in contrast to other contexts the DMFAS debate is not meant to be closed, and the software is not likely to become stabilized. DMFAS, and other global technological systems, are characteristically open, indefinite artifacts that reflect the changing user preferences and the changing beliefs and circumstances of global institutions.

The last examples also point to another aspect of the global character of DMFAS that stands in contrast to the global logic of FOREXS. DMFAS includes not only transnationally negotiated features of which some respond to local user demand, it is also a system that exists in different countries in different languages (it has been translated from English into French, Spanish, Arabic, and Russian). As indicated before, it also exists in different versions, upgrades, and states of maintenance. All recent versions include common core variables and allow users to produce specific standardized reports demanded by global institutions. Nonetheless, one has to imagine the tapestry created by DMFAS systems in user countries as more of a patchwork of pieces of various fabrics, colors, and figures rather than a shiny new cloth of one texture and design. DMFAS, though it is centrally designed and supported by a team of approximately 24 experts located in Geneva, is the work of assimilation, not that of unification. DMFAS embodies the attempt to make things similar on a worldwide basis. But it has done so in leaps and bounds, so to speak, by providing fragmentary solutions for some emerging problems, maintaining flexibility with respect to others, by growing cumulatively over time and then modularizing the growth, and by patching up the technical and taxonomic difficulties that arise. User countries contribute their own share to the patchwork by the ways in which they make use of DMFAS. But debt officers in single countries are also crucially important to the survival of this global technology: they create the fit between the local idiosyncrasies of statistical categories and data collection procedures in individual user countries, and the formal categories and menus provided by the global software. Clearly, these fitting procedures are what allows the "scoping up" of local worlds. What we want to stress then is the assimilated and at times incongruent character of a globally distributed software system that "lives" in national institutional spaces rather than to inhabit, as FOREXS does, an exclusive global sphere.

In emphasizing DMFAS's assimilating globality, we also want to emphasize again its relation to the nation-state and the inclusive strategy of globalization in terms of which we analyzed its development. The system cannot be made to exit this relationship. On one level, it embodies a globality that remains firmly rooted in transnationality and reaches only as far as transnational negotiations and adaptations. Yet on another level, the same system also feeds global institutions (various creditor and debtor groups and conferences, the IMF, the World Bank, and the DMFAS group in Geneva) by means of regularly produced forms and reports, thus sustaining the global character of world economic institutions and an emerging global governance that centrally rests on the harvesting and exchange of knowledge and information.

FOREXS, on the other hand, is the embodiment of a globally exclusive strategy. As a global system and technology, FOREXS sustains, one might say, a separate province of the global world, one of many emerging global forms that do not connect, tie together, or integrate the rest of the planet but co-exist with it. Most of the relevant details indicating this have been given before; they need not be repeated here. But we want to sum up and highlight three aspects of the characteristic exclusiveness of FOREXS: First, FOREXS runs on intranets rather than on the Internet, meaning the infrastructural inter-bank connections on which the systems depend are available only to the accredited institutions that buy the systems. Up to the point when we concluded this research, private, non-institutional actors could not buy the systems, even if they would have been willing and able to finance its enormous cost. The systems are designated for banking institutions, though versions have been available to other institutional market participants. Second, the FOREXS market rests on the establishment of centers of institutional trading in the financial hubs of the three major time zones: in New York, London, Tokyo, and perhaps Zurich, Frankfurt, or Singapore. (For a description of this global urban system, see Sassen 1991, chapter 7.) The centers cover the world by covering the clock. In other words, they cover time zones rather than countries; they provide trading opportunities for banks and institutional investors around the clock in regions of the world of predominant interest to financial service industries (the United States, Southeast Asia, and Europe) during their respective working times. The foreign exchange market does not attempt to penetrate all countries or even continents. Regions of the world are interesting only to the extent that they issue major currencies that can be traded within a speculative regime. This sort of trading can be done with ease, and grace, from the world's time-zone-based financial centers. Third and most importantly, the foreign exchange market is exclusive in the sense of being disembedded from individual nation-states. As indicated before, it has been "set free" deliberately by successive waves of liberalization of capital flows and financial services from the control of individual nation-states over the last 35 years. (See the overview in Swary and Topf 1992.) The removal of barriers between national financial markets, particularly currency markets, enabled a system to emerge in which economists consider frictions and impediments to be minor and that appears in fact beyond the control of any regulatory structure. The uncoupling can also be gleaned from the role currency markets play as an independent power in testing and determining the value of currencies against the authority of central banks and governments. They illustrate the role of the external observer and evaluator of national macroeconomic policies financial markets often take. The market is also disembedded on an action- and institutional level: Traders are oriented to their counterparts in other areas of the world, not to the rest of the banking center in which they work. They often change jobs, move to other places, and can take up trading practically immediately in any financial center of the world. Institutional means of separating trading from the rest of the bank include access limitation to trading floors, different compensation systems for traders, specific management structures, and so on. (See Knorr Cetina and Bruegger 2002 for details.)

Concluding Remarks

Two things should be emphasized in concluding this chapter. First, we used the notion "technology" in a wide sense, including in it "systems" (e.g. Hughes 1987) that incorporate many different elements and information feeds (FOREXS) on the one hand and software packages that are complex in their own right but do not have the same heterogeneous and systemic character (DMFAS) on the other. While it may be worthwhile to draw distinctions between more systemic and more homogeneous forms of technology, these distinctions are not central to the present argument and have been left out of consideration. Second, DMFAS and FOREXS have been described as scoping systems that imply a specific mechanism of coordination which we distinguished from that of networks. But this does not mean that networks have no role to play in the context of these systems' practical instantiation and use. For example, traders in foreign exchange markets cultivate small networks of contacts and friendship with other traders in different countries and time zones. These often include no more than a handful of persons. Traders rely on these contacts for early morning and late evening exchanges of information, they may call upon a "friend" in another time zone to watch their position during the night, they may chat with them during periods of low market activity, and so on. Though traders have indicated that "99.9 percent" of their transactions are anonymous, one assumes that the respective networks keep information alive, provide opportunities for debating and contesting different market stories, and generally provide a kind of social liquidity to financial markets. In the DMFAS case, relationships between the Geneva experts and specific debt officers in single countries were deemed to be important by the experts with a view to the correct implementation and maintenance of the system, with a view to obtaining user feedback, etc. The point here is that the transition to scopic mechanisms of coordination does not eliminate other mechanisms but rather makes the overall system more complex. A specific question then is how different mechanisms of coordination become coupled in specific domains. Knorr Cetina (2003) claims that the FOREXS market was historically a network market but that networks no longer dominate this market today. Rather, scopic systems have provided the basis for a form of Schutzean intersubjectivity in these markets that is based in the synchronicity, immediacy and continuity of market observation (Knorr Cetina and Bruegger 2002). In the DMFAS case, networks and the interpretations that travel through them may well play an important role in securing cooperation between dispersed national interests and in maintaining global order.

Acknowledgments

We are heavily indebted to the managers, traders, salespersons, and analysts in the financial markets studied, and to the IT, financial, and project management experts in the DMFAS group in Geneva, all of whom so generously shared with us the information we collected. We also thank Urs Bruegger, the first author's co-author on other papers, with whose assistance some of the financial market information was collected. Research for the present study is supported by a grant from the Deutsche Forschungsgemeinschaft and by the University of Konstanz.

Notes

1. For descriptions of bond, stock, and other financial markets, see Abolafia 1996a,b, 1998; Smith 1981, 1990, 1999; Hertz 1998. For more general discussions of markets, see Fligstein 2001; White 2002; Swedberg 2003.

2. The study is based on ethnographic research conducted since 1997 on the trading floor of a major global investment bank in Zurich, New York, and London and in several other, for example private and second tier banks. Unlike other financial markets, the foreign exchange market is not organized mainly in centralized exchanges but derives from inter-dealer transactions in a global banking network of institutions; it is what is called an "over-the-counter" market. Over-the-counter transactions are made on the trading floors of major investment and other banks. For a description of this research, see Knorr Cetina and Bruegger 2002. For an extensive description of currency trading in all its aspects, see Bruegger 1999.

3. Institutional currency (foreign exchange) traders use several systems simultaneously, including Reuters 3000 (the most recent version of Reuters's dealing and information systems) and EBS (an electronic brokerage system developed by the banks themselves). For present purposes, we do not differentiate between these systems. We use the shorthand FOREXS to refer to them.

4. To avoid confusion it should be noted that DMFAS uses the notion "projection" as a technical term for different kinds of forecasts (DMFAS User Guide for version 5.3, chapter 26).

5. Scoping systems may of course become coupled with social-authority-based organizational mechanisms, network mechanisms, and other mechanisms.

6. These figures were reported in the *New York Times* (Barringer 2002).

7. For an overview of the "DMFAS Programme" see http://r0.unctad.org/dmfas/. Research on the DMFAS system is based on ethnographic fieldwork and interviews conducted during 2005 and 2006 at the DMFAS program unit in Geneva. For more details, see Grimpe forthcoming.

8. Other funds come from the United Nations and from a trust fund financed mainly by the Netherlands, Norway, and Sweden.

9. According to Enrique Cosio-Pascal, a previous program leader of DMFAS.

10. There is only one other public debt management software that is comparable to DMFAS in the sense that it is used by a number of developing countries: The CS-DRMS of the Commonwealth Secretariat. More detailed information on this system is available at http://www.csdrms.org.

11. The various characteristics of DMFAS are extensively described in chapters 3, 9, 11, and 6 of the DMFAS User Guide 5.3. The following details are based on this description and on personal interviews and field data collected by the second author of this study at the DMFAS unit in Geneva.

12. The users' guide mentioned before described tranches as the "distinct parts of a loan as defined by the creditor in the detailed payment schedules sent to the debtor" (DMFAS User Guide for version 5.3, chapter 9).

13. The degree to which DMFAS participants actually make use of these updating functions (and other functions) varies among countries. In one of the African countries, for example, the system had not been updated for months, and higher level officials in the country appeared not to mind. During a mission to the country, the DMFAS project manager reacted by taking on the role of a traditional development worker: he emphasized the need for updates in user training workshops "for the benefit of the country" and pushed for such updates, but without much success. This situation is not just an exception. DMFAS program activities do occasionally follow the paradoxical patterns of foreign aid initiatives. (For a comprehensive case study of such patterns see Rottenburg 2002.) The analysis of the actual uses of DMFAS software functions and their variations across countries is a study in its own right that has yet to be completed

14. Source: http://publications.worldbank.org.

15. Original communication in German.

16. The full name was Reuters Monitor Money Rates (Read 1992: 301).

17. Read (1992: 290) reports that the general manager of Reuters at the time, Gerald Long, was able to claim by the early 1970s that the firm was "operating the largest and most technically advanced news and information network in the world." This network had two main arteries: one crossed the Atlantic and Pacific (Tatpac), joining London to Singapore and Tokyo through Montreal, Sydney, and Hong Kong; the other, Europlex, linked the main cities of the continent (London, Paris, Geneva, Frankfurt, The Hague, and Brussels) in an 88-channel ring fed by dataspurs and leased teleprinter lines from other European cities. Reuters also leased 18 channels in a new round-the-world Commonwealth cable to Sydney via Canada in 1964. Europlex became operational in 1967. A great deal of Reuters's transformation into a financial service firm has yet to be written.

18. The latest technology was a system that fed tickertape signals from stock exchange and other markets into a master computer in New Jersey that processed the material for feeding it into subsidiary computers, such as one in London, that was in turn connected to the offices of brokers and other subscribers who had small desk units that gave them access to the latest information on stock prices by pressing a button (Read 1992: 296–297). It was perhaps the first attempt, outside the "big sheet" technology, to create a scopic system from tickertape.

19. In fact, Read reports that Reuters proudly pronounced at the time that it was now operating a global stock exchange, since material was being fed into the system from all the main exchanges (1992: 298).

20. The Paris Club met for the first time in 1956. Source: http://www.clubdeparis .org.

21. Source: http://www.unctad.org.

22. Enrique Cosio-Pascal: "My participation as the UNCTAD representative in the meeting of the Paris Club in the early 1980s confirmed that [the] lack of information [of a country concerning its debt situation] applied generally to developing countries as a whole. The question that arose at this stage is, why did countries not develop their own CBDMS [computer-based debt management system]?" (DMFAS 1999/2000: 13).

23. For example, DMFAS has the following characteristics of an inclusive system, too: First, it is principally distributed to government agencies such as ministries of finance or central banks only. Second, many of these financial institutions meticulously keep their debt databases in confidence. For instance, for big data conversions supported by the DMFAS staff in the headquarter some of these user institutions accordingly hesitate to send their databases to Geneva.

24. On the other hand, DMFAS efforts seem to be mired by difficulties having to do with "ineffective" entrenched procedures in some countries. The DMFAS program maintains that institutional restructurings are sometimes necessary to achieve "effective debt management." In one of the African countries, for example, a certain kind of "paper organization" and a certain organizational hierarchy and work flow within the ministry of finance that existed already before the installation of the software seem to persist. The DMFAS project manager responsible for this user country frequently criticizes these organizational features; in his view, these feature stand against effective debt management procedures.

References

Abolafia, M. 1996a. *Making Markets: Opportunism and Restraint on Wall Street*. Harvard University Press.

Abolafia, M. 1996b. Hyper-rational gaming. *Journal of Contemporary Ethnology* 25: 226–250.

Abolafia, M. 1998. Markets as cultures: An ethnographic approach. In *The Laws of the Market*, ed. M. Callon. Blackwell.

Bank for International Settlements. 2007. Triennial Central Bank Survey of Foreign Exchange and Derivatives Market Activity in April 2007. Preliminary Results.

Barringer, F. 2002. Bloomberg, without Bloomberg, faces an industry in retreat. *New York Times*, September 8.

Beunza, D., and D. Stark. 2004. Tools of the trade: The socio-technology of arbitrage in a Wall Street trading room. *Industrial and Corporate Change* 13, no. 1: 369–401.

Bowker, G., and S. Star. 1999. *Sorting Things Out: Classification and Its Consequences*. MIT Press.

Bruegger, U. 1999. Wie handeln Devisenhändler? Eine ethnographische Studie über Akteure in einem globalen Markt. Dissertation, Universität St. Gallen.

Castells, M. 1996. *The Rise of the Network Society*, first edition. Harper & Row.

Castells, M. 2003. *The Internet Galaxy: Reflections on the Internet, Business, and Society*. Oxford University Press.

Clough, P. 2000. *Autoaffection: Unconscious Thought in the Age of Teletechnology*. University of Minnesota Press.

DMFAS. 2004. Participant's Handbook on Production of a Debt Statistical Bulletin. Third Draft.

DMFAS/UNCTAD. 2004. Newsletter 16.

DMFAS/UNCTAD. 1999/2000. Newsletter 12.

Fligstein, N. 2001. *The Architecture of Markets: An Economic Sociology of 21st Century Capitalist Societies*. Princeton University Press.

Grimpe, B. Forthcoming. Ethnography of a Global Information Technology: The Development and Use of a Public Debt Management Software in the Context of International Financial Regulation (working title). Dissertation, University of Constance.

Hamilton, G., and N. Biggart. 1993. Market, culture and authority: A comparative analysis of management and organization in the Far East. In *The Sociology of Economic Life*, ed. M. Granovetter and R. Swedberg. Westview.

Hertz, E. 1998. *The Trading Crowd: An Ethnography of the Shanghai Stock Market*. Cambridge University Press.

Hughes, T. 1987. The evolution of large technological systems. In *The Social Construction of Technological Systems*, ed. W. Bijker et al. MIT Press.

IMF (International Monetary Fund). 2003. External Debt Statistics: Guide for Compilers and Users.

Knorr Cetina, K. 2003. From pipes to scopes: The flow architecture of financial markets. *Distinktion* 7: 7–23.

Knorr Cetina, K., and U. Bruegger. 2002. Global microstructures: The virtual societies of financial markets. *American Journal of Sociology* 107: 905–50.

Knorr Cetina, K., and A. Preda. 2005. *The Sociology of Financial Markets*. Oxford University Press.

Leyshon, A., and N. Thrift. 1997. *Money/Space: Geographies of Monetary Transformation*. Routledge.

Muniesa, F. 2000. Un robot Walrasien. Cotation electronique et justesse de la découverte des prix. *Politix* 13, no. 52: 121–154.

Pinch, T., and W. Bijker. 1987. The social construction of facts and artifacts: or how the sociology of science and the sociology of technology might benefit each other. In *The Social Construction of Technological Systems*, ed. W. Bijker et al. MIT Press.

Preda, A. 2006. Socio-technical agency in financial markets: The case of the stock ticker. *Social Studies of Science* 36, no. 5: 753-782.

Read, D. 1992. *The Power of News: The History of Reuters*. Oxford University Press.

Rottenburg, R. 2002. *Weitergeholte Fakten: Eine Parabel der Entwicklungshilfe*. Lucius & Lucius.

Sassen, S. 2001. *The Global City*, second edition. Princeton University Press.

Schutz, A. 1955. Symbol, reality, and society. In *Symbols and Society*, ed. L. Bryson et al. Harper.

Schutz, A., and T. Luckmann. 1973. *The Structures of the Life-World*. Northwestern University Press.

Smith, C. 1981. *The Mind of the Market: A Study of the Stock Market*. Rowman & Littlefield.

Smith, C. 1990. *Auctions: The Social Construction of Value*. University of California Press.

Smith, C. 1999. *Success and Survival on Wall Street: Understanding the Mind of the Market*. Rowman & Littlefield.

Swary, I., and B. Topf. 1992. *Global Financial Deregulation: Commercial Banking at the Crossroads*. Blackwell Finance.

Swedberg, R. 2003. *Principles of Economic Sociology*. Princeton University Press.

White, H. 2002. *Markets from Networks: Socioeconomic Models of Production*. Princeton University Press.

Zaloom, C. 2003. Ambiguous numbers: Trading technologies and interpretation in financial markets. *American Ethnologist* 30, no. 2: 258–272.

6 The Politics of Patent Law and Its Material Effects: The Changing Relationship between Universities and the Marketplace

Elizabeth Popp Berman

The expansion of intellectual property has been a cornerstone of global economic policy over the past few decades. From *Diamond v. Chakrabarty*, which in 1980 affirmed the patentability of living microorganisms in the United States (Bugos and Kevles 1992; Kevles 1994), to the TRIPs (Trade-Related Aspects of Intellectual Property Rights) Agreement of 1994, which extended and helped unify international intellectual property law (Sell 2003), the reach of private ownership into the world of ideas has been steadily extended, with complex and less than fully understood results.

Research universities have been one important venue for the negotiation of these changes in intellectual property rights. In the United States, universities—which historically took little interest in the patenting of research and often opposed it on principle—have begun patenting the inventions of their faculty in large numbers. Whereas fewer than 500 U.S. patents were being issued to U.S. universities and colleges each year in the early 1980s, since the late 1990s more than 3,000 have been issued each year. (See figure 6.1.) Universities now also bring in substantial amounts of money by licensing their patents to corporations. Net licensing royalties increased from a little under $200 million in 1993 (the first year for which data are available) to well over $800 million in 2003. (See figure 6.2.)

These changes in academia have not taken place without controversy. Introducing an element of ownership and profit into what is still largely a Mertonian and communistic system of science has raised fears and concerns. Some are concerned with the fairness of patenting inventions made with taxpayer dollars, and see patenting as encouraging universities to act in their own self-interest (e.g. in defending their patents through costly litigation) at the expense of the public interest (Leaf 2005). Others raise the specter of an "anticommons" hindering the advancement of science (Heller and Eisenberg 1998). In contrast to the classic "tragedy of the commons," in which common ownership of a resource leads to its overuse,

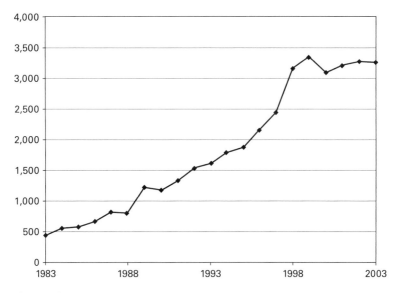

Figure 6.1
Patents issued to U.S. universities and colleges, 1983–2003. Source: NSF 2006.

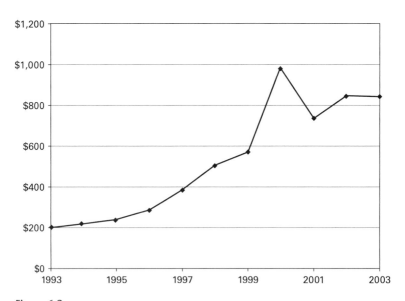

Figure 6.2
Net licensing royalties to U.S. universities and colleges (millions), 1993–2003. Source: NSF 2006.

an anticommons is created when too many claims to ownership exist in a area of knowledge, effectively limiting others' ability to build on that knowledge.

Despite these concerns, the practice of university patenting is now quite entrenched in the university, with organizational structures (technology licensing offices) and normative behaviors (taking patents seriously in tenure decisions) helping to reproduce it. The many supporters of university patenting argue that without the protection of patent rights, firms would lack sufficient incentives to invest in the costly process of bringing an embryonic invention to the marketplace. They suggest that universities, with their close proximity to scientist-inventors, are uniquely positioned to connect those who possess technical know-how with those who have the resources to turn that know-how into marketable products.

Scholars—mostly economists—have produced a large literature on university patenting. Most of it is policy-oriented, and most of it focuses on patenting's causes and consequences: what kinds of universities are more likely to patent, whether university patents have a greater impact than other patents, what kind of university-industry connections facilitate the use of patents. Though some recent efforts have tried to evaluate the effects of university patenting on the cumulative nature of science (Murray and Stern 2005; Walsh et al. 2003), little of this research is framed by the questions of economic sociology—about the creation of economic institutions, social effects within those institutions, or their non-economic outcomes.[1] Even less is oriented toward the questions of technology studies—of the interaction of social and technical factors in the production and use of technology.[2]

This should not be the case. Patents, as legal mechanisms for creating private property, fall clearly within the domain of economic sociology, and as economic phenomena which increasingly extend the scope of formal property, their use and governance can teach us about the ever-growing role of markets in society. The relationship of public and private in the development of science and technology is a question of increasing importance within Science and Technology Studies, and patenting in both a university and an industrial context unquestionably shapes the paths technologies take in their development. Furthermore, because patents are sites at which the socioeconomic intersects with the technical, they are a place where the disciplines jointly can contribute more than either of them could on its own.

This chapter will give a short demonstration of how one might think about patenting from the perspective of economic sociology. It will consider the Bayh-Dole Act, which promoted patenting in universities, as one of a number of ways in which universities were encouraged to adopt the logic of the marketplace beginning in the late 1970s. Through a discussion of the political process behind the Act, the chapter will argue that its effects on universities as organizations were largely unintended, an unanticipated byproduct of an effort to improve the utilization of government-funded technologies that was not initially driven by macroeconomic considerations. Then it will go on to discuss how the methods of technology studies are needed to push this kind of research to the next level, by providing tools for thinking about how political and organizational changes actually interact with the process through which new technologies are constructed—thus moving beyond social abstractions and into material effects. Finally, it will suggest that this kind of joint conversation between economic sociology and technology studies has the potential to make a real contribution to the current debate over patent policy, which is dominated by an economics-of-innovation approach.

University Patenting and the Bayh-Dole Act

The passage of the Bayh-Dole Act in 1980 (more formally, the Patent and Trademark Law Amendments Act, P.L. 96-517) was a landmark event in the development of university patenting. In the popular press, the Act has been lionized as "possibly the most inspired piece of legislation to be enacted in America over the past half-century" ("Innovation's Golden Goose" 2002) and demonized as causing "what used to be a scientific community of free and open debate" to become "a litigious scrum of data-hoarding and suspicion" (Leaf 2005). Recent research has tried to put Bayh-Dole in a longer-term perspective, pointing out the rapid increase in patenting in the decade prior to the Act and arguing that the legislation itself did not cause patenting to increase to the extent that has generally been assumed (Mowery et al. 2001, 2004; Mowery and Ziedonis 2002).[3] Nevertheless, the Act, in which the federal government first fully sanctioned the patenting of taxpayer-funded research by universities, was a symbolic turning point for technology transfer in the United States, and helped introduce a new discourse about universities as economic actors (Krücken, Meier, and Muller 2007).

Universities had been patenting the government-funded inventions of faculty on a limited basis for decades prior to Bayh-Dole. But federal patent policy regulating such behavior was a tangled mess of laws governing different funding agencies, on top of which was laid a complex set of regulations that were applied very differently from agency to agency. Some agencies made it easy for universities to patent their inventions, others allowed it but required the university to jump through a lot of bureaucratic hoops, and yet others made it almost impossible for universities to retain title to faculty inventions. Add in the reality that many university inventions were funded by more than one government agency and that many universities had no full-time person who managed patent activity and it becomes clear why university patenting and licensing did not increase more quickly before 1980.

The Bayh-Dole Act streamlined this complex process, giving universities the patent rights to government-funded inventions by default and unifying this policy across federal agencies. But the Bayh-Dole Act had a second important purpose as well: to encourage commercial activity and industry collaboration in universities. Its stated "Policy and objective" includes the following goals:

to promote collaboration between commercial concerns and nonprofit organizations, including universities; to ensure that inventions made by nonprofit organizations and small business firms are used in a manner to promote free competition and enterprise; to promote the commercialization and public availability of inventions made in the United States by United States industry and labor. (P.L. 96-517: Sec. 200)

The text of the Bayh-Dole Act goes beyond just encouraging universities to patent research so that it can be better utilized. It specifically promotes both university-industry collaboration and active university participation in the commercialization of research. These sorts of roles once would have been considered inappropriate by many universities.[4] But they can be seen as part of a larger shift in the university's orientation toward the economy and its gradual embrace of an explicit economic role for itself.

Why Did the Bayh-Dole Act Become Law? Three Possible Explanations and an Alternative

One could imagine several plausible hypotheses to explain why legislation increasing university interaction with the marketplace might have been passed. I will briefly suggest three possible reasons before proposing a fourth alternative that better fits the empirical data.

We might imagine that the legislation was motivated by a larger project of neoliberalism. The idea that market incentives can make organizations and institutions run more efficiently has prompted a variety of legislative reforms in the United States in recent decades, ranging from the creation of pollution permit markets to welfare reform to the North American Free Trade Agreement. The Act's passage in 1980 would be fairly early in the emergence of neoliberalism in U.S. politics, but it is conceivable that the political forces that helped elect Ronald Reagan also might have helped pass Bayh-Dole.

Such legislation would have been in the interest of industry, so another explanation might be that Bayh-Dole was driven by the business sector. During the 1970s, industry was cutting back on its own research spending, particularly for basic research, as rising international competition placed new pressures on firms to cut costs. In constant dollars, industry spending on basic research peaked in 1966, then declined by 27 percent over the next six years before very slowly beginning to increase again. By 1980, it had only returned to its 1963 level (NSF 1999, table A-26).[5] Places like Bell Labs were in decline, and industry was becoming increasingly reliant on university science. Thus one might imagine that industry would have thrown its support behind legislation that would encourage universities to license inventions to the private sector as well as to collaborate with it.

We could also imagine a straightforward resource dependence explanation for this legislation (Pfeffer and Salancik 1978). Federal funding for university research was stagnant for most of the 1970s (NSF 2004, Appendix table 4–6), and universities were in sorry financial shape as well (see, e.g., Cheit 1971, 1973). Perhaps universities reacted to the end of substantial increases in government support by looking to the marketplace, and saw the activities encouraged by Bayh-Dole as a potential source of revenue.

I will argue that none of these explanations fits the history of the Bayh-Dole Act very well, and that the Act was not passed for any of these reasons. There was no interest group whose goal was to introduce market forces into the university: not industry, not universities, and certainly not the government officials who were the driving force behind the Act. It passed in the form that it did because of the convergence of two unrelated political trends. And it affected universities in particular because of the political circumstances leading up to it. It was not originally created with universities in mind, and thus the effects it might have on them as organizations were not widely considered or discussed.

The original proponents of the Act were aligned with a broader political trend in the 1960s and the early 1970s toward encouraging better utilization of government research. Proponents were not focused on universities, and their effort did not have an explicitly economic component. Instead, they aimed to improve utilization by giving inventors and their organizations the patent rights to all government-funded research contracts and grants, whether the inventors were at MIT or Boeing. They did this because they believed that inventors were more motivated to and capable of developing their inventions than government could ever be. After years of failed effort to garner interest in such legislation, supporters decided to change tactics in two ways. First, the proposed legislation was significantly limited in scope, so that it covered only universities, other nonprofits, and small business, and excluded big business in a concession to liberal concerns with government giveaways. Second, they took advantage of a political opportunity by reframing the bill as helping to solve U.S. economic stagnation by closing the growing "technology gap" with Japan. The legislation as originally conceived would not have affected universities in the same way the legislation that was eventually passed did, and possible effects of Bayh-Dole on the university as an organization were inadvertent and unanticipated.

The evidence for this argument comes from primary historical research, including about twenty volumes of government hearings on patent policy between 1960 and 1980; about twenty volumes of government studies and reports evaluating the patenting of federally funded research and related policy; publications on university patent policy and on the development of a university patenting community (e.g. proceedings of early conferences, surveys of policies, etc.); and many articles on patent policy from the contemporary press, both academic and popular. It also draws on the personal records of Norman J. Latker, former patent counsel to the Department of Health, Education and Welfare and an architect of the Bayh-Dole Act. This historical research was supplemented with a limited number of interviews of people involved in the effort to pass what would become the Bayh-Dole Act.

Creating Government Patent Policy: The Pre-History of the Bayh-Dole Act

The idea behind the Bayh-Dole legislation was not a new one. In fact, a heated on-and-off debate over government patent policy had been going on for almost forty years, beginning almost as soon as the federal government started spending real money on scientific research and development during World War II. The question quickly arose of who would own any inventions that resulted from this kind of research. There were two main policy alternatives for disposing of such inventions, a government-title policy or a government-license policy, along with a host of intermediate possibilities.

A government-title policy would assign title to any such inventions to the government by default. The main argument in favor of this policy was that what the public funds, the public should own—i.e., that whenever possible, publicly funded research should remain in the public domain, accessible to all, but that if it must be patented for some reason, the patent should at least be owned by the government and used in the public interest. So if a university scientist used a federal grant to invent a new hearing aid, the scientist would have to disclose that invention to the government, which would then have the option of either publishing the invention or of patenting it itself. The scientist would not retain rights to the invention.

A government-license policy would keep such rights with the inventor and his or her institution. The government would only reserve for itself the option of a royalty-free non-exclusive license. It could use the invention for public purposes if it wanted, but it would not actually own the invention. In this case, the scientist who invented the hearing aid would decide with the university whether or not to patent the research and try to commercialize it.

Because no one managed to forge an early compromise between these two policies, no uniform federal patent policy was ever created.[6] Instead, federal policy emerged piecemeal from different pieces of legislation affecting different funding agencies. By the time of the Bayh-Dole Act, at least twenty-two different statutes addressed some aspect of federal patent policy (P.L. 96-517: Sec. 210). Some agencies were not guided by statute at all, and instead followed federal regulations, which were themselves complex and ambiguous and thus interpreted very differently from agency to agency. The problem was complicated by the fact that many inventions had received funding from several different agencies, which meant that multiple policies would need to be reconciled. The delay in creating a uniform system of federal patent policy essentially resulted in no coherent policy at all.

In the years following the deployment of the atomic bomb and the end of World War II, science enjoyed a great deal of public support and to a large extent could chart its own course. In 1945, Vannevar Bush, in his influential report, *Science—The Endless Frontier*, proposed that in order to flourish, basic science needed to be free to follow the most intellectually interesting questions, without consideration of whether they were likely to result in "useful" knowledge. Furthermore, he argued, engaging with questions of application would actively *hinder* the development of basic, pure science. If basic science were funded generously and scientists were given free rein to pursue whatever questions they wished, useful applications would flow more or less automatically from this basic research through the process of serendipity.

During the 1950s, this argument became generally accepted by science policymakers, and a wall was established between basic and applied science, with basic science held up as the ideal. Bush helped establish the National Science Foundation to promote basic research, and science funding began to increase dramatically. But by the beginning of the 1960s, problems with this model of organization were beginning to emerge. For one thing, as federal spending on basic research rose tenfold in real terms between 1953 and 1963 (NSF 2005, table C-1), Congress began to pay more attention to science. For another, it was becoming increasingly clear that basic science innovations were not always turning into useful applications automatically, and Congress was beginning to ask whether, in fact, it was getting enough bang for its basic science buck (U.S. Senate 1963; U.S. President's Science Advisory Committee 1963; National Academy of Sciences 1965). Across federal government during the 1960s, a conversation was developing around the issue of how to better utilize government-funded science.

This debate applied broadly to federally funded science. But a piece of it had implications for government patent policy in particular. Some fields of research seemed to be encountering utilization problems precisely because of current government patent policies. One such area was the National Institutes of Health's program in medicinal chemistry, which began to draw negative attention. This relatively small ($8 million annually) grant program funded mostly academic chemists whose work resulted in compounds with potential pharmaceutical applications. Due to NIH patent policy, though, no pharmaceutical companies were willing to become involved in screening and developing the compounds. This was because NIH did not permit exclusive licenses on patents resulting from research it funded. Even then, getting a drug to market was so expensive in comparison to the cost of copying the finished product (i.e., creating a generic version) that there was no incentive for a pharmaceutical company to invest in development unless it were assured of exclusive rights to the original invention. The result of NIH's policy was that promising inventions were withering on the vine because no one was willing to develop them.

In the mid 1960s, Congress and the executive branch both commissioned studies of how federal patent policy was affecting this and other science programs (Harbridge House 1968; U.S. GAO 1968). As the federal government began collecting comprehensive data on patents for the first time, other issues were emerging as well. For instance, relatively few government-held patents were ever licensed—in 1968, for example, of 18,638 patents available for licensing, only 1,661, or 9 percent, were actually licensed (FCST 1968: 62–63).[7] Not surprisingly, in light of the broader problems with utilization that were emerging, a lot of people were becoming concerned that government patent policies might be preventing scientific innovations from being put to use.

By the early 1970s, a political coalition interested in changing government patent policy was developing around this issue. Dominated by administrators in federal science agencies and including both civil servants and political appointees, it extrapolated from the above observations to argue that exclusive patent rights were necessary to encourage the development of government-funded inventions. It also argued that the ongoing involvement of inventors was critical to the successful development of inventions, and that this made government uniquely unsuited to manage inventions. Instead, the organizations in which inventors resided were most likely to be able to get inventions into wider use. The focus was on the invention and the inventor, and no great distinction was made between an invention made in a firm under government contract and an invention made in a university with a government grant. The policy goals of this coalition were (1) to streamline and unify federal patent policy so that it encouraged patenting of government-funded research and (2) to keep patent rights, by default, with the inventor and the inventor's organization rather than giving them to the government.

Changing Federal Patent Policy: The Road to Bayh-Dole

The first efforts in this direction focused on changing federal regulations rather than on legislation (FCST 1965, 1966, 1967, 1968, 1970). While they met with some success, it became apparent fairly quickly that regulatory change alone was not going to accomplish these objectives, since most federal patent policy was governed by statute, not regulation. During the 1970s, proponents of a broad government-license patent policy published draft bills three separate times, the last of which became the Bayh-Dole Act. The legislation that eventually succeeded looked significantly different from that originally proposed.

The first draft bill was published by the congressionally appointed Commission on Government Procurement in 1972. (For the text, see Commission on Government Procurement 1972: 139–146.) It would have given contractors and grantees rights to all inventions except in the unusual situation that the government could show that it was prepared to take active steps to develop an invention. The second bill was drafted by the executive-branch Federal Council for Science and Technology in 1976 (published in FCST 1976: 93–119). It was introduced in 1977 by a member of the House Committee on Science, Space and Technology, Representative Ray Thornton (D-Arkansas), but never made it out of committee (Latker 1977). This version looked fairly similar to the 1972 draft, though it further reduced the limited circumstances under which the government would maintain rights to the inventions of contractors and grantees. The final bill was introduced by Senators Robert Dole (R-Kansas) and Birch Bayh (D-Indiana) in 1978. It looked substantially different from the earlier two bills, and a fundamentally similar version became law in December 1980.

The 1972 and 1976 draft bills had more similarities than differences. Each began with the same four main legislative goals: (1) to encourage the utilization of federally funded research, (2) to encourage contractor participation (that is, to persuade for-profit companies to contract with the government by permitting them to patent inventions resulting from such contracts), (3) to simplify administration of such inventions, and (4) to protect competition (a response to the standard criticism that patents, as a form of monopoly, would discourage competition more than they would encourage the development of inventions).

Another significant feature of the 1972 and 1976 drafts was that neither of them focused on universities. The Bayh-Dole Act would apply specifically to universities, other nonprofit organizations, and small businesses; it did not address most industry contractors. The 1976 bill, in contrast, would have applied to all federal contractors and grantees, for-profit and nonprofit, large and small, and did not single out universities in any way. The 1972 bill also focused on industry contractors as well as university grantees, but actually *excluded* universities and other nonprofits from its provisions unless they had explicitly demonstrated the capability to manage the development of inventions (Commission on Government Procurement 1972: 142). So two characteristics of the early bills were that (1) they were not aimed directly at universities and (2) they focused on the utilization of inventions rather than on macroeconomic goals.

By 1977, after the second draft bill had died in committee, it was becoming clear to the members of the coalition behind these efforts that their approach was not working. Supporters of patent policy reform were still primarily interested in encouraging utilization and in protecting and rewarding inventors, but in order to have a realistic hope of passing legislation they changed two things about the bill.

First, they decided that the only way such a bill might pass was if its scope were narrowed significantly. Supporters therefore decided to focus on universities and small businesses, and to cut big business out of the bill, thus reducing its scope by more than two-thirds.[8] A lot of the criticism of earlier proposed legislation was rooted in the argument that federally funded research was paid for by the public and thus that its results should not be given away to private companies. While allowing Lockheed or Northrop to patent research done under federal contracts seemed like a government giveaway to many, allowing a university or small business to do the same thing raised fewer concerns. A further advantage of this approach was that universities and the small business community had themselves become very supportive of such legislation and could be relied on to help drum up political support.

Second, while supporters' interest in changing federal patent policy grew out of a desire to improve the utilization of scientific research, they decided to reframe their proposal around macroeconomic issues. By the second half of the 1970s, the U.S. economy had been in the doldrums for quite some time. Everyone wanted to know how to turn the situation around. One major fear was that the nation was stagnating economically because the Japanese were doing better at creating and using new technology. The architects of the bill decided it would have a better chance of success if they could argue that it was a partial solution to this "technology gap" and thus to the country's economic woes.

What these decisions meant was that the Bayh-Dole Act looked different from the earlier draft bills, and had a different effect on universities than the earlier bills would have had. Not only did the final legislation apply only to universities, other nonprofit organizations, and small business, but its stated goals had changed. The Bayh-Dole Act had seven goals, as opposed to the earlier bills' four, and while it still included those four (encourage utilization, encourage contractor participation, simplify administration, protect competition), it added three more that were more market-oriented. One of these, encouragement of small business, did not affect universities. But the final two—encouraging universities to collaborate with industry and encouraging not only the utilization of research but its actual commercialization—actively tasked universities with market activities that earlier bills would not have explicitly given them. These latter goals highlighted hopes that the legislation would have a broader economic impact as well as improving the utilization of government-funded research.

What is interesting about this shift is not only that the bill itself evolved for political reasons while the intent of its creators remained the same, but that because people were focused on their original goals they didn't always consider how the bill might affect universities as organizations. Although universities played a significant part in lobbying for the final legislation, universities were not the motivating force behind the bill. The bill's strongest proponents, the people who had been pushing for this legislation for many years, worked in government—mostly for federal agencies.

The lack of attention to potential effects on universities was a point that came up unprompted in several interviews: people I spoke with were surprised that the Bayh-Dole Act ended up affecting universities at all (beyond the obvious intent of encouraging them to patent inventions), and told me that at the time no one really thought that much about what effect patenting and licensing inventions might have on universities as organizations. Instead, proponents of the bill continued to focus on the inventor and the invention, not on the institution that housed the inventor. Congressional hearings preceding the passage of the Act similarly fail to discuss how it might affect universities. Arguments in favor of the legislation emphasize its economic consequences and the way it will help get new discoveries into general use, and infrequently mention the specific role of universities in this process. Opposition to the bill focused on the issue of whether it is a giveaway of a public good, not on whether it will cause the commercialization of universities. In 3,400 pages of congressional hearings immediately preceding the Bayh-Dole Act (U.S. House 1979; U.S. Senate 1978, 1979, 1980), I found no references to the effects such legislation might have on them.

Explaining the Passage of the Bayh-Dole Act

The Bayh-Dole Act was not created with the goal of introducing market forces into universities, or even with much awareness that that would be one of its effects. It took the final form that it did, a form that *would* affect universities, because its proponents, concerned about the utilization of science, took advantage of a political opportunity to reframe it in terms of late-1970s fears about the economy and about falling behind technologically. Because there were so many contingencies, and because its effects on universities were incidental to the intentions behind it, it is easy to come up with counterfactuals in which a slightly different set of circumstances would have led either to no legislation or to legislation that would have had different effects on universities. For example, if such a strong barrier between basic and applied science had not been institutionalized in the 1950s, there might never have been a backlash when it turned out that that barrier was counterproductive. If there had been no backlash, there would have been no broad movement to encourage better utilization of science, which gave proponents of the Act much of their support. Or if political circumstances had been slightly different in 1972—if one or two of the strongest opponents of this kind of legislation had died or left Congress—the bill might have passed in the form in which it was first proposed, in which case it would not have included explicit economic goals, and would have excluded many universities anyway.

Even at the last minute, things almost turned out differently. Bayh-Dole hadn't been voted on by the November 1980 elections, in which Senator Birch Bayh was unexpectedly defeated by Dan Quayle. The bill had to be brought to a vote in the last weeks of a lame-duck Congress, and then had to avoid a pocket veto by President Jimmy Carter.[9] There was no one who could have quickly taken Bayh's place as a sponsor, and the issue might have been tabled for several more years. As a result, the existing situation of regulatory complexity might have persisted, continuing to discourage university patenting—not to mention that university patenting would not have been symbolically legitimized in the same way. This is not to claim that any of these potential scenarios would have meant that universities, in the long run, wouldn't have been affected by the marketplace. But they might not have been affected by market forces at this particular time in this particular way if any one of these things had changed.

Let us return, then, to the question of why the Bayh-Dole Act did pass, and how we can explain the institutionalization of university patenting. Earlier I suggested several possible explanations for this process, none of which fits the story of Bayh-Dole as well as an explanation of politics and contingency does.

The first possible explanation for the passage of the Bayh-Dole Act was neoliberal ideology. In one sense the motivations behind changing government patent policy were compatible with neoliberal ideas. Certainly the promoters of Bayh-Dole thought that inventors and universities would do a better job of managing patents than the government could. But they were not broadly committed to the neoliberal idea that markets always manage things better than government, and they did not see Bayh-Dole as part of a larger project of reducing the size and scope of government. Instead, they were people who saw a specific problem with government administration of one particular thing—federally funded inventions—and proposed a specific solution.

The second explanation proposed was the political interests of industry. But the back story to the legislation makes it clear that this wasn't the case. While industry did need to rely more on collaboration with universities to meet its basic research needs during this period, industry was not involved in the passage of this bill; to the contrary, big business was rather irritated that it had been cut out of legislation originally designed for it. In fact, while industry did not actively oppose the bill, it threw its support behind an alternate bill which would not have simplified university patenting but which it saw as better meeting its own needs.

The resource dependence argument doesn't really explain the passage of Bayh-Dole either. While a handful of universities make a lot of money from licensing patented inventions today, almost no universities received significant income from licensing in the 1970s. Proponents of university patenting, even those from universities, were not terribly focused on the money that licensing might bring in—for the simple reason that it wasn't expected to bring in all that much. It wouldn't have been a drop in the bucket compared to federal R&D funding. University patenting people, like everyone else involved in this, were initially concerned with utilization (National Conference on the Management of University Technology Resources 1974). At the outside, it might be argued that this is a different kind of resource dependence: that universities were focusing on improving their utilization of science to stay in the good graces of their government patrons. But it was not a case of universities actively trying to manage resource dependencies by taking on the patent management role.

Instead, the passage of the Bayh-Dole Act as a bill encouraging an economic role for universities resulted from a political effort to take advantage of two converging policy trends. People who were interested in promoting the utilization of science, but not solely or particularly in universities, decided to reframe their legislation in light of a second emerging trend of increased concerns about technology and the economy. If those two things hadn't happened when they did in the way that they did, patenting and licensing activity at universities would not have been institutionalized in this particular form.

Patenting as a Sociotechnical Institution

The case above demonstrates that patenting can be a useful site at which to study our decisions, as a society, about what will be owned and who it will be owned by—that is, a useful site at which to do economic sociology. But this kind of research could be pushed forward another step by drawing on technology studies as a complementary approach. The story above leaves us with this question: What does it *mean* for universities to begin patenting, or to adopt an market orientation in general? What is the impact of such a change? That is difficult to evaluate without understanding the way technology actually develops, both within the laboratory and outside it. And this is where the question of private and public intersects with the material reality of evolving technology.

For example, consider the biotechnology industry during the late 1970s, in its infancy. The industry was built on techniques to create recombinant DNA and monoclonal antibodies. No one knew if its products would be patentable in the United States or elsewhere. Biotech startup firms waited in eager anticipation for the courts to make critical decisions. Many observers thought that patents were absolutely critical to the industry's future. But the situation was far from clear-cut. Some thought that patents were unnecessary. David Baltimore, a scientist at MIT and later director of the Whitehead Institute, was quoted as saying "There is enough potential in the field that it doesn't need patent protection to stimulate activity." (Clark 1980) Representatives of the smallest firms were concerned that patents might put them at a disadvantage ("Patent Issue" 1980). Some observers worried "They are trying to patent all of molecular biology"; others thought that if changing small aspects of genetically engineered organisms could lead to new patents, the original patent rights would "mean nothing" ("Genetic Patents" 1980). While many thought that universities should not be involved in patenting biotechnology, others thought that should the new products not be patentable "their colleagues [would] resort even further to secrecy" (Clark 1980).

Here is a case where there is clear evidence that academic scientists, as well as the firms and venture capitalists who were funding them, were paying attention to the evolving political reality of patent policy. It seems quite likely that this affected the material development of biotechnology, particularly since by 1980 almost all leading recombinant DNA researchers in academia had some industry involvement. The questions no one has answered are these: What kind of effects did the political and economic environment have on the development of the technology? How did that technological development in turn affect the evolution of that environment?

This is a location where economic sociology and technology studies can be fruitfully brought together. An economic sociologist might look at the reactions of scientists and investors to this uncertain legal situation, and might consider their efforts to shape emerging policies in their own interests or to protect themselves from an undesired outcome. But alone, economic sociology would not have much to say about the new technologies themselves, which matter not only for their own sake but also because their material reality enables, constrains, and shapes decisions made in other social realms.

At the same time, this sort of joint effort would represent a move in a direction already being espoused by some STS scholars. As early as 1996, Trevor Pinch was pointing out that the lack of attention to structural factors in social construction of technology (SCOT) studies was due not to an "inherent limitation" of the SCOT approach but rather to its original focus on technological design (Pinch 1996: 33; see also Klein and Kleinman 2002). Since then some of the most interesting work in technology studies has made organizational, economic, and political factors central (Doing 2004; Hyysalo 2006; Parthasarathy 2005; Vaughan 1999). Some of this research has even focused on the university-industry interface, showing that the interplay between the academic and private sectors is subtle but significant and that it often works in unexpected ways (Kleinman 2003; Mody 2006). Extending this kind of work into questions of patenting would be natural. Here, too, things are more complex than they initially appear: introducing the ideal of commercialization into the academic laboratory would seem to encourage secrecy, yet at the same time the protection of patents can make it possible for more openness rather than less, by decreasing the need for trade secrets.

Ultimately, such studies would have a significant contribution to make to the predominant economics-of-innovation-oriented literature on patenting, as found in journals like *Research Policy*. They could help answer questions that are difficult to get at, or that are simply not asked by those using other methods. Let me propose two possible areas in which this might be the case. One would be in better understanding the very different roles patents play in different scientific fields and industries. These roles are deeply conditioned by the materiality of the innovations they apply to as well as the social environment in which the roles are enacted. Studies sensitive to those factors would certainly provide better explanations for these cross-field differences than we currently have and, by implication, give us ideas about how to better manage these differences. A second focus might be properties of patents that don't easily fit into an economics-of-innovation framework. For example, nearly all research on patenting begins from the assumption that the purpose of patents is to allow inventors to exploit a monopoly for a limited time, in order to reward them for their inventiveness and provide an incentive for firms to invest in the development of inventions. But one recent study draws on a large body of evidence to argue that, with the notable exception of the pharmaceutical industry, "firms use patents not in order to appropriate their innovations and exclude other firms but rather in order to facilitate coordination with the other actors of innovation" (Penin 2005: 648). That is, in most industries, patents are used not as property that can be exploited but as, among other things, "myth and ceremony" (Meyer and Rowan 1977): they signal a firms' scientific competence, for example, and they act as facilitating mechanisms for inter-organizational collaboration. This suggests a need to at least reconsider the microfoundations of the economics of patent policy, an area in which the perspectives of other disciplines would both be pertinent. Of course, it would take a lot of work to bridge the disciplinary gap and find a common language. But the study of patent policy need not be dominated by economists. It should also be a location at which economic sociology and technology studies meet and push each other forward.

Notes

1. A notable exception is the work of Woody Powell and his collaborators (e.g. Colyvas and Powell 2006; Owen-Smith 2003; Owen-Smith and Powell 2001, 2003).

2. Owen-Smith 2005, which examines commensuration work done by technology licensing offices to compare and make decisions about how to handle very different technologies, is an exception that comes to mind.

3. This debate has not been completely resolved, however; see Shane 2004.

4. Though some universities—notably MIT, but other technical institutes and to some extent the land-grant schools as well—always accepted a role for themselves in the technology transfer process.

5. Deflators used are from table 10.1 of Executive Office of the President 2005.

6. For a comprehensive history of the postwar debate over government patent policy, see Kleinman 1995.

7. There are reasons this figure might be misleading, however. First, about two-thirds of these patents were held by the Department of Defense, which routinely turned over rights to inventions that contractors were interested in developing; thus the government only ended up with patents to the less promising inventions. Second, some agencies employed large numbers of patent attorneys who would be out of jobs if no inventions were being patented—thus they had an incentive to patent inventions even if they were not likely to be licensed. Third, the government often

permitted the use of its patented inventions without a formal license. See Eisenberg 1996 for a complete legislative history of federal patent policy on government-funded research starting with World War II, including a discussion of these statistics.

8. Nonprofit organizations other than universities were covered by the bill as well, but they received relatively little federal research funding and played no noticeable political role in this legislative battle. Universities, other nonprofits, and small businesses received less than 32% of extramural federal R&D spending in 1977 (NSF 2002, table 8; NSF 1982, table B-10). This figure excludes the research of federal employees, who would not have been given invention rights under any of these bills. It uses data on R&D funding for businesses with fewer than 1000 employees (about 1.2% of the extramural total) as a (slightly high) proxy for small businesses covered by the Bayh-Dole Act.

9. For an account of the final phase of the legislative battle, see Stevens 2004.

References

Bugos, G., and D. Kevles. 1992. Plants as intellectual property: American practice, law, and policy in world context. *Osiris* 7: 74–104.

Bush, V. 1945. *Science—The Endless Frontier: A Report to the President on a Program for Postwar Scientific Research*. U.S. Government Printing Office.

Cheit, E. 1971. *The New Depression in Higher Education: A Study of Financial Conditions at 41 Colleges and Universities*. McGraw-Hill.

Cheit, E. 1973. *The New Depression in Higher Education—Two Years Later*. Carnegie Commission on Higher Education.

Clark, M. 1980. The miracles of spliced genes. *Newsweek*, March 17: 62.

Colyvas, J., and W. Powell. 2006. Roads to institutionalization. *Research in Organizational Behavior* 27: 305–353.

Commission on Government Procurement. 1972. *Report of the Commission on Government Procurement*, volume 4. U.S. Government Printing Office.

Doing, P. 2004. "Lab hands" and the "scarlet O": Epistemic politics and (scientific) labor. *Social Studies of Science* 34: 299–323.

Eisenberg, R. 1996. Public research and private development: Patents and technology transfer in government-sponsored research. *Virginia Law Review* 82: 1662–1727.

Executive Office of the President. 2005. *Budget of the United States Government: Historical Tables Fiscal Year 2005*. U.S. Government Printing Office.

Federal Council for Science and Technology (FCST). 1965. *Annual Report on Government Patent Policy*. U.S. Government Printing Office.

Federal Council for Science and Technology. 1966. *Annual Report on Government Patent Policy*. U.S. Government Printing Office.

Federal Council for Science and Technology. 1967. *Annual Report on Government Patent Policy*. U.S. Government Printing Office.

Federal Council for Science and Technology. 1968. *Annual Report on Government Patent Policy*. U.S. Government Printing Office.

Federal Council for Science and Technology. 1970. *Annual Report on Government Patent Policy*. U.S. Government Printing Office.

Federal Council for Science and Technology. 1976. *Report on Government Patent Policy, 1973–1976*. U.S. Government Printing Office.

Genetic patents: Less than meets the eye. 1980. *Business Week*, June 30: 48.

Harbridge House, Inc. 1968. *Government Patent Policy Study: Final Report*. U.S. Government Printing Office.

Heller, M., and R. Eisenberg. 1998. Can patents deter innovation? The anticommons in biomedical research. *Science* 280: 698–701.

Hyysalo, S. 2006. Representations of use and practice-bound imaginaries in automating the safety of the elderly. *Social Studies of Science* 36: 599–626.

Innovation's Golden Goose. 2002. *The Economist*, December 14.

Kevles, D. 1994. Ananda Chakrabarty wins a patent: Biotechnology, law, and society, 1972–1980. *Historical Studies in the Physical and Biological Sciences* 25: 111–135.

Klein, H., and D. Kleinman. 2002. The social construction of technology: Structural considerations. *Science, Technology, and Human Values* 27: 28–52.

Kleinman, D. 1995. *Politics on the Endless Frontier: Postwar Research Policy in the United States*. Duke University Press.

Kleinman, D. 2003. *Impure Cultures: University Biology and the World of Commerce*. University of Wisconsin Press.

Krücken, G., F. Meier, and A. Müller. 2007. Information, cooperation, and the blurring of boundaries: Technology transfer in German and American discourses. *Higher Education* 53: 675–696.

Latker, N. 1977. Current Trends in Government Patent Policy. Paper presented at Conference on University Research Management, New York University.

Leaf, C. 2005. The law of unintended consequences. *Fortune*, September 19: 250–260.

Meyer, J., and B. Rowan. 1977. Institutionalized organizations: Formal structure as myth and ceremony. *American Journal of Sociology* 83: 340–363.

Mody, C. 2006. Corporations, universities, and instrumental communities: Commercializing probe microscopy, 1981–1996. *Technology and Culture* 47: 56–80.

Mowery, D., R. Nelson, B. Sampat, and A. Ziedonis. 2001. The growth of patenting and licensing by U.S. universities: An assessment of the effects of the Bayh-Dole Act of 1980. *Research Policy* 30: 99–119.

Mowery, D., R. Nelson, B. Sampat, and A. Ziedonis. 2004. *Ivory Tower and Industrial Innovation: University-Industry Technology Transfer before and after the Bayh-Dole Act*. Stanford University Press.

Mowery, D., and A. Ziedonis. 2002. Academic patent quality and quantity before and after the Bayh-Dole Act in the United States. *Research Policy* 31: 399–418.

Murray, F., and S. Stern. 2005. Do Formal Intellectual Property Rights Hinder the Free Flow of Scientific Knowledge? An Empirical Test of the Anti-Commons Hypothesis. Working paper 11465, National Bureau of Economic Research.

National Academy of Sciences. 1965. *Basic Research and National Goals: A Report to the Committee on Science and Astronautics, U.S. House of Representatives*. U.S. Government Printing Office.

National Conference on the Management of University Technology Resources. 1974. *Technology Transfer: University Opportunities and Responsibilities*. Case Western Reserve University.

National Science Foundation (NSF), Division of Science Resources Statistics. 1982. *Research and Development in Industry: 1980*. NSF 82-317.

National Science Foundation, Division of Science Resources Statistics. 1999. *National Patterns of R&D Resources: 1998*. NSF 99-335.

National Science Foundation, Division of Science Resources Statistics. 2002. Federal Funds for Research and Development: Fiscal Years 2000, 2001, and 2002. NSF 02-321.

National Science Foundation, Division of Science Resources Statistics. 2004. Science and Engineering Indicators 2004. NSB 04-01.

National Science Foundation, Division of Science Resources Statistics. 2005. Survey of Federal Funds for Research and Development: Fiscal Years 2002, 2003, 2004. NSF 05-307.

National Science Foundation, Division of Science Resources Statistics. 2006. Science and Engineering Indicators 2006. NSF 06-01.

Owen-Smith, J. 2003. From separate systems to a hybrid order: Accumulative advantage across public and private science at research one universities. *Research Policy* 32: 1081–1104.

Owen-Smith, J. 2005. Dockets, deals, and sagas: Commensuration and the rationalization of experience in university licensing. *Social Studies of Science* 35: 69–97.

Owen-Smith, J., and W. Powell. 2001. Careers and contradictions: Faculty responses to the transformation of knowledge. *Research in the Sociology of Work* 10: 109–140.

Owen-Smith, J., and W. Powell. 2003. The expanding role of university patenting in the life sciences: Assessing the importance of experience and connectivity. *Research Policy* 32: 1695–1711.

Parthasarathy, S. 2005. Architectures of genetic medicine: Comparing genetic testing for breast cancer in the USA and the UK. *Social Studies of Science* 35: 5–40.

Patent and Trademark Law Amendments Act (Bayh-Dole Act). 1980. U.S. Public Law 517, 96th Congress, 2nd Session (12 December).

Patent issue seen as a spur to DNA products. 1980. *Chemical Week*, May 28: 20.

Penin, J. 2005. Patents versus ex post rewards: A new look. *Research Policy* 34: 641–656.

Pfeffer, J., and G. Salancik. 1978. *The External Control of Organizations: A Resource Dependence Perspective*. Harper and Row.

Pinch, T. 1996. The social construction of technology: A review. In *Technological Change*, ed. R. Fox. Harwood.

Sell, S. 2003. *Private Power, Public Law: The Globalization of Intellectual Property Rights*. Cambridge University Press.

Shane, S. 2004. Encouraging university entrepreneurship? The effect of the Bayh-Dole Act on university patenting in the United States. *Journal of Business Venturing* 19: 127–151.

Stevens, A. 2004. The enactment of Bayh-Dole. *Journal of Technology Transfer* 29: 93–99.

U.S. General Accounting Office (GAO). 1968. *Problem Areas Affecting Usefulness of Results of Government-Sponsored Research in Medicinal Chemistry*. U.S. Government Printing Office.

U.S. House of Representatives. 1979. Committee on Science and Technology. Subcommittee on Science, Research, and Technology. *Government Patent Policy*. U.S. Government Printing Office.

U.S. President's Science Advisory Committee. 1963. *Science, Government and Information: The Responsibilities of the Technical Community and the Government in the Transfer of Information*. U.S. Government Printing Office.

U.S. Senate. 1963. Select Committee on Small Business. *The Role and Effect of Technology on the Nation's Economy: Hearings*. U.S. Government Printing Office.

U.S. Senate. 1978. Select Committee on Small Business. Subcommittee on Monopoly and Anticompetitive Activities. *Government Patent Policies: Institutional Patent Agreements*. U.S. Government Printing Office.

U.S. Senate. 1979. Committee on the Judiciary. *University and Small Business Patent Procedures Act*. U.S. Government Printing Office.

U.S. Senate. 1980. Committee on the Judiciary. Subcommittee on Courts, Civil Liberties, and the Administration of Justice. *Industrial Innovation and Patent and Copyright Law Amendments*. U.S. Government Printing Office.

Vaughan, D. 1999. The role of the organization in the production of techno-scientific knowledge. *Social Studies of Science* 29: 913–943.

Walsh, J., A. Arora, and W. Cohen. 2003. Research tool patenting and licensing and biomedical innovation. In *Patents in the Knowledge-Based Economy*, ed. W. Cohen and S. Merrill. National Academies Press.

III Technology and the Material Arrangements of the Market

7 Technology, Agency, and Financial Price Data

Alex Preda

The popularity of Adam Smith's metaphor of the invisible hand (Smith 1991 [1776]: 351) has been manifest, among others, in fictionalizations that conjure the Scottish philosopher to reflect upon the problems of global markets. Among the more recent invocations is that of the German philosopher Peter Sloterdijk, who imagines Smith giving a dinner toast to Lord North (the British prime minister in the 1770s). In this toast, the powers of the invisible hand that coordinates the interests of market actors are related to an artifact: the cloth[1] woven by Penelope, Ulysses' wife, while she waited for her husband to return. Smith toasts as follows:

We have it much better today, gentlemen, since we have the privilege to observe how an invisible hand produces the same thing day and night, a cloth which is many thousand times bigger, knottier, richer in threads and patterns than the wedding shirt of Ithaca, and much more useful too, because, as you know, that wedding shirt should never be worn, since Ulysses came home at last. And how bewildered must we be, many times more than the troop of impudent guests who wooed the favors of a matron? While Penelope's own hand undid what she had woven during the day, weaves the world market behind our backs according to its own obscure laws exactly that, which we disbanded when we entrusted our fate into the division of labor and trade. (Sloterdijk 2006: 317–318)

The metaphor of the market as an invisible hand (i.e., coordinating device) is supplemented here by that of the cloth. Penelope's cloth can be seen as mediating between her, the unwanted suitors, and the absent husband: as such, it maintains a web of social relationships. This provides a good starting point for investigating the role of artifacts and technologies in weaving market relationships. One of the domains where this can probably be best investigated is financial markets. Massively relying on and investing in data collection, processing, and transmission technologies, financial markets are now in the process of a global transition to "click and trade" procedures, whereby face-to-face transactions are replaced by face-to-screen ones (Zaloom 2006; Knorr Cetina and Bruegger 2002a,b).

Technology has been seen as a mediating agent (Pinch 2003: 248; Callon 2004: 121; Callon and Muniesa 2003; Callon 1998: 15; Barry and Slater 2002: 177) that aligns the positions and interests of heterogeneous, dispersed actors (e.g., of traders and investors) by producing and distributing standardized information, which contributes to the rationalization of economic action. Technology standardizes financial data, allowing economic actors to rationalize their future courses of action and to project the outcomes of these actions, a process that establishes boundaries between efficient and non-efficient actions. Nevertheless, can we identify and analyze features of technology that, while allowing for standardizing routines, open up supplementary paths of institutional invention?

I suggest here the concept of 'generator' as a way of grasping such features, and apply this concept to the study of price-recording technologies (e.g., the stock ticker). These technologies are relevant because transactions depend on price data, because interpretations of market events are grounded in collecting and processing price data as an explanandum, and because formal models of price behavior (a central tool in derivatives markets—see MacKenzie and Millo 2003) cannot work without price data.

In the first step of the argument, I discuss the ways in which the relationship between technology and (financial) markets has been conceptualized. In the second step of the argument, I present the concept of generator, defined through mutually reinforcing (1) temporal structures, (2) visualization modes, (3) representational and interpretive languages, (4) cognitive tools and categories, and (5) group boundaries. I argue that these five features constitute agential aspects of financial technologies. In the third step of the argument, I apply the concept of generator to a case study of the stock ticker, the first custom-tailored financial technology. Given the fact that there have been relatively few historical studies of these technologies, the ticker case can contribute to a better understanding of how markets are shaped by technology.

The stock ticker was invented in 1867 by Edward A. Calahan, an engineer associated with the American Telegraph Company. It was a printing telegraph with two independent type wheels, placed under a glass bell jar (to keep off dust) and powered by a battery (Jenkins et al. 1989: 153). The wheels were mounted face to face on two shafts and revolved under the action of an electromagnet. The first wheel had the letters of the alphabet on it; the second wheel had numerals, fractions, and some letters. The inked wheels printed on a paper tape divided into two strips: the security's

Figure 7.1
Cartoon of an investor consulting the ticker tape. Source: Harper 1926.

name was printed on the upper strip and the price quote on the lower one, beneath the name. In the 1870s, the ticker also began to record the traded volume, printing it on the tape immediately before the price. The machine could be manned with only one (expensive) Morse operator (at the recording end), instead of two (with one at each end). In December 1867, ticker operators and machines were installed on the floor of the New York Stock Exchange and in the brokerage offices of David Groesbeck; Work, Davis & Barton; Greenleaf, Norris & Co.; and Lockwood & Co. (Calahan 1901: 237). The ticker has been in operation ever since, being continuously upgraded. The mechanical version (figure 7.1) was replaced in 1960 by an electronic one.

The primary data I use are provided by US investor manuals, brochures, newspaper articles, reports, investors' diaries, and the reminiscences and correspondence of stockbrokers covering a period from about 1868 to 1910. This time span coincides with the period when the ticker was introduced and enthusiastically adopted, first in the United States and later in Great Britain. I examine not only public representations and comments on this technology, but also how individual users perceived and combined it with other technologies like the telegraph and written correspondence. My approach is that of a historical case study, grounded in a reconstruction of knowledge processes from the documents of the financial marketplace.

Technology as a Market Agent

The notion of agent sends us to the capacity of (social) action to transform given structures and to reconfigure its own context (Emirbayer and Mische 1998: 970). The projective dimension is essential to the concept: we speak of agency when the iteration of past actions is accompanied (or replaced) by the generation of new paths of action, leading to structural change and to contextual reconfiguration (Emirbayer and Mische 1998: 971; Giddens 1987: 204). A reduction of agency to intentional human action is plagued by conceptual regress and circularity (Pickering 2001: 164; Schatzki 2002: 190; Lynch 1992: 251–252). Part of the solution to these problems is considering artifacts and technologies as "endowed with powers of determination that either render these entities as potent as social phenomena or make materiality and sociality codetermining" (Schatzki 2002: 108). A large body of research has examined how technology contributes to projecting new paths of action and to structural change (e.g., Pickering 1995; Woolgar 1991; MacKenzie and Wajcman,1985; Grint and Woolgar 1995; Bijker et al. 1987), while cautioning against treating all possible future courses of action as pre-determined by technological structures.

   Against this general background, the concept of calculative agency has been put forward as a way of capturing how technologies shape markets. Operating with a broad definition of technology, which includes theoretical and disciplinary aspects (Callon 2004: 123; Barry and Slater 2002: 181), calculative agency is defined by framing, disentanglement, and performativity. Framing designates the process through which technology creates calculable objects (Callon 1998: 15; Barry and Slater 2002: 181) and separates them from non-calculable ones. Disentanglement is the marking of boundaries between what is relevant or non-relevant with respect to calculability (Callon 1998: 16). Performativity designates the status of technology (including economic theories) as a set of intervention tools in market transactions.

   At the core of this approach is the notion of technology as a standardizer.[2] Making objects calculable requires treating them as abstract, homogeneous entities characterized by a restricted set of properties, to which a set of context-independent operations is applied. Standardization implies boundary marking (e.g., between relevant and irrelevant properties) as well as the projection of similar paths of action across various contexts. Economic theories, as well as material artifacts, provide both the tools and the criteria for standardization.

Standardization is different from the compression of information into data (e.g., Tarr, Finholt, and Goodman 1987: 79; Yates 1986: 150; Rousseau and Sylla 2001: 35; O'Rourke and Williamson 1999: 215). Standardization, which involves a set of rules, conventions, and tools, transfers data across contexts. It plays a major part in the rationalization of decision making and in the constitution of financial transactions as separated from broader social ties and obligations. Trust and authority are dissociated from individuals and transferred to technology: trustworthy data are data produced or recorded by an authoritative technology, which can be transferred across heterogeneous contexts without losing their properties.

Standardizers can create new boundaries and/or shift existing ones with respect to professional jurisdictions (e.g., Abbott 1988: 219–220), gender (e.g., Bertinotti 1985; Fischer 1992; Siegert 1998; Bakke 1996), and time and space (Stein 2001: 115; Flichy 1995: 10–11). The agential force of standardizers thus consists mainly in opening predictable and reliable paths of economic action across heterogeneous contexts. In doing this, they expand the sphere of economic transactions and make market actors to adopt distinctions and operations incorporated in technology. By promoting order and efficiency, standardizers introduce routines which may preclude further reconfigurations of action contexts. Once a mode of calculation (i.e., a set of routines) has been established, unexpected paths of action are discouraged. There are situations, however, in which the agential force of technology might include opening up domains of institutional invention, along with the introduction of routine-related constraints. Standardizers invent routines, but can we conceive of technology as inventing "invention" (Mumford 1967: 255) alongside routines?

Financial Technology as a Generator

Sociologically speaking, technology can be seen as a set of rules, conventions, and tools (e.g., MacKenzie and Wajcman 1985: 3; Bowker and Star 1999). This view is grounded in Karl Marx's notion of machines as "crystals of social substance" and "implements of labor" (1996 [1867]: 48, 389) and in Émile Durkheim's insight that (practical) knowledge is incorporated in the artifacts with which we operate (1995 [1915]: 440). Technology has an iterative dimension (skills and routines), as well as a projective one.

Once we accept that technology is social action, we have to treat it as having its own temporal structures. Action that is directed toward other actors (users) and toward the future (the agency condition) is defined through an internal time structure (Schutz 1967: 68–69). For instance, a technology that produces data sporadically and at irregular intervals differs from one that produces data continuously and at regular intervals. Data perceived as representing past transactions differ from data representing current transactions. Moreover, action that projects its own temporal structures toward other actors (users) elicits responses that also are temporally structured. The rhythm of price data requires temporally specific actions (as manifested in observation, attention, and interpretation). Thus, at a first level, technology generates temporal structures visible in the rhythm of data and in the rhythm of the users' responses to them.

Furthermore, the temporal structure of action is visually articulated and presented *as* action to other actors. These articulations can manifest themselves as working action or as performed action (Schutz 1967: 214). Working action is generated in a continuous flow, in the present tense, whereas 'performed action' refers to the past. Working action is visualized as a continuous flow of data; performed action is articulated in closed visual arrangements. Data presented as a continuous flow differ from a table or a list which refers to the past. Thus, technology displays its own temporal structures, eliciting specific responses from its users. These responses include adequate language tools for designating, describing, and interpreting both (price) data and the users' reactions to them. Taken together, visual arrangements, time structures, and language tools project avenues for interpreting and processing price data as transaction-relevant. This requires that users regard data not only as an object of contemplation, but also as a "manipulative area" (ibid.: 223), acted upon with the help of interpretive tools.

Therefore, once we conceive technology as social action, we are compelled to take into account, as intrinsic agential features, its temporal structures, visualizations of those structures, and the tools with which such visualizations are processed. If they are to be transferred across contexts, the temporal structures of technology have to be endowed with authority and tied to users' responses. This requires distinctions between users who are entitled and able to respond and those who are not, and between users who are entitled to take up further tools and those who are not. Such distinctions, in their turn, are relevant to issues of access and control: who is entitled to own the technology, to observe price data, to interpret them, to use them in transactions. A price-recording technology would thus be tied to status and access issues. It could reinforce existing status boundaries, but also create new, access-based ones.

Issues of status and access are related to distinctions like the one between public and private transactions. When the authority of price data combines reliability (iteration across contexts) with charismatic features (status and prestige), a double movement emerges: authoritative data are kept in the sphere of public transactions, but access to them is controlled. One means of control would be restricting access to data according to status; another would be restricting access to the tools which help interpret these data. Both would imply the emergence of a group that controls the tools (data analysts) and a reorganization of the activities related to their use.

To reiterate: The concept of generator emphasizes that, when we conceive technology as social action, we are compelled to deal with its intrinsic temporal structures and with the modes for visualizing those structures. Temporal structures open up means of creative intervention in financial transactions, means that go beyond and embed data standardization. Using this concept, I will turn now to the analysis of a specific case: the stock ticker as the first technology specifically designed to be used in financial markets.[3] I will start from the distinction between the price-recording and the price-transmitting features of technology, on the premise that the two will not necessarily overlap. I will focus on the user side of the stock ticker and examine how it generated and visualized temporal structures, representational languages, and boundaries which changed the structural conditions of financial action. I will first specify the methodology employed here.

Methodology

I use the case study method, which allows the reconstruction and analysis of interrelated yet heterogeneous variables (Stake 2000: 24) constituting user practices, variables which otherwise cannot be grasped with quantitative tools. This method also allows an investigation of the temporal structures brought about by a new price-recording technology, as well as the changes triggered by them.

Available historical data also can enhance a qualitative case study. Before the 1870s, many brokerages did not keep transaction ledgers, and so it is hard to evaluate how the number of investors evolved before and after the introduction of the ticker. Stockbrokers operated with paper slips, which are now lost. Many "kept accounts in their heads" and "considered the whole paraphernalia of book keeping a confounded fraud" (Clews 1888: 152, 154). Quite a few brokerage houses were ephemeral; others had

various lines of trade, financial securities being just one of them. Brokers'
archives have been lost or destroyed. To the best of my knowledge, there
is no aggregate data on individual or institutional investors, and brokerage
houses for the period before World War I, even for New York City.

Although this case study cannot be seen as representative for all stock
exchanges, stockbrokers, or investors in the United States, it shows how a
pattern of change took place on the New York Stock Exchange—at the
time, one of the most important in the world.

The data I examine consist of investment manuals, journal articles,
descriptions of the New York Stock Exchange, reminiscences of stockbrokers, diaries of investors, and letters from brokerage houses, covering
the period from 1868 to about 1910.[4] I closely examined more than 340
original documents. In accordance with the principles of archival research
(Hill 1993: 64), I cross-checked data from different archives and combed
archival holdings for new data. Combing was repeated until no new relevant data could be found. I verified the reliability of the sources by cross-
checking documents from different years and sources for each of the
aspects analyzed and by checking printed documents against manuscripts
(Kirk and Miller 1986: 42).

Equipped with this apparatus, I will now turn to the ticker, starting before the advent of the ticker with the constitution of securities prices as
data.

The Constitution of Price Data before the Ticker

In spite of all its apparent benefits, the telegraph did not automatically induce investors to send price information by telegram. When we examine
the correspondence of investors and stockbrokers, we can see that, even
after the inauguration of the transatlantic cable in 1865 and the introduction of the first telephones to Wall Street in 1878 (Anonymous 1927: 753),
brokers still used letters extensively. The business correspondence of
Richard Irvine & Co. (a major New York brokerage house with an international clientele) with its British clients shows that in most cases orders were
placed by letter. In other cases, Irvine & Co. included price quotations in
letters to investors, and asked them to order back by cablegram. In 1868,
Irvine was still providing twelve-day-old quotations to some of his clients.
Of course, cablegrams were rather expensive, and investors used them par-
simoniously. At the same time, the fact that price quotations were circulated by letters between New York and Europe indicates two aspects. On
the one hand, information was enmeshed with narrative structures gener-

ated in letters. When writing to clients about a successful shipment of fruit, Irvine & Co. offered some attractive stocks too, together with the latest New York quotations:

We have shipped to you care of Messrs Lampart and Holt, by this steamer, the apples you ordered in your favor of the 20th September last. We are assured the peaches and oysters are of the best quality, and trust they will prove so. Below we give you memo of their cost to your debit. We think it well to mention that 1st Mortgage 6% gold Chesapeake and Ohio Railroad bonds can now be bought here to a limited amount at 86% and accrued interest. They are well thought of by investors, and were originally marketed by the company's agents as high as 14% and interest. We enclose today's stock quotations. (letter of Richard Irvine, New York, to J. A. Wiggins, London, 1872; New York Historical Society)

This, among other documents, makes clear why brokers and investors cherished letters. They were a more efficient means of distributing information, networking, and deal making—in short, of producing knowledge and relationships at the same time. In this perspective, brokers were nodes in a network in which knowledge, deals, and private services overlapped. For all these practical purposes, letters worked very well. In this example, relevant information cannot be separated from a complex narrative structure evocative of deep social ties, of an economy of favors, and full of allusions impossible to render by means of a telegram.

On the other hand, financial actors needed accurate, timely information about price variations. It is more or less irrelevant to know that the price of, say, the Susquehanna Railroad Co. is at $53\frac{1}{8}$. What is really relevant is whether it is higher or lower than 30 minutes ago, or an hour ago, or yesterday.

First and foremost, this kind of information required that prices be recorded in an adequate fashion. What the public got to see were price lists, published in the commercial and general press. In fact, it was impossible to determine what kinds of prices were being published in the lists and how they were recorded in the first place. The practice of publishing closing prices was not common everywhere. In New York City, publications like the *Wall Street Journal* began publishing closing quotations only in 1868. The New York Stock Exchange got an official quotation list on February 28, 1872. 'Official' did not mean, however, that the NYSE guaranteed price data. In the 1860s in London, only the published quotations of consols (a form of British government bond) were closing prices. A closer inquiry into what 'closing prices' meant provides us with the following specification: [Until 1868] "there were no official closing quotations. The newspapers would publish such late quotations as some broker, who remained

late, saw fit to furnish." (Eames 1894: 51) Besides, some prices were compiled from the floor of the Exchange, while others were compiled from private auctions, which ran parallel to those on the exchange floor (Martin
1886: iv).

There was also a long history of forging price lists. Typographic errors
were all but infrequent. Stock quotations published in newspapers were accordingly not always perceived as reliable; in fact, stock price lists had a singularly bad reputation in the United States (and not only there) for being
unreliable and prone to manipulation (e.g., Anonymous 1854: 10).

Thus, data on prices, not data on price variations data, were regarded as
relevant. Technologies of private communication (letters) were used in
order to confer credibility upon price data. These communicative technologies emphasized personal authority and trust over accuracy and timeliness.
The boundary between private and public price data was tilted in favor of
the private domain.

Temporal Structures, Price Data Boundaries, and Status Groups in Financial Markets

This brings us to the question of the organization of knowledge production
underlying financial transactions. In the 1860s, just what did the words
'Wall Street' designate? Trade in securities was in fact carried out by two
wholly distinct classes of brokers. One comprised members of the Regular
Board, who inherited or paid hefty sums for their seats and who traded in
tailcoats and tall hats, sitting on their personal chairs, from a fixed place
in the room. The other class comprised brokers of the Open Board, who
did not inherit any seats, paid much lower membership fees (under a tenth
of what Regular Board members paid), and traded *in the street* (standing, of
course). The public could not join the sessions of the Regular Board, but
constantly mixed with the Open Board (Anonymous 1848: 8–9). The trading volume of the Open Board was estimated by some contemporaries to be
ten times that of the Regular Board (Medbery 1870: 39).[5]

The Regular Board traded by calls: securities were called out loudly, one
by one. For each call, stockbrokers bought and sold according to orders
received in advance; afterwards, trading was interrupted. The vice-president
of the Board repeated the price to his assistant secretary, who repeated it to
a clerk, who wrote it down on a blackboard. Then the next call was traded,
its price was ceremoniously repeated, and so forth. The Open Board took
the ground floor, the entrance to the building (the doors stayed open),
and the street. The public was in and out all the time. The Open Board

traded uninterruptedly and moved in the evening to the Fifth Avenue Hotel, so that its market was open for about 12–14 hours a day. By contrast, the Regular Board traded between about 10 a.m. and 2 p.m., and its members made sure they did not miss their lunch breaks (Eames 1894: 51–57; Smith 1871: 76–77).

We encounter here a closed status group operating a discontinuous market (the Regular Board) and a relatively open group operating a continuous market (the Open Board). There were multiple prices for one and the same security. The prices of the Regular Board were recorded after each call, but were discontinuous. The prices of the Open Board were—according to contemporaries—more or less continuous, but not all of them were necessarily (and certainly not accurately) recorded.[6]

Brokerage offices employed messenger boys to record prices from Broad Street. In an article published in 1901, Edward A. Calahan (the ticker's inventor) reminisced that each office employed 12–15 boys (rarely over the age of 17), who had to ascertain prices from the street and from the building (Downey 2000: 132–133). Each boy focused on certain stocks and yelled prices to other boys, who wrote them down on paper slips, which were lost, misread, misdirected, or forged on a daily basis (Stedman 1905: 433). About 30 or 40 brokerage houses sent their armies to 10–12 Broad Street (Calahan 1901: 236):

... intermittent messenger-boys twist in and out, carrying hurried whispers back to offices, or dashing forward with emergent orders for brokers whose names are shouted by the page boys in shrillest treble. The roar from the cock-pit rolls up denser and denser. The President plies his gravel, the Assistant Secretaries scratch across the paper, registering bids and offers as for dear life. The black tablet slides up second by second with ever-fresh figures evolved from the chaos below. Every tongue in every head of this multiform concourse of excited or expectant humanity billowing hither and tither between the walls, is adding its contribution to the general bedlam. (Medbery 1870: 30)

On the first floor of 10–12 Broad Street, agents listened for the prices of the Regular Board, wrote them down, and sold the information in the street. This privilege was sold for $100 a week (Clews 1888: 8)[7]. Seen up close, this was "perfect bedlam" (Smith 1871: 76). Seen from afar, this technology was well adapted to multiple prices. Messenger boys were highly mobile, and the tools they employed (paper slips, pencils), though neither accurate nor forgery-proof, could be easily carried in their pockets.

Under these circumstances, it was understandable that even big brokerage houses did not bother much with telegrams: the telegraph did not solve a basic problem of the marketplace, namely that of tying price data directly to floor transactions. In between there were paper slips littering the floor, crowds of courier boys running in all directions, shouts and yells, and, not infrequently, forgers. In the Open Board market, contracts were written down by back office clerks at the end of the working day. In the Regular Board market, stockbrokers stopped the market to record transactions. In this arrangement, the time at which prices were recorded could be estimated with some accuracy, but this did not help much, since on the ground floor, where the Open Board reigned, prices changed continuously. The existence of two markets in the same building—one continuous, the other discontinuous—contributed to multiple prices and to parallel, heterogeneous time structures. The technology obscured any direct relationship between the published price data and the interaction side of financial transactions. These transactions were, then as in the eighteenth century (Preda 2001b) and as today (Knorr Cetina and Bruegger 2002b), conversational exchanges: securities prices were set by conversational turns.

All this meant that the interactional price-setting mechanism of the marketplace was the speech act. Speech acts had to fulfill specific felicity conditions in order to be valid, of course: participants knew one another, had legitimate access to the floor, and had a transactional record, among other things. But this does not obscure the fact that it was the perlocutionary force of a speech act (Austin 1976 [1962]) that set the price. Paper slips fixed and visualized this conversational outcome post hoc and only for momentary needs. They were an ephemeral trace left by conversations which, if observed from the visitors' gallery, appeared as a cacophonous jumble of shouts and wild gestures. This spectacle was intriguing enough for tour operators to routinely include a visit to Wall Street in their "visit New York" packages, marketed to the middle classes of provincial cities. A special gallery was built on the first floor of 10–12 Broad Street so that tourists could contemplate the "mad house" (Hickling 1875: 12). Conversations could not be directly and individually witnessed; the paper slips, sole proof of their ever having taken place, were less long-lived than a fruit fly. This is why all commentators of the financial marketplace, from the eighteenth century on, emphasized the importance of honor as an unspoken condition for the felicity of transactions.

In the same way in which the speech act's felicity conditions required that the broker was honorable and known to the other participants on the floor, the paper slip had to be handwritten, signed, and certified as original and emanating from its author. This, of course, made the price dependent on the individual, context-bound features of conversational exchanges.

The higher-status brokers had a monopoly on trust and credibility. The Regular Board, through its elaborated rituals of recording price data, invested it with authority. The lower-status brokers, working continuously in the street, did not invest its data with similar features. The boundaries between authoritative and less authoritative price data and between higher and lower social statuses were not affected by paper slips and messenger boys.

To sum up: We encounter here a technology with a ragged temporal structure (irregular intervals, parallel times, and holes), a technology which neither presents actual transactions to observers nor represents (accurately) every transaction. The assemblage of price data in lists is separated from the generation of data. Market interactions are not made visible to the public. Existing boundaries between public and private transactions and between status groups are kept in place.

The Ticker and Price Data Monopolies

With the advent of the ticker, the brokerage office was directly connected to the floor of the exchange and had access to real-time prices. However, technical problems were soon to arise: on the one hand, the stiletto blurred and mixed up letters and numbers, instead of keeping them in two distinct lines. On the other hand, tickers required batteries, which consisted of four large glass jars with zinc and copper plates in them, filled with sulphuric acid; this, together with non-insulated wires, made accidents very frequent in the tumult of a brokerage office.

These problems were solved in the 1870s by Henry van Hoveberg's invention of automatic unison adjustment and by the construction of special buildings for batteries, respectively; brokerage offices were connected now to a central power source (the battery building) and to the floor of the Stock Exchange. The Gold Exchange received a similar instrument, the gold indicator; a clocklike indicator was placed on the facade of the Exchange so that the crowds could follow price variations directly (Stedman 1905: 436).

After technical problems had been dealt with, several companies competed for the favors of brokerage houses and investors alike. Samuel Laws's Gold and Stock Reporting Telegraph Company (which reported gold prices) competed with Calahan's Gold and Stock Telegraph Company until 1869, when they merged. The outcome, the Gold and Stock Telegraph Company, merged in its turn with Western Union in 1871. In the early 1870s, Western Union's competition was the Gallaher Gold and Stock Telegraph Co.

(Jenkins et al. 1989: 357). Technical improvements were accompanied by conflicts about patents: Thomas Alva Edison set out to circumvent Calahan's patent by developing his own "cotton ticker," with the type wheels mounted on the same shaft. In 1869, he developed a one-wire transmission technology, which competed with the three-wire technology of the Gold and Stock Telegraph Co.

According to Calahan, Samuel Law's company had secured an exclusive contract with the New York Stock Exchange even before the technology was developed. When Calahan appeared with his ticker, he replaced Law as the exclusive supplier of the stock exchange. Official brokers competed for being first on Calahan's delivery list (Calahan 1901: 237).

In the 1870s, for instance, the New York firm of Ward & Co. paid a monthly rent of $25 for their "instrument." Other firms rented tickers at a rate of $1 per day. While the figures about the number of tickers in use at the turn of the twentieth century are contradictory, it is clear that the ticker was present in provincial towns, at least between the Midwest and the East Coast. Around 1900, there were 1,750 tickers in Manhattan, Brooklyn, and New Jersey (Pratt 1903: 139). Edmund Stedman (1905: 441) claimed that 23,000 US offices subscribed to ticker services. *The Magazine of Wall Street* (Anonymous 1927: 753) stated that in 1890 there were about 400 tickers installed in the United States; in 1900 there were more than 900, and in 1902 the number reached 1,200. Other publications (Gibson 1889: 82) claimed that by 1882 there were 1,000 tickers in New York City alone, rented to offices at a rate of $10 per month. Peter Wyckoff (1972: 40, 46) estimates the number of tickers in use at 837 in 1900 and 1,278 in 1906. Bond tickers were introduced in 1919. Contemporary observers thought that price-recording over the ticker was less prone to errors than it was over the phone (Pratt 1903: 142); the ticker thus expanded even after the introduction of telephone services.[8]

Edward A. Calahan, who had worked as a messenger boy in his youth (Calahan 1901: 236), wrote of the necessity of quelling the noise and the confusion emanating from the stock exchange. It appears that his motivation for developing the stock ticker was not tied to issues of increased efficiency or speed in disseminating the price data. The main issue was eliminating the disorder that affected the working of the Regular Board. Another issue was paying only one skilled, expensive Morse operator. As I will detail below, the users, on their part, were not motivated by ideas of efficiency or of increased access to price data. They were driven by their desire to consolidate their status. They wanted a monopoly over authoritative data and sought to have exclusive deals with telegraph companies.

The technology brought together engineers, who developed competing machines, the telegraph companies, which sought to secure exclusive contracts, and the official stock brokers, who wanted exclusivity of use. The ticker was not wanted for efficient, accurate, and broad diffusion of price data. It was wanted because it helped reinforce social status and a monopoly over authoritative price data.

The Ticker as a Time Generator

In 1907, Edwin Lefevre represented the effect of the ticker upon speculators as follows in his novel *Sampson Rock of Wall Street*:

He [Sampson Rock] approached the ticker and gazed intently on the printed letters and numbers of the tape—so intently that they ceased to be numerals and became living figures. Williams was ten million leagues away, and Rock's vision leaped from New York to Richmond, from Richmond to Biddleboro, from Biddleboro back to the glittering marble and gold Board Room of the Stock Exchange. The tape characters were like little soldier-ants, bringing precious loads to this New York office, tiny gold nuggets from a thousand stock holders, men and women and children, rich and poor, to the feet of Sampson Rock. (Lefevre 1907: 16)

The binding of the observer to the rhythms of the price data is apparent not only in individual diaries, but also in warnings about its pernicious, if not outright addictive effects. An investor like Edward Neufville Tailer often inserted a lapidary yet telling note in his diary—for example: "Erie going into the hands of a receiver, was the cause of the great decline in stocks in Wall Street to day & I passed the entire day in the office of Mssrs. Webb & Prall watching the ticker." (Tailer 1893, July 26) But another author, Henry Harper, felt that the public had to be warned:

The individual who trades or invests in stocks will do well to keep away from the stock ticker; for the victim of "tickeritis" is no more capable of reasonable and self-composed action than one who is in the delirium of typhoid fever. The gyroscopic action of the prices recorded on the ticker tape produces a sort of mental intoxication, which foreshortens the vision by involuntary submissiveness to momentary influences. It also produces on some minds an effect somewhat similar to that which one feels after standing for a considerable time intently watching the water as it flows over Niagara Falls. (Harper 1926: 10–11)

The ticker endowed price data with a new temporal structure, visualized in a new way. It made visible price variations, which now flowed without interruption. Irregular, large time gaps were eliminated, or made imperceptible. Shrunken time intervals required more attention and coordination on the part of market actors. In the morning, actors set their watches before the tickers began to work (Selden 1917: 160), so that individual schedules were coordinated with the schedule of the machine. The time for the delivery of stocks, for example, was marked on the tape by 'time' printed twice, after which the ticker gave fifteen distinct beats (Pratt 1903: 139). The ticker worked as a device for reciprocal temporal coordination. From the perspective of the brokerage house, it helped orient the participants' time and rhythm to that of Wall Street. In this sense, the ticker appears as a networking technology (Latour 1999: 28–29, 306) that allows the transfer of temporal patterns across various contexts and the coordination of future paths of action.

The flow of price variations visualized the results of ongoing conversational exchanges, and disassociated their results from the individual authority of the participants in those conversations. This flow linked the results to each other, made the ties that bound them visible as the tape unfolded, and made the market in its turn visible as an abstract, faceless, yet very lively whole. All the felicity conditions which made the speech act valid (intonation, attitude, look, wording, pitch of voice, etc.) were blanked out. The flow of figures and letters on the ticker tape became an appresentation (Husserl 1977 [1912]: 112, 124–125) of market transactions: perception (of price rhythms) and representation (of floor transactions) fused together.

While stock list compilations separated the process of composition from the process of inscription (putting together the prices of various securities came after each price was recorded), the ticker integrated them (Collins and Kusch 1998: 109). At the same time, lists became more sophisticated and began to show, not only opening and closing prices, but also prices at different times of the day—a fact mirrored by more detailed market reports in the New York and provincial newspapers. In 1884, with the ticker a solid market fixture, Charles Dow began publishing average closing prices of active representative stocks (Wyckoff 1972: 31), thus initiating the first stock market index.

The ways in which brokers and investors worded their decisions had to be adapted to continuous data about price variations. This made brokerage houses adopt and distribute telegraphic codes to their clients, fitting the language of financial transactions to the new temporal structure. For example, the house of Haight & Freese (known in Street parlance as "Hate & Freeze") could telegraph a client "army event bandit calmly" instead of "Cannot sell Canada Southern at your limit. Please reduce limit to 23."

(Anonymous 1898: 385, 396). Here, 'army' stood for 'cannot sell at your limit', 'event' for 'Canada Southern', 'bandit' for 'reduce limit to', and 'calmly' for '23'. Had the investor wanted to sell, say, 150 shares Pacific Mail, he would have telegraphed back "alpine [sell 150 shares] expulsion [Pacific Mail]." This language was exclusively centered on representing the world of finance: one could build sentences using the word bandit, but it was impossible to formulate sentences about bandits.

Some code books reached mammoth dimensions: in 1905, the Hartfield Telegraphic Code Publishing Co. published *Hartfield's Wall Street Code*, containing about 467,000 cipher words, all related to securities and financial transactions in Wall Street. Others, like the *Ticker Book and Manual of the Tape*, published in the same year, included not only codes, but also maintenance instructions for the machine and tips for tape reading.

An effect of the new transactional language was that investors were bound to their brokers even more closely than before: as an investor, one had to learn a special telegraphic code from the broker's manual, spend as much time as possible in his office, and read his chart analyses. Brokerage houses advertised their codes as a sign of seriousness and reliability. Some distributed them to their clients for free, while others charged a fee. This was a means of asserting the prestige of the brokerage house, but also an attempt to keep a record of the investors using the code and thus limit forged orders. On the other hand, free telegraph codes were a means of attracting more clients and increasing business (Wyckoff 1930: 21).

The ticker abbreviations of the quoted companies became nicknames, widely used by market actors:

> ... because MP stands on the tape for Missouri Pacific, that stock is generally called 'Mop.' NP stands for Northern Pacific, which goes by the name 'Nipper,' the common being called 'little' and the preferred 'big.' PO standing for People's Gas Light and Coke Company, that stock is often called 'Post-office.' The same law of economy in the use of words applies to all the active stocks. (Pratt 1903: 136)

Not only was the new language standardized and adapted to the rapidity of transactions; it was tied to the visibility of price variations. The observer of financial markets could no longer be equated with the confused tourist standing in the visitors' gallery. The observer of the market was the observer of the tape. In this sense, the ticker contributed to a radical abstraction and reconfiguration of the visual experience of the market. This is perhaps best illustrated by subsequent developments. Due to its physical size, the ticker tape could be observed directly by only a few people at once. An operator, or tape reader, sat by the ticker and read the data. In

1923, a device called the Trans-Lux Movie Ticker was developed and tested on the floor of the NYSE. It projected the image of the ticker tape onto a translucent screen in real time, so that the flow of quotations was visible at once to all those present. Built according to NYSE's specifications, the Trans-Lux machine had a projection bandwidth of at least twenty quotations. Its success was so great that the parent company also developed the Trans-Lux Movie Flash Ticker and the Movie News Ticker. They were installed in banks, brokerage offices, and at different locations in the NYSE building. The Movie Flash Ticker projected a single, flowing line of business and political news onto the screen, while the Movie News Ticker could project a block of eight lines of text. The Trans-Lux company claimed that in 1929 there were 1,500 Movie Tickers in operation in 211 trading centers (Burton 1929: 14).

While the use of the ticker provided the same data everywhere, these were contingent upon temporal structures and the related mode of visualizing the market. The ragged time structure of paper slips was replaced by the smooth, uninterrupted, unique time of the ticker tape. The visualization of this structure replaced the rhythm of conversational transactions, but at the same time was equivalent with it. The visualization of price rhythms promoted observational and sociolinguistic principles superimposed on (and equivalent to) economic transactions. This should not imply that the ticker was built according to abstract economic principles. Quite the contrary: its way of working generated temporal structures directing investors and brokers along specific paths of action, orienting them toward price variations, re-entangling authority with human actors, coordinating individual schedules, changing the language. The ticker was neither a tool nor a proxy for human action (Collins and Martin Kusch 1998: 119). It did things that market actors could never have done without it.

From Tapes to Financial Charts: The Generation of Cognitive Tools

Perhaps not entirely by accident, the ticker was invented at a time when US psychologists were engaged in heated debates about constant attention as a fundamental condition of knowledge (Crary 1990: 21–23). The ticker firmly bound investors and brokers to its ticks. Constant presence, attention, and observation were explicitly required by manuals of the time. The duty of the stockbroker was simply to be always by his "instrument," which "is never dumb" and which ensures that the United States is "a nation of speculators" (Anonymous 1881). Investors too felt motivated to spend more time in their brokers' offices, watching the quotations and socializing. In his reminiscences, Richard D. Wyckoff, a stock operator and pioneer of chart analysis, wrote that in 1905 friends of his could sit and watch the tape for an hour and a half without any interruption (Wyckoff 1934a: 37). Wyckoff himself had trained hard so that he could watch the ticker tape for up to an hour. He remembered how, in 1907, James R. Keene, a financial speculator, fell into a "ticker trance":

I used to stand facing him, my left elbow on his ticker while talking to him. He would hold the tape in his left hand and his eye-glasses in his right as he listened to me, then on went the glasses astride his nose as he bent close to the tape in a scrutiny of the length that had passed meanwhile. I might be talking at the moment his eye began to pick up the tape again, but until he finished he was a person in a trance. If, reading the tape, he observed something that stimulated his mental machinery, I might go on talking indefinitely; he wouldn't get a word of it.... He appeared to absorb a certain length of tape, and to devote to its analysis a specified interval, measured by paces. Sometimes he returned to the ribbon for another examination, followed by more pacing. (Wyckoff 1930: 148)

The ability to watch and be in touch all the time was a key condition of playing the investing game (Wyckoff 1934a: 38; 1933: 26). Some brokerage offices tried to restrict access to the customers' rooms, where the tickers were placed, on account of the great number of "chair warmers, just sitting there and watching the ticker and talking, who repel the better class of business men" (Selden 1917: 106). Not only was one's presence in the stockbroker's office a must, if one was to be au courant with the latest price variations; it was also a must for the investor eager to hear "scientific" interpretations and analyses of price variations.

Financial charts had been in use in England and France since the 1830s (Preda 2001a). Traditional procedures of data collection, however, allowed the visualization of price variations over years or months, but not over a couple of days or over a single day. Price data were not collected daily, much less hourly, a difficulty mentioned by many chart compilers. The source of prices and the way in which they had been recorded were often highly uncertain. With the ticker, it was possible to visualize (and analyze) minute price fluctuations over hours, days, or months. Many commentators became aware of minute time differences, complaining about the fact that during peak hours the ticker might fall behind for up to four minutes, or that it would not record the entire volume of traded securities (Pratt 1903: 136; Selden 1917: 160; Wyckoff 1933: 24). This indicates an awareness of real-time coordination between the tape flow and market transactions, which in its turn requires coordination between the investor-cum-observer and the ticker tape:

The most expert type of tape-reader carries no memorandums, and seldom refers to fluctuation records. The tape whispers to him, talks to him, and, as Mr. Lawson puts it, "screams" at him. Every one is not fitted to become an expert tape-reader, any more than in the musical world can every one be a Paderewski. (Rollo Tape 1908: 34)

After the introduction of the Trans-Lux machine, a new time lag appeared, because the speed of the ticker tape was greater than the movie ticker's projection speed. By increasing spacing between quotations, this minute time lag was eliminated (Wyckoff 1934b: 20).

By 1900, detailed, by-the-hour financial charts began to be published in investor magazines. However, their importance had already been acknowledged in the mid 1870s: ". . . it becomes almost a necessity that the broker or operator should consult a table of fluctuations of stocks before he can form an intelligent opinion of future prices" (Anonymous 1875: 11). Since the data now had to be recorded much more rapidly, new cognitive skills were required: those of the "tape reader," a trained clerk or a stockbroker who stood by the ticker and picked out and recorded the price variations of a single security in a diagram, so that at the end of the trading day a chart was already available. All this required a great deal of concentration, not to mention agility, good eyesight, and a well-trained memory. The chart analyst became an established presence in brokerage offices, as well as in investor magazines:

There was a chart-fiend in our office—a wise-looking party, who traveled about with a chart book under his arm, jotting down fluctuations, and disposing in an authoritative way, of all questions relating to 'new tops,' 'double bottoms,' etc. Now, whatever may be claimed for or brought against stock market charts, I'll say this in their favor, they do unquestionably show when accumulation and distribution of stocks is in progress. So I asked my expert friend to let me see his 'fluctuation pictures,' my thought being that no bull market could take place till the big insiders had taken on their lines of stock. Sure enough, the charts showed, unmistakably, that accumulation had been going on at the very bottom. (Anonymous 1907: 2, 4)

In his thinly disguised biography of Jesse Livermore, Edwin Lefevre (1998 [1923]: 18–19, 22) stressed the importance of judgment by the chart, of having a proper system of assessing the meaning of fluctuations on the basis of charts alone, together with the strong desire for constant action. Commentators saw the technical chart as "the bird's eye view of the stock market" (Pratt 1903: 138). "Reading charts is like reading music, in which you endeavor to interpret correctly the composer's ideas and the expression of his art. Just so a chart of the averages, or of a single stock, reflects the ideas, hopes, ambitions and purposes of the mass mind operating in the market, or of a manipulator handling a single stock." (Wyckoff 1934c: 10)

The ticker tape is "the recorded history of the market" (Wyckoff 1934d: 16). It is the resultant force of fundamental statistics, economic changes, and political developments (Wyckoff 1934e: 12) and requires a type of analysis fundamentally different from statistics (Wyckoff 1934f: 23). The chart was the market, as well as the means of understanding the market.

At the same time, the establishment of the stock analyst as a distinct profession—one whose purpose was to ensure the impartial distribution of meaningful information to investors and help them make their decisions—was being loudly urged in journal articles. A stock analyst would be on the same plane as a physician who recommends a medicine solely on its curative merits; he "would have to stand on a plane with George Washington and Caesar's wife. He must have no connection with any bond house or brokerage establishment, and must permit nothing whatever to, in any way, warp his judgment. He must know all securities and keep actual records of earnings and statistics which show not only whether a security is safe, but whether it is advancing or declining in point of safety." (Anonymous 1908b: 35)

The new financial charts—unlike the older ones—came with their own metaphorical luggage and discursive modes: there were now "points of resistance," "double bottoms," "tops," and "shoulders" to enrich the analyst's arsenal. What's more, the chart continued the process initiated by the tape: it was the visualization of concatenated representations of conversational outcomes. Correspondingly, the analytical language is full of visual metaphors: we do not need any references to the bricks, furnaces, tracks, and machinery of stock companies any more. Price variations suffice. Discursive modes supported the chart as a cognitive instrument, which in its turn conferred authority on the stock analyst as the only one skilled enough to discover the truth of the market in the dotted lines. The analyst promoted the charts, which required a special language.

We encounter here a double game of authority and trust: on the one hand, authority and trust are transferred to the ticker. On the other hand, the rhythm of price data generates interpretive tools which require personal skills and thus invests authority in groups able to interpret charts. Instead of completely disentangling authority from human actors, the ticker gives rise to new status groups in financial transactions.

The Ticker as a Device for the Organization of Knowledge

The rhythm of price variations required a temporally structured response from users, which led to rearrangement of (1) the relationship between the brokerage office and the trading floor and (2) activities on the trading floor.

The ticker transformed the stockbroker's office into a kind of community-cum-communications center, where investors could spend the whole day watching quotations, talking to each other, and placing orders. In the customers' room, rows of tickers (attended by clerks) worked uninterruptedly, while clients seated on several rows of seats watched other clerks updating the quotations board. The modern brokerage office had a separate telegraph room, an order desk, a ticker room, and a back office. It is not evident from the available evidence that brokerage offices had been organized in this way before the introduction of the ticker. At the center of this spatial arrangement was the ticker room. (See figure 7.2.)

In the more important brokerage houses, the ticker room was often elegant, but noisy; in it, the "ever-changing position of the great markets" is "like a kaleidoscope" (Pratt 1903: 162). Advertising brochures praised the stockbroker's office as a model of efficient communication, accuracy, and machine-inspired modernity:

Figure 7.2
Trading on the Regular Board before the introduction of the ticker, with brokers seated and wearing tall hats. Source: Eames 1894.

A passenger standing on the observation platform in the engine-room of a modern ocean-liner will observe great masses of steel, some stationary, some whirling at terrific speed; he may go down into the boiler-room where is generated the power with which the great ship is driven, but all this will give him only a crude idea of the actual workings of the machinery of propulsion.... So it is with the machinery of a large banking and brokerage house. A client may spend many days in the customers' room, from which vantage point he will observe much, but his knowledge of the inner workings of the machine, built to handle orders in the various markets, must still be superficial.... Everything is run with clock-like precision. No matter how large a business is being done, there is no confusion, the plant being designed to handle the maximum volume of orders. (Anonymous 1908a: 7)[9]

The emphasis on observation and attentiveness fitted in very well with the overall discourse of the "science of financial markets," popular in late nineteenth century. It required of investors precisely those qualities preached by manuals: attention, vigilance, and constant observation of financial transactions and of price variations. For the investor, it is only reasonable to follow the market movements and to try to be efficient.

Moreover, the broker's office could influence the market by bringing the ticker into action. Richard Wyckoff recalled that in 1905 some friends of his were sitting in the New York office of Eddie Wasserman. Noticing that "the tape was barely moving," Wasserman

said to the clients in the office: 'Let's make up a little pool in Southern Railway and start a move in it. I'll buy thousand if you will.' Eddie went over on the floor and bought a few thousand shares all at one price. 'It came easily,' he said. Then he called up friends and told them there was going to be a move in Southern Railway. When all these trades appeared on the tape and in such an absolutely dead market, it did look as though something had started. Here was a chance for some of the thousands of people sitting around hundreds of tickers all over the country to get a little action. Outside buying orders began to come into the crowd; in a few minutes Southern Railway was up a point and a half. Eddie and his friends quickly took their profits. The evening papers said Morgan had been buying Southern Railway. (Wyckoff 1934a: 37)

In this account, "moving the tape" and "moving the market" are treated as synonymous; the idleness of the tape requires action and intervention. At the same time, this episode is indicative of the possibilities for manipulating a market that is perceived as the flow of letters and figures on a tape. On yet another level, this manipulation could now be performed anonymously: one did not have to show up in the marketplace in order to corner the market. It became more difficult to know who was actually "moving the tape."

Figure 7.3
The organization of the trading floor with specialized trading posts and rows of tickers in the background. Source: Eames 1894.

The floor of the exchange was reshaped too. In 1868, members of the Open Board began adopting a particular place for each security traded. In November 1870, the Regular Board moved downstairs and began trading in the same room with the Open Board, mixing up with its members. The two were officially merged. The practice of trading by Calls was abandoned. The market became a continuous, single-price entity (Eames 1894: 69; Smith 1871: 71; Selden 1917: 90). The merger created a larger status group, but excluded from access to data all those who were not members of the former Open and Regular Boards. The boundary between authoritative and less authoritative price data was redrawn by absorbing old competitors and by keeping out any new ones.

The stock exchange floor was organized in specialized "crowds"—brokers and market makers trading a single security around a ticker, under a street-lamp-like signpost (Nelson 1900: 19; Anonymous 1875: 9; Anonymous 1893). Each trading post had an indicator showing the latest quotation. Investors were relegated to a special, enclosed area on the floor, access to which cost $100 a year (Cornwallis 1879: 27). Tourists had access only to the gallery. The public could no longer mingle freely with brokers. Stock-brokers were issued with personal identification numbers; an electric panel was installed, and every time a stockbroker was called, his number was flashed in a color corresponding to the category of the caller—client, back office, and the like (Selden 1917: 90).

Re-drawing the Boundaries: The Fights around Ticker Use

I have argued that official stockbrokers adopted the ticker in order to main-tain their monopoly over credible, authoritative price data. Yet, since the technology brought together social groups with heterogeneous interests (developers, operators, users), the boundary between credible and less cred-ible data soon became fragile. These groups engaged in both competition and cooperation: ticker operators, for instance, cooperated with stock-brokers, but competed with each other. Official stockbrokers wanted to keep unofficial brokers away from the ticker, while the brokers sought ac-cess to ticker operators. Because operators wanted to expand their business (after all, there was only a limited number of official brokers), they started selling the technology to unofficial brokers. Struggles arose for control of ticker machines and tapes. Engineers fought about patents, founded ticker companies, and got involved in mergers. By the early 1870s, Western Union emerged as the dominant (though not the sole) provider of price quotations. Patents on existing machines were circumvented by patenting new models, with minor modifications; the result was that in the early 1870s there were gold, stock, and cotton tickers, which differed only slightly from each other.

Bucket shops[10] sought access to the ticker and wanted to compete with official brokerage houses. The willingness of operators to sell their services to unofficial brokers threatened the monopoly of the official brokerage houses. Power struggles raged over who should have access to price data; in 1889, in a short-lived attempt to drive bucket shops out and restore the value of the stockbrokers' seats, the NYSE banned all stock tickers (Wyckoff 1972: 33). In 1887, in another short-lived attempt, the President of the Chicago Board of Trade destroyed the tickers on the floor of the exchange and cut all the electric wires (Hochfelder 2001: 1). There were multiple reasons for these actions: on the one hand, bucket shops[11] dealt primarily in derivatives, taking thus a chunk of the business from the more estab-lished brokerage houses. On the other hand, trading in derivatives was still associated with gambling; the big houses argued that the combination of shaky bucket shops and derivatives was ruining the reputation of financial investments.

Together with the NYSE, official brokerage houses tried hard to control the flow of price information. In 1890, the New York Stock Exchange bought a controlling interest in the New York Quotation Co., which received the exclusive right to furnish quotations to the members of the exchange (Pratt 1903: 134). The public was serviced by other quotation companies, under certain restrictions. Tickers could be installed only in brokerage offices approved by the NYSE (Selden 1917: 158). This triggered bitter litigation between the NYSE and brokerage houses, which ended in 1892 in favor of the NYSE. In the late 1880s, the New York Stock Exchange tried (unsuccessfully) to deprive the Consolidated Stock and Petroleum Exchange (a much smaller exchange located in Lower Manhattan) of its ticker service, so as to restrict the Consolidated to unlisted securities (Wyckoff 1930: 23). The latter fought back, and was able to retain its tickers for a time.

On yet another level, the ticker greatly stimulated arbitrage activities (and competition too) between the New York and the London stock exchanges. In the 1890s, it took only minutes for New York prices to reach London and vice versa. The NYSE saw arbitrage as damaging to its interests and to its control over price information, and in 1894 it temporarily withdrew all tickers (Eames 1894: 65, 91). At that time, the London Stock Exchange was actively trying to win business away from the New York Stock Exchange and build up an "American Market" in London (Michie 1999: 79). Banning the ticker was seen as a means of discouraging arbitrage, maintaining a tight hold on price information, and keeping business in New York. For its part, the London Stock Exchange protected its business by furnishing incomplete price information over the ticker, much to the distress of American brokers. The tickers of the London Stock Exchange did not print the price and the volume of each trade (Gibson 1889: 83); this made "an American stock trader in London feel as though he had no information about the market worth mentioning with only these meager figures to go upon" (Selden 1917: 162). At the same time, in London prices were recorded by the ticker not from "crowds," but from inscriptions made by brokers on a blackboard on the exchange floor (Gibson 1889: 35). Incomplete inscriptions, made in haste, led to incomplete price data.

The enlarged Regular Board continued to have a monopoly over authoritative, continuous price data. By re-drawing the boundary between authoritative and less authoritative data, the ticker rather reinforced the social position of the Regular Board. It wiped out whatever informational advantages (i.e., continuity) unofficial brokers may have had before. Deprived of them, many unofficial brokers re-oriented themselves toward unlisted securities, which were outside the NYSE's monopoly. At the beginning of the 1920s, unofficial brokers dealing in unlisted stocks founded the American Stock Exchange.

Differences in the organization of knowledge and in price data were thus perceived as critical for access to financial transactions.[12] The organization of the stock exchange was synonymous with the "high organization of knowledge" (Anonymous, 1876: 25). How to monitor and collect quotations efficiently, how to transmit them further without missing anything, were topics regularly repeated by brokerage houses in their advertising brochures and manuals. In this sense, the ticker triggered a process of self-monitoring at the institutional level and generated principles of organization superimposed on the sociolinguistic view it disseminated. Both the organization and the individual had to pay more attention to what they did and to weigh courses of action.

Conclusion

I will come back now to the question of Penelope's cloth, the artifact which intervenes in the organization of human relationships. How does technology intervene in shaping market transactions? The generator concept, proposed here, takes into account that the agential features of financial technologies are inseparable from temporal structures, visualizations, representational languages, and cognitive tools. In the present case, a ragged temporal structure was replaced by a smooth one, with the consequence that price variations became visualizations of market transactions and objects of symbolic interpretation. The ticker made market exchanges visible as they happened, disentangled them from local conversations, and transformed them into something which is both abstract and visible in several forms to everybody at once. They are visible in the flow of names and prices on the paper strip, but also in the financial charts, which are nowadays also produced in real time. The quality of price data changed: instead of multiple, discontinuous, heterogeneous, and unsystematically recorded prices, we now have single, continuous, homogeneous, nearly real-time price variations. This does not mean, of course, that financial fraud and manipulation were eradicated forever: the evidence points to the contrary. But when moving the market means moving the tape, the possibilities for (anonymous) manipulation all but disappear.

Price-recording technologies are not neutral with respect to what constitutes price data. The quality of the data depends, among other things, on how they are generated: before the ticker, minute, continuous price variations could not constitute any real information for actors who were not permanently present in the marketplace; traders did not have the memory of a whole herd of elephants and the computational capabilities of an army of accountants. Financial economics postulates that direct observation of securities prices reduces uncertainty (Biais 1993). This postulate has to be sociologically complemented: forms of price observation in the marketplace are constituted by price-recording technologies. It is this constitution that processes cognitive and informational uncertainties.

The change in the quality of price data has not been without consequences for modern financial theory. Today's ultra-high-frequency research—a notion from financial economics, not from particle physics— is grounded in recording all financial transactions as they happen and analyzing (among other things) the spacing of data (e.g., Engle 2000: 1–2). Such speed would have been unimaginable without the appropriate technology. Sociologists have recently argued that financial theory has a performative character: it changes the market processes it claims to describe (MacKenzie and Millo 2003). Yet financial theory itself depends upon price data, which are impossible to obtain without complex technical systems.

In recent years, economic sociology has emphasized that prices are determined not only by efficiency criteria and computational rules, but also by factors such as social networks, interests, or status (see e.g. Swedberg 2003: 129). This perspective has become known as the social constitution (or construction) of prices. The generation and recording of price data, together with modes of data observation and analysis, are intrinsic to the social processes through which (securities) prices are constituted. In many markets, price data are generated and/ or recorded by technological systems. The investigation of these systems appears as a fruitful way of bringing together various disciplinary interests and of deepening our understanding about how markets work.

Acknowledgment

This is a revised version of a paper published in *Social Studies of Science* 36 (2006), no. 5.

Notes

1. Though this cloth was actually a burial shroud, Sloterdijk treats it as a wedding shirt.

2. This aspect is also emphasized by organization sociologists (e.g., Perrow 1967; Blau et al. 1976; Carruthers and Stinchcombe 2001), historians (e.g., Porter 1995), accountants (Miller 2001: 385), and sociologists of technology (e.g., Hughes, Rounce-field, and Toulmie 2002) who show that financial technologies standardize informa-tional outputs and make them transferable across various contexts.

3. Two years before the introduction of the ticker in New York City, there was a short-lived attempt to introduce an alternative technology in Paris, called the pante-legraph (Preda 2003). This technology, however, was quickly abandoned and was never used on financial exchanges other than the Paris Bourse. Since it does not fit the theoretical question about socio-technical agency, I will not examine here the successful adoption of a technology vs. the failure of another.

4. The data have been obtained from the following archives: the New York Historical Society, the New York Public Library, the Library Company of Philadelphia, the Guildhall Library, the British Library, the Archives of the Bank of England, and the Bibliothèque Nationale de France. These archives have extensive holdings of public and private financial documents, not to be found elsewhere.

5. I am examining here only the organization of trading on the New York Stock Ex-change. For reasons of space, the other stock and commodities exchanges which existed in Lower Manhattan in the second half of the nineteenth century will not be discussed here.

6. The Paris Bourse was organized in a similar way. There was a status group of sixty stockbrokers who were practically civil servants. They operated indoors and inherited seats. A much larger group of unofficial stockbrokers dealt in the street. Paris also had multiple prices and several quotation lists for the same securities (Vidal 1910). There were at least three official daily quotations, plus a number of unofficial ones (Maddi-son 1877: 17).

7. In the 1840s and 1850s, the Regular Board traded behind closed doors and under a veil of secrecy. Quotations were not disclosed during trade (Warshow 1929: 70). Agents would listen at the keyhole and sell the price information in the street.

8. Some New York restaurants had tickers in their dining rooms. At Miller's and Del-monico's, investors could follow the price variations in real time, ordering a meal and some stocks at the same time (Babson 1908: 47). Private stock auctioneers also installed tickers on the exchange floor. Not only was the ticker present in places where the upper middle classes congregated, it was also present on the fringes of the marketplace, in the cramped, badly lit bucket shops where poorer people came to invest their few dollars. So strong was the influence of the ticker and the prestige associated with it that some bucket shop operators felt compelled to install fake tick-ers and wires going only to the edge of the rug, together with additional parapherna-lia like mock quotation tables and fake newsletters (Fabian 1990: 191; Cowing 1965: 103). The ticker was praised by some contemporaries as 'the only God' of the market

(Lefevre 1901: 115) and as a wonder of the outgoing nineteenth century (King 1897: 108). Successful stock operators were fictionalized as 'Von Moltke of the ticker' and as 'masters of the ticker' (Lefevre 1907: 83, 122). It became a prized possession, to be kept until a speculator's last breath: when Daniel Drew, the famed speculator of the 1850s and early 1860s died in 1879, his only possessions were a Bible, a sealskin coat, a watch, and a ticker (Wyckoff 1972: 28).

9. See also Nelson 1900: 27–28 and Anonymous 1893. For period photographs of brokerage offices see, for example, *Guide to Investors* (Anonymous 1899: 62, 78).

10. The expression 'bucket shop' generally designates an unofficial brokerage office. There were, however, considerable differences among bucket shops, with regard to their wealth, as well as to their reputation and to the financial securities they traded.

11. A contemporary observer estimated that in 1898 there were about 7,000 bucket shops and 10,000–15,000 brokerage houses in the United States (Hoyle 1898: 17). The difference between bucket shops and brokerage houses was not only a matter of financial muscle. Some bucket shops were very prosperous and upmarket, while others were quite modest. Bucket shops specialized mainly in derivatives, while brokerage houses did more "classical" securities trading.

12. A comparison with the Paris Bourse can show best the knowledge-organizational changes induced by the ticker. In the 1860s, the NYSE and the Paris Bourse had a similar organization of knowledge: both were multiple (discontinuous *and* continuous) markets with multiple prices (Vidal 1910; Walker 2001). For specific reasons, the Paris Bourse did not introduce the ticker at all and continued to work without a price-recording technology until at least 1920. There continued to be two classes of stockbrokers: one was a high-status group; the other an illegal, yet tolerated, class that greatly outnumbered the first. Multiple prices and parallel price lists were the rule until 1920 (Parker 1920: 37), and even later. In New York, by contrast, the two boards of brokers merged and worked in the same room from late 1870 on; multiple quotations disappeared and price data became both continuous and standardized. Instead of adopting a price-recording technology, some Paris stockbrokers developed an interest in abstract models of price variations (Preda 2004), seen as a corrective to inaccurate price information.

References

Abbott, A. 1988. *The System of Professions: An Essay on the Division of Expert Labor*. University of Chicago Press.

Anonymous. 1848. *Stocks and Stock-Jobbing in Wall Street*. New York Publishing Co.

Anonymous. 1854. The Bulls and Bears or, Wall Street Squib No. I.

Anonymous. 1875. *Secret of Success in Wall Street*. Tumbridge.

Anonymous. 1876. *The Rationale of Market Fluctuations*. Effingham Wilson.

Anonymous. 1881. Stocks and the big operators of the streets: The science of speculation as studied by a leading operator. *Cincinnati Enquirer*, April 13.

Anonymous. 1893. *The Boston Stock Exchange*. Hunt & Bell.

Anonymous. 1898. *'Guide to Investors,'* Haight & Freese's *Information to Investors and Operators in Stocks, Grain and Cotton*. Haight & Freese.

Anonymous. 1907. A Method of Forecasting the Stock Market. *The Ticker* 1, no. 1.

Anonymous. 1908a. Modern brokerage establishments: Organization and machinery of the House of J.S. Bache & Company. *The Ticker* 1, no. 4: 7.

Anonymous. 1908b. Why not investment experts? Demand for advice and opinions on investments, suggests the establishment of a new profession. *The Ticker* 1, no. 6: 35.

Anonymous. 1927. *The Magazine of Wall Street* 40, no. 9: 753.

Austin, J. 1976 [1962]. *How to Do Things with Words*. Oxford University Press.

Babson, R. 1908. The theory of financial statistics. *The Ticker* 1, no. 3.

Bakke, J. 1996. Technologies and interpretations: The case of the telephone. *Knowledge and Society* 10: 87–107.

Barry, A., and D. Slater. 2002. Introduction: The technological economy. *Economy and Society* 31, no. 2: 175–193.

Bertinotti, D. 1985. Carrières féminines et carrières masculines dans l'administration des postes et télégraphes à la fin du XIXe siècle. *Annales ESC* 3: 625–640.

Biais, B. 1993. Price formation and equilibrium liquidity in fragmented and centralized markets. *Journal of Finance* 48, no. 1: 157–185.

Bijker, W., T. Hughes, and T. Pinch, eds. 1987. *The Social Construction of Technological Systems: New Directions in the Sociology and History of Technology*. MIT Press.

Blau, P., C. Falbe, W. McKinley, and P. Tracy. 1976. Technology and organization in manufacturing. *Administrative Science Quarterly* 21: 20–40.

Bowker, G., and S. Star. 1999. *Sorting Things Out: Classification and Its Consequences*. MIT Press.

Burton, J. 1929. *The Story of Trans-Lux*. Trans-Lux Daylight Picture Screen Corp.

Calahan, E. 1901. The evolution of the stock ticker. *The Electrical World and Engineer* 37, no. 6: 236–238.

Callon, M. 1998. Introduction: The embeddedness of economic markets in economics. In *The Laws of the Markets*, ed. M. Callon. Blackwell.

Callon, M. 2004. Europe wrestling with technology. *Economy and Society* 33, no. 1: 121–134.

Callon, M., and F. Muniesa. 2003. Les marchés économiques comme dispositifs collectives de calcul. *Réseaux* 21, no. 122: 189–234.

Carruthers, B., and A. Stinchcombe. 2001. The social structure of liquidity: Flexibility in markets, states, and organizations. In A. Stinchcombe, *When Formality Works*. University of Chicago Press.

Clews, H. 1888. *Twenty-Eight Years in Wall Street*. Irving.

Collins, H., and M. Kusch. 1998. *The Shape of Actions: What Humans and Machines Can Do*. MIT Press.

Cornwallis, K. 1879. *The Gold Room and the New York Stock Exchange and Clearing House*. A. S. Barnes.

Cowing, C. 1965. *Populists, Plungers, and Progressives: A Social History of Stock and Commodity Speculation*. Princeton University Press.

Crary, J. 1990. *Techniques of the Observer: On Vision and Modernity in the Nineteenth Century*. MIT Press.

Downey, G. 2000. Running somewhere between men and women: Gender in the construction of the telegraph messenger boy. *Knowledge and Society* 12: 129–152.

Durkheim, É. 1995 [1915]. *The Elementary Forms of the Religious Life*. Free Press.

Eames, F. 1894. *The New York Stock Exchange*. Thomas G. Hall.

Emirbayer, M., and A. Mische. 1998. What is agency? *American Journal of Sociology* 103, no. 4: 962–1023.

Engle, R. 2000. The econometrics of ultra high frequency data. *Econometrica* 68, no. 1: 1–22.

Fabian, A. 1990. *Card Sharps, Dream Books, and Bucket Shops: Gambling in 19th-Century America*. Cornell University Press.

Fischer, C. 1994. *America Calling: A Social History of the Telephone to 1940*. University of California Press.

Flichy, Patrice. 1995. *Dynamics of Modern Communication: The Shaping and Impact of New Communication Technologies*. Sage.

Gibson, George Rutledge. 1889. *The Stock Exchanges of London, Paris, and New York: A Comparison*. Putnam.

Giddens, A. 1987. *Social Theory and Modern Sociology*. Polity.

Grint, K., and S. Woolgar. 1995. On some failures of nerve in constructivist and feminist analyses of technology. *Science, Technology and Human Values* 20, no. 3: 286–310.

Harper, H. 1926. *The Psychology of Speculation: The Human Element in Stock Market Transactions.* Privately printed.

Hickling, J. 1875. *Men and Idioms of Wall Street: Explaining the Daily Operations in Stocks, Bonds, and Gold.* John Hickling & Co.

Hill, M. 1993. *Archival Strategies and Techniques.* Sage.

Hochfelder, D. 2001. Partners in Crime: The Telegraph Industry, Finance Capitalism, and Organized Gambling, 1870–1920. Paper presented at 2001 IEEE Conference on the History of Telecommunications.

Hoyle [pseudonym]. 1898. *The Game in Wall Street, and How to Play It Successfully.* J. S. Ogilvie.

Hughes, J., M. Rouncefield, and P. Toulmie. 2002. The day-to-day work of standardization: A sceptical note on the reliance on IT in a retail bank. In *Virtual Society?* ed. S. Woolgar. Oxford University Press.

Husserl, E. 1977 [1912]. *Cartesianische Meditationen.* Felix Meiner.

Jenkins, R., et al., eds. 1989. *The Papers of Thomas Alva Edison: The Making of an Inventor, February 1847–June 1873.* Johns Hopkins University Press.

King, M., ed. 1897. *King's Views of the New York Stock Exchange: A History and Description with Articles on Financial Topics.* King's Handbooks.

Kirk, J., and M. Miller. 1986. *Reliability and Validity in Qualitative Research.* Sage.

Knorr Cetina, K., and U. Bruegger. 2002a. Traders' engagement with markets: A postsocial relationship. *Theory, Culture and Society* 19, no. 5–6: 161–187.

Knorr Cetina, K., and U. Bruegger. 2002b. Global microstructures: The virtual societies of financial markets. *American Journal of Sociology* 107, no. 4: 905–950.

Latour, B. 1999. *Pandora's Hope: Essays on the Reality of Science Studies.* Harvard University Press.

Lefevre, E. 1901. *Wall Street Stories.* McClure, Phillips & Co.

Lefevre, E. 1907. *Sampson Rock of Wall Street.* Harper & Brothers.

Lefevre, E. 1998 [1923]. *Reminiscences of a Stock Operator.* Wiley.

Lynch, M. 1992. Extending Wittgenstein: The pivotal move from epistemology to the sociology of science. In *Science as Practice and Culture,* ed. A. Pickering. University of Chicago Press.

MacKenzie, D., and Y. Millo. 2003. Negotiating a market, performing theory: The historical sociology of a financial derivatives exchange. *American Journal of Sociology* 109, no. 1: 107–146.

MacKenzie, D., and J. Wajcman. 1985. Introductory essay: The social shaping of technology. In *The Social Shaping of Technology*, ed. D. MacKenzie and J. Wajcman. Open University Press.

Maddison, E. 1877. *The Paris Bourse and the London Stock Exchange*. Effingham Wilson.

Martin, Joseph G. 1886. *Martin's Boston Stock Market*. Published by the author.

Marx, K. 1996 [1867]. *Capital*, volume 1. Lawrence & Wishart.

Medbery, James. 1870. *Men and Mysteries of Wall Street*. Fields, Osgood & Co.

Michie, R. 1999. *The London Stock Exchange: A History*. Oxford University Press.

Miller, P. 2001. Governing by numbers: Why calculative practices matter. *Social Research* 68, no. 2: 379–395.

Mumford, L. 1967. *The Myth of the Machine*. Secker & Warburg.

Nelson, S., ed. 1900. *The ABC of Wall Street*. S. A. Nelson.

O'Rourke, K., and J. Williamson. 1999. *Globalization and History: The Evolution of a Nineteenth-Century Atlantic Economy*. MIT Press.

Parker, W. 1920. *The Paris Bourse and French Finance. With Reference to Organized Speculation in New York*. Columbia University.

Perrow, C. 1967. A framework for the comparative analysis of organizations. *American Sociological Review* 32, no. 2: 194–208.

Pickering, A. 1995. *The Mangle of Practice: Time, Agency, and Science*. University of Chicago Press.

Pickering, A. 2001. Practice and posthumanism: Social theory and a history of agency. In *The Practice Turn in Contemporary Theory*, ed. T. Schatzki et al. Routledge.

Pinch, T. 2003. Giving birth to new users: How the Minimoog was sold to rock and roll. In *How Users Matter*, ed. N. Oudshoorn and T. Pinch. MIT Press.

Porter, T. 1995. *Trust in Numbers: The Pursuit of Objectivity in Science and Public Life*. Princeton University Press.

Pratt, S. 1903. *The Work of Wall Street*. Appleton.

Preda, A. 2001a. The rise of the popular investor: Financial knowledge and investing in England and France, 1840–1880. *Sociological Quarterly* 42, no. 2: 205–232.

Preda, A. 2001b. In the enchanted grove: Financial conversations and the marketplace in England and France in the 18th century. *Journal of Historical Sociology* 14, no. 3: 276–307.

Preda, A. 2003. Les hommes de la Bourse et leurs instruments merveilleux: Technologies de transmission des cours et origins de l'organisation des marchés modernes. *Réseaux* 21, no. 122: 137–166.

Preda, A. 2004. Informative prices, rational investors: The emergence of the random walk hypothesis and the nineteenth century "science of financial investments." *History of Political Economy* 36, no. 2: 351–386.

Rollo Tape [Richard D. Wyckoff]. 1908. Market Lecture. Manipulation—Tape Reading—Charts. *The Ticker* 1, no. 4: 33–35.

Rousseau, P., and R. Sylla. 2001. Financial Systems, Economic Growth, and Globalization. Working paper 8323, National Bureau of Economic Research.

Schatzki, T. 2002. *The Site of the Social: A Philosophical Account of the Constitution of Social Life and Change.* Pennsylvania State University Press.

Schutz, A. 1967. *Collected Papers I: The Problem of Social Reality.* Martinus Nijhoff.

Selden, G. 1917. *The Machinery of Wall Street.* The Magazine of Wall Street.

Sloterdijk, P. 2006. *Im Weltinnenraum des Kapitalismus.* Suhrkamp.

Smith, A. 1991 [1776]. *Wealth of Nations.* Prometheus Books.

Smith, M. 1871. *Twenty Years Among the Bulls and Bears of Wall Street.* J. B. Burr & Hyde.

Stake, R. 2000. The case study method in social inquiry. In *Case Study Method*, ed. R. Gomm et al. Sage.

Stedman, E., ed. 1905. *The New York Stock Exchange: Its History, Its Contribution to the National Prosperity, and Its Relation to American Finance at the Outset of the Twentieth Century.* New York Stock Exchange Historical Company.

Stein, J. 2001. Reflections on time, time-space compression and technology in the nineteenth century. In *Timespace*, ed. J. May and N. Thrift. Routledge.

Swedberg, R. 2003. *Principles of Economic Sociology.* Princeton University Press.

Tailer, E. 1893. Diaries. Manuscript, New York Historical Society.

Tarr, J., T. Finholt, and D. Goodman. 1987. The city and the telegraph: Urban telecommunications in the pre-telephone era. *Journal of Urban History* 14, no. 1: 38–80.

Vidal, E. 1910. *The History and Methods of the Paris Bourse.* US Government Printing Office.

Walker, D. 2001. A factual account of the functioning of the 19th century Paris Bourse. *European Journal for the History of Economic Thought* 8, no. 2: 186–207.

Woolgar, S. 1991. The turn to technology in social studies of science. *Science, Technology and Human Values* 16, no. 1: 20–50.

Wyckoff, P. 1972. *Wall Street and the Stock Markets: A Chronology (1644–1971)*. Chilton.

Wyckoff, R. 1930. *Wall Street Ventures and Adventures*. Harper and Brothers.

Wyckoff, R. 1933. 1901: Inside Information. *Stock Market Technique* 2, no. 5: 24–26.

Wyckoff, R. 1934a. 1905: Studying the big fellows. *Stock Market Technique* 3, no. 2: 37–39.

Wyckoff, R. 1934b. Why the dots on the tape? *Stock Market Technique* 3, no. 2: 20.

Wyckoff, R. 1934c. It's not the *kind* of a chart but your ability to interpret that counts. *Stock Market Technique* 3, no. 2: 10.

Wyckoff, R. 1934d. Why you should use charts. *Stock Market Techniques* 2, no. 6: 16.

Wyckoff, R. 1934e. The tape is the best guide. *Stock Market Technique* 3, no. 2: 12.

Wyckoff, R. 1934f. 1902–1903 Experiences in the brokerage business. *Stock Market Technique* 2, no. 6: 23–25.

Yates, J. 1986. The telegraph's effect on nineteenth-century markets and firms. *Business and Economic History* 15: 149–163.

Zaloom, C. 2006. *Out of the Pits: Traders and Technology from Chicago to London*. University of Chicago Press.

8 Tools of the Trade: The Socio-Technology of Arbitrage in a Wall Street Trading Room

Daniel Beunza and David Stark

What counts? This question expresses most succinctly the challenge facing securities traders in the era of quantitative finance. As for other industries where actors are immersed in a virtual flood of information, the challenge for traders is not faster, higher, stronger—as if the problem of the volume of data could be solved by gathering yet more—but selecting what counts and making sense of the selection. The more information is available to many simultaneously, the more advantage shifts to those with superior means of interpretation. How is a trading room organized for making sense of what is to be taken into account?

What counts? This question also expresses most succinctly a challenge for economic sociology. What is valuable, and by what metrics of value and what performance criteria? In its contemporary form, economic sociology arguably began when Talcott Parsons made a pact with economics. You, the economists, study value; we sociologists study values. You study the economy; we study the social relations in which economies are embedded. This paper is part of a research agenda that breaks with that pact by turning to problems of valuation and calculation (Boltanski and Thevenot 1991; White 1981, 2001; Thévenot 2001; Stark 2000; Girard and Stark 2002; Callon and Muniesa 2002; Callon et al. 2002). Just as post-Mertonian studies of science moved from studying the institutions in which scientists were embedded to analyze the actual practices of scientists in the laboratory, so a post-Parsonsian economic sociology must move from studying the institutions in which economic activity is embedded to analyze the actual calculative practices of actors at work.

Our analysis starts with the fundamental theme that network analysis shares with other schools of economic sociology—the conception that markets are social (Granovetter 1985; Fligstein 1990; Uzzi 1997). But we extend and deepen that perspective by arguing that social network analysis should not be limited to studying ties among persons. Because the social consists of humans and their non-humans (artifacts), in place of studying "society" we must construct a science of associations—an analysis that examines not only links among persons but also among persons and instruments (Callon 1998; Latour 1988, 1991; Hutchins 1995). What counts? Tools count. Instrumentation must be brought into the accounts of economic sociologists. Calculation, as we shall see, is not simply embedded in social relations. Calculative practices are distributed across persons and instruments.

Studying Quantitative Finance

To analyze the organization of trading in the era of quantitative finance we conduct an ethnography of arbitrage, studying how traders recognize opportunities in the trading room of a major international investment bank. We focus on arbitrage because it is the trading strategy that best represents the distinctive combination of connectivity, knowledge, and computing that we regard as the defining feature of the quantitative revolution in finance. With the creation of the NASDAQ in 1971, Wall Street had an electronic market long before any other industry. With the development of Bloomberg data terminals in 1980, traders in investment banks were connected to each other in an all-inclusive computer network long before other professionals. With the development of formulas for pricing derivatives such as the Black-Scholes formula in 1973, traders gained powerful mathematical tools. And with the dramatic growth in computing power traders were able to combine these equations with powerful computational engines. This mix of formulas, data to plug into them, computers to calculate them, and electronic networks to connect them was explosive, leading to a decisive shift to "quantitative finance" (Lewis 1999; Dunbar 2000).

   To date, the leading analytic strategy by sociologists studying modern finance has been to focus on one or another of the key components of the quantitative revolution. Exemplary, in this light, is a recent paper by Knorr Cetina and Bruegger (2002) that analyzes one of the key trends of the quantitative revolution, the rise of electronic markets, arguing that electronic trading has altered the relationship between market participants and physical space. Their work is pathbreaking for the insight that the numbers on the screens of the electronic traders do not *represent* a market that is elsewhere; instead, the market is said to be *appresented* (p. 4). Like a conversation taking place through instant messaging (but unlike, say, a movie or a television show), electronic markets constitute an on-screen reality that lacks an off-screen counterpart. This has important implications for the practice of quantitative finance. Just as the eyes of traders in a commodities pit are glued to the gestures of other traders, Knorr Cetina and Bruegger found that the eyes of their currency traders are glued to the screen—because in both cases that is where the market is. Electronic markets, they assert, have brought the marketplace to the trader's screen, prompting the traders to shift from a "face-to-face world" to a "face-to-screen world" and bringing about the "diminishing relevance of the physical setting" (p. 23).

While Knorr Cetina and Bruegger focus on the rise of connectivity in finance, MacKenzie and Millo (2001) focus on another leg of the quantitative revolution, the rise of mathematical formulas and their consequences for trading (see also MacKenzie 2002).[1] The mathematical formulas of modern finance, they argue, do not *represent markets* so much as constitute part of a network (also made up of people, computers, ideas, etc.) that *performs the market* in the sense developed by Callon (1998). As an example of such a "performative" that does not just mirror a reality but is constitutive of it, they point to the role of the Black-Scholes formula in predicting and later setting option prices on the Chicago Board Options Exchange.

The two studies are nicely complementary: Knorr Cetina and Bruegger examine the network connectivity of electronic trading, but ignore formulas entirely; MacKenzie and Millo address the role of formulas but ignore the connectivity of electronic trading. But if we are to understand the organization of trading in the era of modern finance, we must examine all three pillars of the quantitative revolution: network connectivity, mathematical formulas, and computing. It is precisely this combination that gives the study of modern arbitrage—as the trading strategy that most powerfully (and, to date, most profitably) exploits the mathematics and the machines of modern market instruments—such analytic leverage.

In taking the limitations of these studies as our point of departure, the opportunity we seize, however, is not just to examine as an ensemble the pieces they had begun to analyze separately. Amidst the circulating information of Knorr Cetina and the diffusing equations of MacKenzie, we find little about the core problem facing any trader—how to recognize an opportunity? We will argue that traders do so by making of their trading room a laboratory, by conducting experiments, by deploying an array of instruments to test the market. In the practices through which value is calculated, equivalencies are constructed, and opportunities are realized, tools count. Calculation is distributed across the human and non-human agents and instruments enacting the trade. But if calculation involves both the mathematics and the machines of quantitative finance, as we shall see, even when it is automated it is far from mechanical: at this level of performance, calculation involves judgment. Moreover, calculation is not detached: whereas the trader is emotionally distant from any particular trade, to be able to take a position, the trader must be strongly attached to an evaluative principle and its affiliated instruments.[2] In the field of arbitrage, to be opportunistic you must be principled, that is, you must commit to an evaluative metric.

Our focus on the problem of identifying value leads us to take into account the dynamics identified by Knorr Cetina, MacKenzie, and their co-authors but to draw radically different analytic conclusions. For Knorr Cetina and Bruegger (2002), the displacement of physical locale in favor of the "global microstructures" on the screen is explained by the ever-increasing rapidity of the circulation of information. We, too, initially approached our research setting as a world of globally instantaneous information. By studying sophisticated derivative traders, able to produce formulas that quantify unknown magnitudes, we hoped to demarcate a world of pure information that could stand as a benchmark against which we could differentiate other calculative settings. And, yes, we encountered a world abundant in information, delivered with dazzling, dizzying speed. But after months of fieldwork, we realized that, as increasingly more information is almost instantaneously available to nearly every market actor, the more strategic advantage shifts from economies of information to socio-cognitive process of interpretation (Weick 1979; Brown and Duguid 2000; Grabher 2002b). Precisely because all the relevant alters have the same information as ego, this particular trading room makes profits (considerably higher than industry-average profits) not by access to better or timelier information but by fostering interpretive communities in the trading room.

Similarly, learning from MacKenzie and Millo (2003) about how the diffusion of formulas shapes markets, we go on to ask the next question. If everyone is using the same formulas, how can you profit? The more that formulas diffuse to perform the market, the more one's profits depend on an original performance. That is, the premium shifts to innovation. As with information (which you must have, but which in itself will not give advantage) so with formulas: the more widely diffused, the more you must innovate.

What then facilitates interpretation and fosters innovation? The answer came only when we stopped regarding the trading room simply as a "setting" and began to regard the spatial configurations of this particular locale as an additional dimension alongside the combination of equations, connectivity, and computing. In analyzing the *modus operandi* of modern finance, we came to see that its *locus operandi* could not be ignored. That is,

whereas Knorr Cetina and Bruegger dismiss physical locale in favor of interactions *in cyberspace*, we found that trading practices are intimately tied to
the deployment of traders and instruments *in the room*.

Arbitrage trading can be seen as an economy of information and speed.
So is flying a fighter aircraft in warfare. Without the requisite information
and the requisite speed neither trader nor pilot could do the job. But
maneuvering in the uncertain environment of markets, like maneuvering
in the fog of battle, requires situated awareness.[3] As we shall see, the configuration of the trading room, as a specific locale, provides the socio-spatial
resources for this sense making. A trading room is an engine for generating
equivalencies. Such associations are made *in situ*, that is, they entail the use
of financial formulas that result from associations among people working
in the same physical place.

The cognitive challenge facing our arbitrage traders—a challenge central
to the process of innovation—is the problem of recognition. On one hand,
they must, of course, be adept at *pattern recognition* (e.g., matching data to
models, etc). But if they only recognize patterns familiar within their existing categories, they would not be innovative (Clippinger 1999). Innovation
requires another cognitive process that we can think of as *re-cognition* (making unanticipated associations, reconceptualizing the situation, breaking
out of lock-in). It involves a distinctive type of search—not like those
searches that yield the coordinates of a known target or retrieve a phone
number, product code, or document locator for a pre-identified entity or
category—but the search where you don't know what you're looking for
but will recognize it when you find it.

The organization of the trading room, as we shall see, is equipped (quite
literally) to meet this twin challenge of exploiting knowledge (pattern recognition) while simultaneously exploring for new knowledge (practices
of re-cognition).[4] Each desk (e.g., merger arbitrage, index arbitrage, etc.) is
organized around a distinctive evaluative principle and its corresponding
cognitive frames, metrics, "optics," and other specialized instrumentation
for pattern recognition. That is, the trading room is the site of diverse, indeed rivalrous, principles of valuation. And it is the interaction across this
heterogeneity that generates innovation.

To explore the socio-cognitive, socio-technical practices of arbitrage,
we conducted ethnographic field research in the Wall Street trading room
of a major international investment bank. Pseudonymous International
Securities is a global bank with headquarters outside the United States. It
has a large office in New York, located in a financial complex in Lower
Manhattan that includes the offices of Merrill Lynch and other major investment banks. With permission from the manager of the trading room

we had access to observe trading and interview traders. We found an environment extremely congenial to academic inquiry. In our conversations with arbitrageurs, who are the intellectual elite of Wall Street, it was not unusual for us to hear timely references to economic history, French cinema of the 1960s, books on network analysis, and even the philosophy of Richard Rorty or Martin Heidegger. More importantly, the traders relished reflecting on the nature of their work, and were correspondingly generous with their time. Our observations extended to sixty half-day visits across more than two years. During that time, we conducted detailed observations at three of the room's ten trading desks, sitting in the tight space between traders, following trades as they unfolded and sharing lunch and jokes with the traders. We complemented this direct observation with in-depth interviews. In the final year of our investigation, we were more formally integrated into the trading room—provided with a place at a desk, a computer, and a telephone. The time span of our research embraced the periods before and after the September 11th attack on the World Trade Center. (For accounts of the trading room's response and recovery, see Beunza and Stark 2003, 2005.)

In the following section we make the case that arbitrage constitutes a distinctive trading strategy that operates by making associations among securities. In contrast to value and momentum investing, we argue, arbitrage involves an art of association—the construction of equivalence (comparability) of properties across different assets. In place of essential or relational characteristics, the peculiar valuation that takes place in arbitrage is based on an operation that makes something the measure of something else—associating securities to each other. Subsequent sections analyze how the trading room is organized to recognize opportunities. We first observe how the spatial organization of the room facilitates general sociability among traders. Second, we examine how these traders are grouped into specialized desks, each deploying distinctive financial instruments and evaluative metrics for pattern recognition. Next, we examine the trading room as an ensemble of multiple desks, exploring how this ecology of diverse evaluative principles facilitates practices of re-cognition; and finally, we examine the room as an assemblage of instrumentation, exploring how the socio-cognitive and the socio-technical are intertwined.

Arbitrage, or Quantitative Finance in the Search for Qualities

Arbitrage is defined in finance textbooks as "locking in a profit by simultaneously entering into transactions in two or more markets" (Hull 1996,

p. 4). If, for instance, the prices of gold in New York and London differed by more than the transportation costs, an arbitrageur could realize an easy profit by buying in the market where gold is cheap and selling it in the market where it is expensive. As such, classical arbitrage lacks sociological as well as economic interest: it relates markets that are the same in every dimension except for an obvious one such, as in this case, the geographical. Reducing arbitrage to an unproblematic operation that links the obvious (gold in London, gold in New York), as textbook treatments do, is doubly misleading, for modern arbitrage is neither obvious nor unproblematic. It provides profit opportunities by associating the unexpected, and it entails real exposure to substantial losses.

Arbitrage is a distinctive form of entrepreneurial activity that exploits not only gaps across markets but also the overlaps among multiple evaluative principles. Arbitrageurs profit not by having developed a superior way of deriving value but by exploiting opportunities exposed when different evaluative devices yield discrepant pricings at myriad points throughout the economy.

As a first step to understanding modern arbitrage, consider the two traditional trading strategies, *value* and *momentum* investing, that arbitrage has come to challenge. Value investing is the traditional "buy low, sell high" approach in which investors look for opportunities by identifying companies whose "intrinsic" value differs from its current market value. They do so by studying a company's annual reports, financial results, products, and executives; they then compare the intrinsic value that emerges from this analysis with the market price of the company (Graham and Dodd 1934). Value investors are essentialists: they believe that property has a true, intrinsic, essential value independent of other investors' assessments, and that they can attain a superior grasp of that value through careful perusal of the information about a company. Value investors map the many aspects of a company by translating them into abstract variables—e.g., return, growth, risk—and collapsing them into a single number ("value") with the use of formulas such as discounted cash flow. They proceed with the belief that mispricings will be eventually corrected—that is, that enough investors will eventually "catch up" with the intrinsic value and drive the price towards it, producing a profit for those who saw it first.

In contrast to value investors, momentum traders (also called chartists, see Smith 2001) turn away from scrutinizing companies towards monitoring the activities of other actors on the market (Malkiel 1973). Like value investors, their goal is to find a profit opportunity. However, momentum traders are not interested in discovering the intrinsic value of a stock.

Instead of focusing on features of the asset itself, they turn their attention to whether other market actors are bidding the value of a security up or down. Alert to trends, they believe in the existence of "momentum," a self-sustaining social process amenable to discovery by studying patterns in the time series of the stock—its chart. In contrast with value investing, a momentum strategy can involve buying when the price is extremely high, as long as the patterns in the chart suggest that it is getting higher. Preoccupied with vectors and directionality, momentum traders plot trajectories. Like the fashion-conscious or like night-life socialites scouting the trendiest clubs, they derive their strength from obsessively asking "Where is everyone going?" in hopes of anticipating the hot spots and leaving just when things get crowded.

As with value and momentum investors, arbitrageurs also need to find an opportunity, an instance of disagreement with the market's pricing of a security. They find it by making associations. Instead of claiming a superior ability to process and aggregate information about intrinsic assets (as value investors do) or better information on what other investors are doing (as momentum traders do), the arbitrage trader tests ideas about the correspondence between two securities. Confronted by a stock with a market price, the arbitrageur seeks some other security—or bond, or synthetic security such as an index composed of a group of stocks, etc.—that can be related to it, and prices one in terms of the other. The two securities have to be similar enough so that their prices change in related ways, but different enough so that other traders have not perceived the correspondence before. As we shall see, the posited relationship can be highly abstract. The tenuous or uncertain strength of the posited similarity or co-variation reduces the number of traders that can play a trade, hence increasing its potential profitability.

Arbitrage, then, is a distinct trading strategy. Whereas value investment is essentialist and momentum trading is extrinsic, arbitrage is associational. Whereas the value investor pegs value on intrinsic worth, and the momentum trader tracks the value assessments assigned by other investors, arbitrage traders locate value by making associations between particular properties or qualities of one security and those of other previously unrelated or tenuously related securities.

Arbitrage hinges on the possibility of interpreting securities in multiple ways. By associating one security to another, the trader highlights different properties (qualities) of the property he is dealing with.[5] In contrast to value investors who distill the bundled attributes of a company to a single number, arbitrageurs reject exposure to a whole company. But in contrast to corporate raiders, who buy companies for the purpose of breaking them up to sell as separate properties, the work of arbitrage traders is yet more radically deconstructionist. The unbundling they attempt is to isolate, in the first instance, categorical attributes. For example, they do not see the Boeing Company as a monolithic asset or property, but as having several properties (traits, qualities) such as being a technology stock, an aviation stock, a consumer-travel stock, an American stock, a stock that is included in a given index, and so on. Even more abstractionist, they attempt to isolate such qualities as the volatility of a security, or its liquidity, its convertibility, its indexability, and so on.

Thus, whereas corporate raiders break up parts of a company, modern arbitrageurs carve up abstract qualities of a security. In our field research, we find our arbitrageurs actively shaping trades. Dealing with the multiple qualities of securities as narrow specialists, they position themselves with respect to one or two of these qualities, but never all. Their strategy is to use the tools of financial engineering to shape a trade so that exposure[6] is limited only to those equivalency principles in which the trader has confidence. Derivatives such as swaps,[7] options,[8] and other financial instruments play an important role in the process of separating the desired qualities from the purchased security. Traders use them to slice and dice their exposure, wielding them in effect like a surgeon's tools—scalpels, scissors, proteases—to give the patient (the trader's exposure) the desired contours.

Paradoxically, much of the associative work of arbitrage is therefore for the purpose of "disentangling" (see Callon 1998 for a related usage)—selecting out of the trade those qualities to which the arbitrageur is not committed. The strategy is just as much not betting on what you don't know as betting on what you do know. In merger arbitrage, for example, this strategy of highly specialized risk exposure requires that traders associate the markets for stocks of the two merging companies and dissociate from the stocks everything that does not involve the merger. Consider a situation in which two firms have announced their intention to merge. One of the firms, say the acquirer, is a biotech firm and belongs to an index, such as the Dow Jones (DJ) biotech index. If a merger arbitrage specialist wanted to shape a trade such that the "biotechness" of the acquirer would not be an aspect of his/her positioned exposure, the arbitrageur would long the index. That is, to dissociate this quality from the trader's exposure, the arbitrageur associates the trade with a synthetic security ("the index") that stands for the "biotechness." Less categorical, more complex qualities require more complex instruments.

When, as in some forms of merger arbitrage, the process of dissociating is taken to the extreme, we could say that merger arbitrageurs trade in securities in order to bet on events. By hedging against all qualities of the stock other than the merger itself, the merger arbitrageur, in effect, is betting about the likelihood of a discrete event. You cannot go to a betting window to wager that two companies will merge (or not) on January 3. But with enough sophisticated instruments, you can shape your exposure to something very close to such a position.

Arbitrageurs do not narrow their exposure for lack of courage. Despite all the trimmings, hedging, and cutting, this is not a trading strategy for the faint-hearted. Arbitrage is about tailoring the trader's exposure to the market, biting what they can chew, betting on what they know best, and avoiding risking their money on what they don't know. Traders expose themselves profusely—precisely because their exposure is custom-tailored to the relevant deal. Their sharp focus and specialized instruments gives them a clearer view of the deals they examine than the rest of the market. Thus, the more the traders hedge, the more boldly they can position themselves.

Arbitrageurs can reduce or eliminate exposure along many dimensions but they cannot make a profit on a trade unless they are exposed on at least one. In fact, they cut entanglements along some dimensions precisely to focus exposure where they are most confidently attached. As Callon argues (Callon and Muniesa 2002; Callon et al. 2002), calculation and attachment are not mutually exclusive. To be sure, the trader's attachment is distanced and disciplined; but however emotionally detached, and however fleeting, to hold a position is to hold a conviction.

How do unexpected and tenuous associations become recognized as opportunities? In the following sections we enter the trading room to see how cognition is distributed and diversity is organized. Before examining the instruments that mediate the markets, we look first at the deployment of the traders themselves within the room. After examining the spatialized sociability of the trading room, we examine the *equipment*—the teams and the tools—of arbitrage.

The Trading Room as an Associative Space

The architecture and the ambiance of the trading room would be unfamiliar to someone who retired from trading several decades ago. To appreciate the changes, consider the following description of a typical Wall Street trading room in the 1980s:

No sooner did you pass the fake fireplace than you heard an ungodly roar, like the roar of a mob ... the bond trading room of Pierce & Pierce. It was a vast space, perhaps sixty by eighty feet, but with the same eight-foot ceiling bearing down on your head. It was an oppressive space with a ferocious glare, writhing silhouettes ... the arms and torsos of young men ... moving in an agitated manner and sweating early in the morning and shouting, which created the roar. (Wolfe 1987, p. 58)

This boiler-room imagery is absent from the Wall Street trading room of International Securities, the global investment bank where we have been conducting ethnographic research. Entering the trading room is like entering the lobby of a luxury hotel. Instead of a low ceiling, the observer finds high ceilings and a huge open space occupying almost the entire twentieth floor of a skyscraper in Lower Manhattan filled with rows of desks, computers, and traders. Instead of a roar, the observer hears a hushed buzz among the traders immersed in the flickering numbers on hundreds of flat-panel screens. Instead of an oppressive space, the observer finds generous corridors, elegant watercolors on the walls, and a dramatic view of Manhattan. Instead of agitated employees, the observer finds relaxed traders in business-casual wear leisurely circulating about the trading room, coffee in hand. Instead of writhing arms and torsos, we see equations and formulas scribbled hurriedly on a large white board located prominently near the center of the trading room. And instead of a fake fireplace, the room is populated by non-human "intelligent agents," the computer programs executing automated trades, referred to by the traders themselves as "robots."

In the traditional corporate office, space is used to emphasize status differences as the hierarchy of concentric rings effectively isolates the highest-ranking employees. At International Securities, by contrast, space is used to create an atmosphere conducive to association. The open plan, not unlike the layout of a newsroom or a new media design studio, contains no cubicles or partitions. There is even a strict "low-monitor" policy enforced by Bob, the manager of the room, that prevents traders from stacking their Bloomberg monitors two or three high. "We try," he says, "to keep the PCs at a low level so that they can see the rest of the room."

Moreover, the social composition of the room promotes association among disparate communities of practice: the room not only accommodates traders and their assistants, but a diversity of employees, including salesmen, analysts, operation officers, and computer programmers. Flouting an industry-wide trend of relegating these latter employees to a back office, International Securities has kept programmers and operations officers in its money-making core. They not only stay in the trading room but are given

desks as large as the traders,' and their area of the room has the same privileged feel as the rest. The objective, Bob states, is to prevent differences in
professional status from undermining interaction among these groups. If
placed in a different building, says Bob, "they might as well be in a different
planet."

At 160 people, the trading room is small by current Wall Street standards.
But this small number and the open plan layout were deliberately chosen
to allow the type of low-key interaction that encourages experimentation
and intellectual risk-taking. Bob says: "Managers, they'll tell you, 'communication, communication,' but you wonder." To make the contrast, he
pointed us to the trading room of another international bank located in
Connecticut:

It's the size of three aircraft carriers. And the reason for it is that it is a source of pride
to the manager. It is difficult to see how traders can communicate shouting at each
other across two aircraft carriers. At [name of bank], what you'll find is chaos that
looks grand.

Instead, at the trading room of International Securities "the key is [to
avoid] social awkwardness":

Two traders are talking to each other. A third needs a piece of information. He has to
interrupt. 'Can I interrupt? Can I interrupt?' The key there is the social cost of the
interruption. Part of my job is to keep those costs down.

Promoting sociability among traders is not an easy task. Tom Wolfe's
"Masters of the Universe" were gregarious to the point of bullying. In the
age of mathematical finance, arbitrageurs are intellectually over-confident
but socially inept:

A trader is like an engineer type. Difficult when they think they're right. Abrasive.
And not very social. Not socially adept. I can easily find you ten traders in the room
who would be miserable at a cocktail party.

If such individualism is not addressed, it can result in fragmented territoriality in the trading room. For example, a trader recalls his experience in
another bank years ago where he began his career as follows:

For years, there were areas of the trading floor I would never venture onto. People I
never, absolutely never, talked to. There was no reason why I should go there, since
we traded completely different things. Being there felt strange. There were these cold
looks.

International Securities avoids this territoriality in the trading room by
moving traders around. "I rotate people as much as I can," Bob says, "because sitting near each other is the best rule of thumb to predict that they

will talk to each other." However, Bob is careful not to displace them too disruptively. He describes his approach as "not really shifting, more like drifting." He continues:

Once two traders have been sitting together, even if they don't like each other, they'll cooperate, like roommates. So, everyone gets moved every six months on average. But not everyone at a time. It's like those puzzles with one empty space in which you move only one piece at a time.

This emphasis on cooperative interaction underscores that the cognitive tasks of the arbitrage trader are not those of some isolated contemplative, pondering mathematical equations and connected only to a screen-world. Cognition at International Securities is a distributed cognition. The formulas of new trading patterns are formulated in association with other traders. Truly innovative ideas, as one senior trader observed, are slowly developed through successions of discreet one-to-one conversations:

First you talk to others. You tell someone else, 'I've got this great idea,' and if he tells you 'I read it yesterday in Barron's,' you drop it. If you get a positive take, then you work it around.

An idea is given form by trying it out, testing it on others, talking about it with the "math guys," who, significantly, are not kept apart (as in some other trading rooms), and discussing its technical intricacies with the programmers (also immediately present). Because they have been stirred up by the subtle churning of the room, traders can test the ideas on those with whom they were once "like roommates" but who might now be sitting in different parts of the room. Appropriately, the end of this process of formulation (and the beginning of the next stage of material instrumentation, see below) is known as a "victory lap"—a movement around the room in and through which the idea was generated. Place facilitates sociability to make associations.

And where is Bob, the trading room manager? He sits in the middle of the room despite the fact that he has a very well-appointed office in one corner, complete with designer furniture, a small conference table, and a home cinema-sized Bloomberg screen to watch the markets that can be controlled from a wireless mouse and keyboard. But he prefers to sit in a regular trader's desk in the middle of the room:

I have that office over there—you just saw it. But I like this place better [referring to his desk]. Here, I am more connected. No one would come to tell me stories if they had to come into my office. Also, here I get a feel for how the market is doing. I have to know this, because the atmosphere definitely influences the way these guys trade.

In this way, the trading room at International Securities overturns the traditional concentric circles of status. Rather than enjoying less accessibility, the trading room manager is the most accessible. He is most easily reached; and he is best positioned to observe, indeed to sense, what is happening in the room.

What is happening is more than exchange of information. To be sure, traders must have access to the most timely and complete array of information; but this is not enough. In addition to being a nexus of data flows, the trading room is a room of bodies. Taking its collective "pulse" is a means to take the pulse of the markets. Whereas Knorr Cetina and Bruegger find their foreign currency traders "viscerally plugged into the screen reality of the global sphere" (2002: 15), our arbitrage traders are reflective about how they are acutely attuned to the social reality of the local sphere:

The phone and online communication are inefficient. It takes longer for people to tell each other what they want. You miss body language. Body language and facial expressions are really important. You're not conscious of body language and so it's another channel of communication, and it's one that's not deliberate. So it's a good source for what's happening. I don't try to get too conscious of how I'm reading body language and facial expressions. I just let it work its way to where it's useful.

Bob's observations (and those of many other traders with whom we spoke) highlight that cognition in the trading room is not simply distributed. It is also a situated calculation. A trader needs tools—the financial instruments of derivatives and the material instruments to execute a trade. But in addition to these calculative instruments, the trader also needs a "sense of the market." Knowing how to use the tools combines with knowing how to read the situation. This situated awareness is achieved by drawing on the multiple sensors (both human and instrumental) present within the room.

The trading room thus shows a particular instance of Castells's paradox: As more information flows through networked connectivity, the more important become the kinds of interactions grounded in a physical locale. New information technologies, Castells (2000) argues, create the possibility for social interaction without physical contiguity. The downside is that such interactions can become repetitive and programmed in advance. Given this change, Castells argues that as distanced, purposeful, machine-like interactions multiply, the value of less-directed, spontaneous, and unexpected interactions that take place with physical contiguity will become greater (see also Thrift 1994; Brown and Duguid 2000; Grabher 2002). Thus, for example, as surgical techniques develop together with telecommunications technology, the surgeons who are intervening remotely on patients in distant locations are disproportionately clustering in two or three neighborhoods of Manhattan where they can socialize with each other and learn about new techniques, etc.[9]

From the perspective of arbitrage as association, trading rooms can be seen as the "space of place" where novel associations emerge. One exemplary passage from our field notes finds a senior trader formulating an arbitrageur's version of Castells's paradox:

It's hard to say what percentage of time people spend on the phone vs. talking to others in the room. But I can tell you the more electronic the market goes, the more time people spend communicating with others inside the room.

The Trading Room as an Ecology of Evaluative Principles

Pattern Recognition at the Desk

From looking at the trading room as a simple society of individuals, we now turn to examine the teams that compose the trading room as a more complex organization of diversity. This organization of diversity begins by demarcating specialized functions. The basic organizational unit, "team," has a specific equipment, "desk." The term "desk" not only denotes the actual piece of furniture where traders sit, but also the actual team of traders—as in "Tim from the equity loan desk." Such identification of the animate with the inanimate is due to the fact that a team is never scattered across different desks. In this localization, the different traders in the room are divided into teams according to the financial instrument they use to create equivalencies in arbitrage: the merger arbitrage team trades stocks in companies in the process of consolidating, the options arbitrage team trades in "puts" and "calls,"[10] the derivatives that lend the desk its name, and so on. The desk is an intensely social place. The extreme proximity of the workstations enables traders to talk to each other without lifting their eyes from the screen or interrupting their work. Lunch is at the desk, even if the sandwich comes from a high-end specialty deli. Jokes are at the desk, a never-ending undercurrent of camaraderie that resurfaces as soon as the market gives a respite.

Each desk has developed its own way of looking at the market, based on the principle of equivalence that it uses to calculate value and the financial instrument that enacts its particular style of arbitrage trade. For example, traders at the merger arbitrage desk value companies that are being acquired in terms of the price of the acquiring firm and specialize in asking, "how solid is company X's commitment to merge. For merger arbitrage traders, the companies in the S&P 500 index are little more than a set of potential acquirers and acquisition targets. In contrast, traders at the index arbitrage desk exploit discrepancies between the price of index securities (e.g., futures on the S&P 500 index) and the actual average price of the companies that constitute such indexes. Given the minuscule and rapidly vanishing nature of the misalignments among these two, they need to trade in high volume and at a high speed. Traders at the convertible bond arbitrage desk look at stocks as bonds, and specialize in information about stocks that would typically interest bondholders such as their liquidity and likelihood of default. The traders at the customer sales desk, meanwhile, take and propose orders to customers outside the confines of the room. Although not specialized in a distinct financial instrument, this most sociable team in the room provides a window on the anxiety level of their customers and thus of the market at large by the sound of their voices on the phone and the banging of headsets against their desks in frustration.

A desk generates its own form of pattern recognition. For example, merger arbitrage traders, keen on finding out the degree of commitment of two merging companies, look for patterns of companies' progressive approximation in stock prices. They probe commitment to a merger by plotting the "spread" (difference in price) between acquiring and target companies over time. As with marriages between persons, mergers between companies are scattered with regular rituals of engagement intended to persuade others of the seriousness of their intent. As time passes, arbitrage traders look for a pattern of gradual decay in the spread as corporate bride and groom come together. A similar correspondence of tools and concepts can be found at other desks.

Such joint focus on visual and economic patterns creates, at each desk, a distinctive community of practice around an evaluative principle with its own tacit knowledge. Traders at a desk develop a common sense of purpose, a real need to know what each other knows, a highly specialized language, and idiosyncratic ways of signaling to each other. This sense of joint membership translates into friendly rivalry toward other desks. A customer sales trader, for example, took us aside to denounce statistical arbitrage as "like playing video games": "If you figure out what the other guy's program is, you can destroy him. That's why we don't do program trades," he explained, referring to his own desk. Conversely, one of the statistical arbitrage traders, told us, in veiled dismissal of manual trading, that the more he looked at his data (as opposed to letting his robot trade) the more biased he becomes.

Within each desk, there is a marked consistency between the trading strategy, mathematical formulas, and tools for pattern recognition that

Table 8.1
The valuation principles, formulas, and tools of arbitrage strategies.

Desk	Valuation principle	Typical formula	Tools
Merger arbitrage	The value of an all-stock acquisition target will converge to the price of the acquirer.	$P_T = P_A \cdot r \cdot p_M$[a]	Index plots, spread plots
Index arbitrage	The price of the index futures contract will converge to the spot price of the constituent stocks.	$F_0 = S_0 e^{(r-q)^T}$[b]	High-bandwidth connections to market data
Convertible bond arbitrage	The value of convertible bond can be expressed as the value of a bond and an option to convert into stock.	NA	Bloomberg valuation model, proprietary valuation model
Statistical arbitrage	The ten-day moving average of stock prices reverts to the mean.	$E_T(X) = \dfrac{1}{T} \displaystyle\sum_{t=t_0}^{t=t_0+T} X_t$ $E_T(X) \to 0$ when $T \to \infty$[c]	Robot, atomic clock, order traffic speed indicator
Customer sales	Execute client's order. No sales on downtick trades.	Orders given by clients	Telephone, market indices, magnifying glass, footprints, active cells

a. P_A = stock price of acquirer, P_T = stock price of target, r = exchange ratio, p_M = probability of merger. Source: Reverre 2001.

b. F_0 = price of the futures contract, q = dividend yield rate, r = risk-free interest rate, T = maturity date. Source: Reverre 2001.

c. X = stock price, T = time, $E_T(X)$ = ten-day moving average of X. Source: Reverre 2001.

traders use. Merger arbitrage traders, as table 8.1 shows, plot spreads on their screens but do not use convertible bond valuation models; neither do they employ Black-Scholes equations or draw on principles of mean-reversion. Convertible arbitrage traders, by contrast, use bond valuation models but do not obsess about whether the spread between two merging companies is widening or narrowing. Customer sales traders are more keen on executing their clients' orders on the day they receive them than on following for months the evolution of the spread between two merging stocks. The complex trades that are characteristic of our trading room, however, seldom involve a single desk/team in isolation from others. It is to these collaborations that we now turn.

Connect to Cut, Co-Locate to Dissociate

The desk, in our view, is a unit organized around a dominant evaluative principle and its arrayed financial instruments (devices for measuring, testing, probing, cutting). This principle is its coin; if you like, its specie. But the trading room is composed of multiple species. It is an ecology of evaluative principles. Complex trades take advantage of the interaction among these species. To be able to commit to what counts, to be true to your principle of evaluation, each desk must take into account the principles and tools of other desks. Recall that shaping a trade involves disassociating some qualities in order to give salience to the ones to which your desk is attached. To identify the relevant categories along which exposure will be limited, shaping a trade therefore involves active association among desks. Co-location, the proximity of desks, facilitates the connections needed to do the cutting. Figure 8.1 illustrates the spatial positioning of the various desks in the trading room at International Securities.

Whereas in most textbook examples of arbitrage the equivalence-creating property is easy to isolate, in practice, it is difficult to fully disassociate. Because of these difficulties, even after deliberate slicing and dicing, traders can still end up dangerously exposed along dimensions of the company that differ from the principles of the desired focused exposure. We found that traders take into account unintended exposure in their calculations in the same way as they achieve association: through co-location. Physical proximity in the room allows traders to survey the financial instruments around them and assess which additional variables they should take into account in their calculations.

For example, the stock loan desk can help the merger arbitrageurs on matters of liquidity. Merger arbitrage traders lend and borrow stock as if they could reverse the operation at any moment. However, if the company is small and not often traded, its stock may be difficult to borrow, and traders may find themselves unable to hedge. In this case, according to Max, senior trader at the merger arbitrage desk, "the stock loan desk helps us by telling us how difficult it is to borrow a certain stock." Similarly, index arbitrageurs can help merger arbitrageurs trade companies with several classes of shares. Listed companies often have two types of shares, one called "A-class" and one called "K-class." The two carry different voting rights, but only one of the types allows traders to hedge their exposure. The existence of these two types facilitates the work of merger arbitrageurs, who can exe-

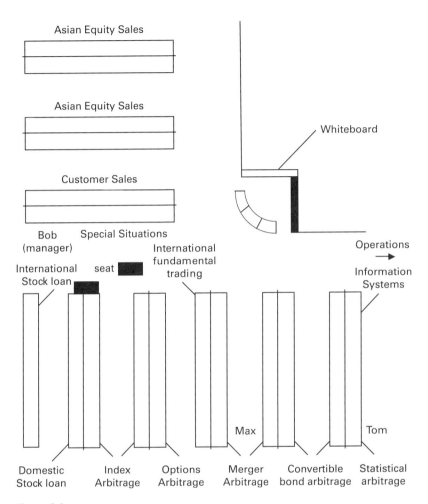

Figure 8.1
Schematic of the trading room at International Securities.

cute trades with the more liquid of the two classes and then transform the stock into the class necessary for the hedge. But such transformation can be prohibitively expensive if one of the two classes is illiquid. To find out, merger arbitrageurs turn to the index arbitrage team, which exploits price differences between the two types.

In other cases, one of the parties may have a convert provision (that is, its bonds can be converted into stocks if there is a merger) to protect the bondholder, leaving merger arbitrage with questions about how this might affect the deal. In this case, it is the convertible bond arbitrage desk that helps merger arbitrage traders clarify the ways in which a convertibility provision should be taken into account. "The market in converts is not organized," says Max, in the sense that there is no single screen representation of the prices of convertible bonds. For this reason,

We don't know how the prices are fluctuating, but it would be useful to know it because the price movements in converts impacts mergers. Being near the converts desk gives us useful information.

In any case, according to Max, "even when you don't learn anything, you learn there's nothing major to worry about." This is invaluable because "what matters is having a degree of confidence."

By putting in close proximity teams that trade in the different financial instruments involved in a deal, the bank is thereby able to associate different markets into a single trade. One senior trader observed:

While the routine work is done within teams, most of the value we add comes from the exchange of information between teams. This is necessary in events that are unique and non-routine, transactions that cross markets, and when information is time-sensitive.

Thus, whereas a given desk is organized around a relatively homogeneous principle of evaluation, a given trade is not. Because it involves hedging exposure across different properties along different principles of evaluation, any given trade can involve heterogeneous principles and heterogeneous actors across desks. If a desk involves simple teamwork, a (complex) trade involves collaboration. This collaboration can be as formalized as a meeting (extraordinarily rare at International Securities) that brings together actors from the different desks. Or it might be as primitive as an undirected expletive from the stock loan desk which, overheard, is read as a signal by the merger arbitrage desk that there might be problems with a given deal.

Practices of Re-cognition

How do the creativity, vitality, and serendipity stemming from the trading room yield new interpretations? By interpretation we refer to processes of categorization, as when traders answer the question "What is this a case of?" but also processes of re-categorization such as making a case *for* something. Both work by association—of people to people, but also of people to things, things to things, things to ideas, etc.

We saw such processes of re-cognition at work in the following case of an announced merger between two financial firms. The trade was created by the "special situations desk," its name denoting its stated aim of cutting through the existing categories of financial instruments and derivatives.

Through close contact with the merger arbitrage desk and the equity loan desk, the special situations desk was able to construct a new arbitrage trade, an "election trade," that recombined in an innovative way two previously existing strategies: merger arbitrage and equity loan.

The facts of the merger were as follows: on January 25, 2001, Investors Group announced its intention to acquire MacKenzie Financial. The announcement immediately set off a rush of trades from merger arbitrage desks in trading rooms all over Wall Street. Following established practice, the acquiring company, Investors Group, made an offer for the shares of the stockholders in the target company. It offered them a choice of cash or stock in Investors Group as means of payment. The offer favored the cash option. Despite this, the head of the special situations desk (Josh) and his traders reasoned that a few investors would never be able to sell their shares. For example, board members and upper management of the target company are paid stocks in order to have an incentive to maximize profit. As a consequence, "it would look wrong if they sold them" John said. In other words, their reasoning included "symbolic" value, as opposed to a purely financial profit-maximizing calculus.

The presence of symbolic investors created, in effect, two different payoffs: cash and stock. The symbolic investors only had access to the smaller payoff. As with any other situation of markets with diverging local valuations, this could open up an opportunity for arbitrage. But how to connect the two payoffs?

In developing an idea for arbitraging between the two options on election day, the special situations desk benefited crucially from social interaction across the desks. The special situations traders sit between the stock loan desk and the merger arbitrage desk. Their closeness to the stock loan desk, which specialized in lending and borrowing stocks to other banks, suggested to the special situations traders the possibility of lending and borrowing stocks on election day. They also benefited from being near the merger arbitrage desk, as it helped them understand how to construct an equivalency between cash and stock. According to Josh, the idea was generated by "looking at the existing business out there and looking at it in a new way. Are there different ways of looking at merger arb? . . . We imagined ourselves sitting in the stock loan desk, and then in the merger arbitrage desk. We asked, is there a way to arbitrage the two choices, to put one choice in terms of another?" The traders found a way. Symbolic investors did not want to be seen exchanging their stock for cash, but nothing prevented another actor such as International Securities from doing so directly. What if the special situation traders were to borrow the shares of the symbolic investors at the market price, exchange them for cash on election day (i.e., get the more favorable terms option), buy back stock with that cash and return it to symbolic investors? That way, the latter would be able to bridge the divide that separated them from the cash option.

Once the special situation traders had constructed the bridge that separated the two choices in the election trade, they still faced a problem. The possibilities for a new equivalency imagined by Josh and his traders were still tenuous and untried. But it was this very uncertainty—and the fact that no one had acted upon them before—that made them potentially so profitable. The uncertainty resided in the small print of the offer made by the acquiring company, Investors Group: how many total investors would elect cash over stock on election day? The answer to that question would determine the profitability of the trade: the loan and buy-back strategy developed by the special situations traders would not work if few investors chose cash over stocks. IG, the acquiring company, intended to devote a limited amount of cash to the election offer. If most investors elected cash, IG would prorate its available cash (i.e., distribute it equally) and complete the payment to stockholders with shares, even to those stockholders who elected the "cash" option. This was the preferred scenario for the special situation traders, for then they would receive some shares back and be able to use them to return the shares they had previously borrowed from the "symbolic" investors. But if, in an alternative scenario, most investors elected stock, the special situations desk would find itself with losses. In that scenario, IG would not run out of cash on election day, investors who elected cash such as the special situations traders would obtain cash (not stocks), and the traders would find themselves without stock in IG to return to the original investors who lent it to them. Josh and his traders would then be forced to buy the stock of IG on the market at a prohibitively high price.

The profitability of the trade, then, hinged on a simple question: Would most investors elect cash over stock? Uncertainty about what investors would do on election day posed a problem for the traders. Answering the question "What will others do?" entailed a highly complex search problem, as stock ownership is typically fragmented over diverse actors in various locations applying different logics. Given the impossibility of monitoring all the actors in the market, what could the special situation traders do?

As a first step, Josh used his Bloomberg terminal to list the names of the twenty major shareholders in the target company, MacKenzie Financial. Then he discussed the list with his team to determine their likely action. As he recalls,

What we did is, we [would] meet together and try to determine what they're going to do. Are they rational, in the sense that they maximize the money they get?

For some shareholders, the answer was straightforward: they were large and well-known companies with predictable strategies. For example, Josh would note:

See ... the major owner is Fidelity, with 13 percent. They will take cash, since they have a fiduciary obligation to maximize the returns to their shareholders.

But this approach ran into difficulties in trying to anticipate the moves of the more sophisticated companies. The strategies of the hedge funds engaged in merger arbitrage were particularly complex. Would they take cash or stock? Leaning over, without even leaving his seat or standing up, Josh posed the question to the local merger arbitrage traders:

"Cash or stock?" I shouted the question to the merger arbitrage team here who were working [a different angle] on the same deal right across from me. "Cash! We're taking cash," they answered.

From their answer, the special situations traders concluded that hedge funds across the market would tend to elect cash. They turned out to be right.

The election trade illustrates the ways in which co-location helps traders innovate and take advantage of the existence of multiple rationalities among market actors. In some ways, the election trade can be seen as a recombination of the strategies developed by the desks around special situations. Proximity to the stock loan desk allowed them to see an election day as a stock loan operation, and proximity to risk arbitrage allowed them to read institutional shareholders as profit maximizers, likely to take cash over stock.

The trade also shows that connectivity and electronic markets play a role that is complementary to place. With easy and automatic access to timely data on prices and transactions, the special situations traders were able to see two payoffs that could be connected in the election trade. The Bloomberg terminals subsequently allowed them to find out the identity of major shareholders. Finally, co-location in the trading room gave them confidence to make a tenuous and uncertain equivalency.

The Trading Room as a Laboratory

In the previous section we showed how calculation is not individual and asocial, but instead is distributed across various desks in the trading room.

In this section we argue that calculation is distributed across socio-technical networks of tangible tools that include computer programs, screens, dials, robots, telephones, mirrors, cable connections, etc.

Although financial instruments (derivatives such as futures, options, swaps, etc) are deemed worthy of study in the *Journal of Finance*, these material instruments supposedly belong to the province of handymen, contractors, and electricians. But traders know they are important, if only because they spend so much time acquiring skills to use, construct and maintain these instruments. Without instruments for visualizing properties of the market, they could not see opportunities; and without instruments for executing their trades, they could not intervene in markets. No tools, no trade.

To see opportunities, traders put on the financial equivalent of infrared goggles that provide them with the trader's equivalent of night-vision. They also delegate calculation to robots that single-mindedly execute their programmed theories, and they scan the room for clues that alert them to the limits in the applicability of these theories.

One cannot appreciate the degree to which quantitative finance is knowledge-intensive without considering the complexity of the trader's tools. According to Knorr Cetina and Bruegger (2002), traders do not quite match up to scientists: when compared to high-energy physicists and their twenty-year long experiments, traders appear as having flat production functions that instead of transforming data merely transpose it onto the screen. By contrast, we found our traders' tools remarkably close to Latour's (1987) definition of scientific instruments as inscription devices that shape a view. Scientific instruments, whether a radio telescope, a Geiger counter, or a Petri dish, display phenomena that are often not visible to the naked eye. They reveal objects in space, radiation waves, or minuscule bacteria that could otherwise not be discerned. Similarly, the trader's tools reveal opportunities that are not immediately apparent. Both scientists and traders derive their strengths—persuasiveness in the former, profits in the latter—from original instrumentation.[11]

Perhaps the most salient instruments at International Securities are the traders' Bloomberg workstations and their individually customized screens. These dramatic extra-wide high-contrast Bloomberg flat panel monitors serve as their workbench. Bloomberg terminals include a specialized monitor, color-coded keyboard and a direct Intranet cable connection to Bloomberg L.P. Even more expensive than the physical terminals is the software that comes with them, structured around five areas that include data (prices, volume, etc.), analytics for parsing and visualizing the data, news (from 1,000 journals around the world), trading support, and information on trade execution. Just as traders are on the look-out for specialized software, they individually tailor their digital workbenches in ways as elaborate as they are diverse: At International Securities, no two screens are the same. Screen instruments are not mere transporters of data, but select, modify, and present data in ways that shape what the trader sees.

Take, for example, the case of Stanley H., a junior trader at the customer trading desk. Like others at his desk, Stan executes arbitrage trades for clients. He does not need to come up with new trades himself, but only to find out the points in time in which he can execute the client's orders. For this purpose, he needs to know the general direction of the market, current developments regarding the companies he is trading, and whether he can trade. His is a world of the here and now. To grapple with it, Stan has arranged on his screens instruments such as a "magnifying glass," trading "baskets," and "active links."

Stan's point of departure is the baseline information that everyone has: a Bloomberg window that graphs the Dow Industrials and the NASDAQ market indexes to give him information on the market's general direction. Next to it, another instrument provides a more personalized perspective. A window that he calls his "magnifying glass" displays 60 crucial stocks that he considers representative of different sectors such as chips, oil, or broadband. Visually, the numbers in this window momentarily increase in size when an order is received, resembling a pulsating meter of live market activity. Stan complements the magnifying glass with the "footprints" of his competitors in tables that display rival banks' orders in the stocks that he trades.

Stan's screens include a clipboard for his operations, an arrangement that simplifies and automates part of the cognitive work involved in making the trades. This is composed of several "trading baskets," windows that show the trades that he has already done. An additional instrument shows pending work. This is contained in an Excel spreadsheet in which Stan introduces entries with "active links" to stock prices, that is, cells that are automatically updated in real time. In the cells next to the links Stan has programmed the conditions that the clients give to him (e.g., "set the spread at 80"). As a result, another cell changes color depending on whether the conditions are met. (Cyan means they are met; dark green means they are not.) The computer, then, does part of the calculation work for Stan. Instead of having to verify whether the conditions hold to execute each of the trades, he follows a much simpler rule: trade if the cell is cyan, do not trade if it is dark green.

Stan is a toolmaker as much as a "trade maker," a craftsman of tools as much as a processor of information. He devotes considerable deliberation to the conscious inscription of his screens. Every day, one hour before the markets open, he arrives to the trading room to prepare his setup; part of that preparation is readying the screens. One by one, Stan opens his windows and places them in their customary places, ensures each has its own color and size, creates new active links as customers order new trades, and discusses possible technical issues with the computer programmers.

Two desks away, at the convertible bond arbitrage desk, Richard looks at stocks from a very different perspective—as if they were bonds. As was noted above, traders in convertible bond arbitrage such as Richard seek to exploit the value of the "convertibility" option that is sometimes included in bonds. These allow the bondholder to convert the bond into a stock, in effect morphing from one type of security into another. To assess the value of the option to convert, Richard uses Bloomberg's proprietary "Convertible Bond Valuation" model, which returns an estimated value of the bond given basic parameters such as volatility of the stock, its delta, and its gamma. Richard's models can be seen as a pair of goggles that highlight the hidden value of convertibility options.

Close to the bond arbitrage desk, Max Sharper at the merger arbitrage desk exploits profit opportunities when companies merge. As has been noted, merger arbitrage traders long the company that is the acquisition target and short the acquirer. Their trades thus become a bet on the probability that the merger will take place. To decide whether to bet on a merger, Max plots the "spread" or price difference between the companies in merger talks. If two companies merge, they will be worth the same, and their spread will be zero. As the merger unfolds, a small spread denotes market confidence in the merger, and a large spread denotes skepticism. Max plots the spread in time to read back from it the "implied probability" that the market assigns to the merger. As with the other traders, Max's spread plots serve as an optical device that brings into focus actors' confidence about a given merger.

The visualization techniques of on-screen instruments, then, are as varied as the principles of arbitrage that guide each desk. Stan's desk executes trades, and the magnifying glasses, trading baskets, rivals' footprints, and active links on his screens display momentary instances of open windows of opportunity in a geometric array of white, green, blue, and cyan squares with numbers dancing in them, lending it the appearance of an animated painting by Piet Mondrian. Richard's desk buys and sells convertible bonds, and the bond valuation models on his screens display a more conventional text interface, a boxy black-on-white combination suggestive of 1980s-style minicomputer screens. The spread plots for betting on mergers on Max's screens show narrow white lines that zigzag in a snake-like manner from left to right over the soothing blue background of his monitor.

The traders' reliance on such goggles, however, entails a serious risk. In bringing some information into sharp attention, the software and the graphic representations on their screens also obscure. In order to be devices that magnify and focus, they are also blinders. According to one trader, "Bloomberg shows the prices of normal stocks; but sometimes, normal stocks morph into new ones." One such occasion is in instances of mergers or bond conversions. If a stock in Stan's magnifying glass—say, an airline that he finds representative of the airline sector—were to go through a merger or bond conversion, it would no longer stand for the sector.

An even more serious risk for the traders is that distributing calculation across their instruments amounts to inscribing their sensors with their own beliefs. As we have seen, in order to recognize opportunities, the trader needs special tools that allow him to see what others cannot. But the fact that the tool has been shaped by his theories means that his sharpened perceptions can sometimes be highly magnified misperceptions, perhaps disastrously so. For an academic economist who presents his models as accurate representations of the world, a faulty model might prove an embarrassment at a conference or seminar. For the trader, however, a faulty model can lead to massive losses. There is, however, no option not to model: no tools, no trade. What the layout of the trading room—with its interactions of different kinds of traders and its juxtaposition of different principles of trading—accomplishes is the continual, almost minute-by-minute, reminder that the trader should never confuse representation for reality.

Instead of reducing the importance of social interaction in the room, the highly specialized instruments actually provide a rationale for it. "We all have different kinds of information," Stan says, referring to other traders, "so I sometimes check with them." How often? "All the time."

Hence, just as Latour (1987) defined a laboratory as "a place that gathers one or several instruments together," trading rooms can be understood as places that gather diverse market instruments together. Seen in this light, the move from traditional to modern finance can be considered as an enlargement in the number of instruments in the room, from one to several. The best scientific laboratories maximize cross-fertilization across disciplines and instruments. For example, the Radar Lab at MIT made breakthroughs in the 1940s by bringing together the competing principles of physicists and engineers (Galison 1996).[12] Similarly, the best trading rooms bring together heterogeneous value frameworks for creative recombination.

Monitoring the Price Mechanism

Another example of distributed calculation can be found in "robots," computer programs used by statistical arbitrage traders that automate the process of buying and selling stocks. As with the other market instruments of the trading room, robots bring benefits but also pose new challenges that are solved by intermingling the social, the cognitive, and the artifactual.

Robots are representations as well as tools for automation. Inscribed with the trader's beliefs, they execute only the trading strategy they were programmed to perform. For example, in deciding whether to buy or sell stocks, a mean reversion robot only takes into account whether the prices are close or distant from their historic average price, while an earnings robot only considers the companies' earnings. Robots enact a complex set of assumptions about the market, and they process an active selection of the available data that are consistent with it.

Sociability in the room is crucial from the moment of the robot's inception, a process of codifying tacit knowledge into algorithms and computer code. This takes place at the whiteboard in meetings of heterogeneous perspectives that might include an index arbitrage trader, a computer programmer, and a merger arbitrage trader. Starting from the whiteboard, an idea for a trade mutates in form from a trader's utterances, to graphs on the board, to abstract models, to mathematical equations, and finally into computer code. The robot is quite literally codified knowledge.

Once codified into a computer program, the robot goes to work with traders specialized in implementing computer programs such as the statistical arbitrage desk. But the story does not end here. Piloting a robot requires inputs from a kind of emergent traffic control—cues and signals from other parts of the room.

Consider the case of Tom, a trader at the statistical arbitrage desk. Instead of trading manually, Tom uses and maintains a robot. Automated trading poses the same challenge as driving a car at a high speed: any mistake can lead to disaster very quickly. "I have," Tom says, "a coin that comes up heads 55 percent of the time." With margins as low as 0.05, the only route to high returns is trading a very high volume—as Tom, "the point is to flip [the coin] a lot." As with Formula 1 car racing or high-speed boating, traders need excellent instrumentation. Indeed, they have navigation instruments as complex as an airplane cockpit. Yet, as it turns out, these are not enough. The price mechanism has to be monitored, and calibrated; and for that purpose Tom obtains crucial cues from the social interactions at the desks around him.

To illustrate the sensitivity of results to timely data (in which the units of measurement are frequently seconds rather than minutes), Tom recounts an instance in which a slight time delay lost millions of dollars for a competing bank—and earned them for International Securities. On that specific day, some banks had been receiving price information with a delay because of problems with the Reuters server. Price movements had been large all through the day, and the market index had risen very quickly. In a rising market, a delay makes the index appear consistently below its real level. In contrast to spot prices, prices for futures contracts were arriving at the banks with no delay. As a result, index arbitrage traders at one bank (traders who exploit differences between spot and S&P 500 futures) perceived as inexpensive securities that were in fact very expensive, and bought extensively. Tom and others at International Securities, in contrast, were getting timely information on both spot and futures prices. Tom recounts:

While they were buying, we were selling . . . the traders here were writing tickets until their fingers were bleeding. We made $2 million in an hour, until they realized what was happening.

The episode illustrates the challenges of working with robots. When trading at Formula 1 speed, "the future" is only seconds away. When the speed of trading amplifies second-by-second delays, the statistical arbitrage trader must be attuned to a new kind of problem: By how many seconds are the data delayed? That is, traders have to remind themselves of the time lag that elapses between what they see—the numbers on their screens—and actual prices. The prices that matter are those that reside in the computer servers of the market exchange, be it the NASDAQ or the New York Stock Exchange, for that is where the trades are ultimately executed. What traders see on their screens are bits and bytes that have been transported from the exchange to the trading room in a long and sometimes difficult path of possible delays. If traders mistakenly take delayed data for real-time data, losses will pile up quickly. In that situation, delegating the trading decisions to the robot could lead to disaster. How do the statistical arbitrage traders prevent these disasters from taking place?

The first line of defense against the risks of high-volume, high-speed automated trading is more technology. Tom's robot provides him with as many dials as a cockpit in an airplane. He trades with three screens in front of him. Two of them correspond to powerful Unix workstations; the third is a Bloomberg terminal. One Unix terminal has real-time information about his trades. Across the top of one, a slash sign rotates and moves from side to side. It is a "pulse meter" to gauge the "price feed," i.e., the speed with which information on prices is arriving at him. The character stops moving when prices stop arriving. It is very important to be aware when this happens, because the price robot can get confused. According to Tom, "it thinks that prices aren't changing and it imagines false opportunities, while in reality prices are moving but not arriving to it."

Tom benefits from numerous additional dials. On the right side of the screen of Tom's second Unix station, there are has five squares. Each of them is a speedometer that indicates how quickly the orders are getting through the servers of the specialists or electronic communication networks. If they are green, everything is fine. If they are yellow, the network is congested and deals are delayed. If they are red, servers are clogged. The clocks in the Unix workstations are synchronized to the atomic clock every day. In addition to a large display of an analog clock in his computer, Tom has two "CPU-meters" which measure congestion in the bank's order flow. When it is engaged for long periods, orders take longer to execute. Thus, to monitor prices in the market, traders must monitor the price mechanism—literally, they must monitor the machines that bring and make prices.

Technology, however, is not the only answer to the problem of execution, for the dials that measure the accuracy of the technology are a representation themselves. Technology, in other words, answers one question ("Is the robot getting the data?") but raises another ("Is the robot right in what it says?"). We call this infinite-regress problem *the "calibration" problem*.

The 1986 nuclear accident at Chernobyl illustrates the calibration problem. Radiation was so high that the dials of the Geiger counters of the control room of the Soviet nuclear power station did not register any abnormal level of radiation even at the peak of the escape. The dials, calibrated to register nuances, failed to detect the sharp increase in radiation levels. Technology permits the execution of automated tasks, but it requires appropriate calibration.

How to solve the calibration problem? Tom solves it by drawing on the social and spatial resources of the trading room. He sits in between the merger arbitrage desk and the systems desk. According to Tom,

When you hear screams of agony around you, it indicates that perhaps it is not a good time to trade. If I hear more screams, maybe I should not use the system even if it's green.

Similarly, price feed in stocks and futures has to come at the same speed. By sitting near the futures arbitrage desk, the stat arb trader can remain alert to any anomaly in the data feed. In addition to getting a sense of when to turn off their robots, statistical arbitrage traders interpret cues from nearby desks to gauge when to take a particular security out of automated trading. The instruments of representation that make up the technology of finance retain their value only so long as they remain entangled in the social relations that spawned them. A trader's tools are socio-technical.

This socio-technical character, finally, governs the placement of the robots in the trading room. While promoting association through proximity, the trading room also uses distance to preserve the requisite measure of variety among the robots. Instead of minimizing differences to produce a "one right way" to calculate, the trading room actively organizes diversity. Of the four statistical arbitrage robots, a senior trader observed:

> We don't encourage the four traders in statistical arb to talk to each other. They sit apart in the room. The reason is we have to keep diversity. We could really get hammered if the different robots would have the same P&L [profit and loss] patterns and the same risk profiles.

Seemingly at odds with the policy of putting all the traders of the same function at the same desk, the statistical arbitrage traders and their robots are scattered around the room. Why? Because the robots, as the traders say, are partly "alive"—they evolve. That is, they mutate as they are maintained, re-tooled, and re-fitted to changes in the market. They are kept separated to reduce the possibility that their evolution will converge (thereby resulting in a loss of diversity in the room). But they are, of course, not pushed out of the room entirely because a given stat arb unit cannot be too far from the other types of arbitrage desks—proximity to which provides the cues about when to turn off the robots.

Conclusion

In the preface to *Novum Organum*, one of the founding documents of modern science, Francis Bacon wrote: "In every great work to be done by the hand of man it is manifestly impossible, without instrumentation and machinery, either for the strength of each to be exerted or the strength of all to be united." (1620/1960: 35) These observations about the importance of instrumentation were crucial to Bacon's broader goal of outlining a new course of discovery. Writing in an age when the exploration, conquest, and settlement of territory was enriching European sovereigns, Bacon proposed an alternative strategy of exploration. In place of the quest for property, for territory, Bacon urged a search for properties, the properties of nature, arguing that this knowledge, produced at the workbench of science, would prove a nearly inexhaustible source of wealth.[13]

Just as Bacon's experimentalists at the beginnings of modern science were in search of new properties, so our arbitrage traders at the beginnings of quantitative finance are in search of new properties—as different from the old notions of property of value investors or momentum traders as Bacon's was from the conquest of territory. And just as Bacon, in the more standard reading, was advocating a program of inductive, experimentalist science in contrast to logical deduction, so our arbitrage traders, in contrast to the deductive stance of neo-classical economists, are actively experimenting to uncover properties of the economy. But whereas Bacon's New Instrument was part of a program for "The Interpretation of Nature,"[14] the new instruments of quantitative finance—connectivity, equations, and computing—visualize, cut, probe, and dissect ephemeral properties in the project of interpreting markets. In the practice of their trading room laboratories, our arbitrage traders are acutely aware that the reality "out there" is a social construct consisting of other traders and other interconnected instruments continuously reshaping, in feverish innovation, the properties of that recursive world. In this co-production, in which the products of their interventions become a part of the phenomenon they are monitoring, such reflexivity is an invaluable component of their tools of the trade.

Economic sociologists, we have argued, need to make the study of technology a part of the tools of our trade. When economists or sociologists study technology, it is most often as a specialized subfield, e.g., the social studies of science or the economics of technological innovation. Such research is invaluable. But we should also incorporate the study of technology in the core subfields of our disciplines. In our epoch, organizational design, for example, is inseparable from design of the digital interface. Similarly, to understand not only the mathematics but also the machines that make up the sophisticated market instruments of quantitative finance we need to analyze the entanglements of actors and instruments in the sociotechnology of the trading room laboratory.

Acknowledgments

This work was originally published in *Industrial and Corporate Change* (13, 2004: 369–400). Our thanks to Pablo Boczkowski, Michael Burawoy,

Michel Callon, Karin Knorr Cetina, Paul Duguid, Geoff Fougere, Istvan Gabor, Raghu Garud, William Guth, Vincent Lepinay, Frances Milliken, Fabian Muniesa, Alex Preda, Harrison White, Sidney Winter, Amy Wrzesniewski, and especially Monique Girard for helpful comments and suggestions on a previous draft. We are also grateful to the Russell Sage Foundation for providing a stimulating and collegial community during Stark's stay as a Visiting Fellow in 2002–03.

Notes

1. For a large-sample approach to the organization of trading rooms, see Zaheer and Mosakowski 1997.

2. Zaloom (2004) correctly emphasizes that, to speculate, a trader must be disciplined. In addition to this psychological, almost bodily, disciplining, however, we shall see that the arbitrage trader's ability to take a risky position depends as well on yet another discipline—grounding in a body of knowledge.

3. For an application of interpretive theories of organization to the military, see Weick and Roberts 1993.

4. We are re-interpreting March's (1991) exploitation/exploration problem of organizational learning through the lens of the problem of recognition. See also Stark 2001; Girard and Stark 2002.

5. At the outset of our investigation, quantitative finance seemed an improbable setting to find actors preoccupied with qualities. On the qualification of goods in other settings and for theoretical discussions of economies of qualities, see Eymard-Duvernay 1994; Thevenot 1996; Favereau 2001; White 2001; Callon et al. 2002.

6. The exposure created by a trade is given by the impact that a change in some variable (such as the price of an asset) can have on the wealth of the trader. Following the quantitative revolution in finance, traders think about their own work in terms of *exposure*, not in terms of *transactions*. Hence, for example, they do not use the expression "buy IBM," but say "to be long on IBM" which means that a trader stands to profit when the price of IBM rises. Similarly, they do not say "sell," but "be short on." The reason for this change in terminology is that, through the use of derivatives, traders can attain a given exposure in different ways.

7. A swap is an agreement to exchange rights or obligations.

8. A stock option is a derivative security that gives its holder the right to buy or sell a stock at a certain price within a given time in the future.

9. Castells's observations are consistent with findings in much of the Computer-Supported Cooperative Work literature on automated control rooms (see, e.g., Heath et al. 1995).

10. A put is a financial option that gives its holder the right to sell. A call gives the right to buy.

11. For insightful treatments of the interaction between valuation and technology in the field of finance, see Preda's (2002) historical study of the ticker and its effects on investor behavior, and Muniesa's (2002) study of the use of telephones in trading rooms.

12. On the architecture of science, see Galison and Thompson 1999.

13. We owe this insightful reading of Bacon's writings, including *Novum Organum* and his (often unsolicited) "advices" to his sovereigns, Elizabeth I and James I, to Monique Girard.

14. *Novum Organum* translates as "New Instrument." Bacon contrasts the deductive method of "Anticipation of the Mind" to his own method of "Interpretation of Nature" (1620/1960: 37).

References

Abolafia, M., and M. Kilduff. 1988. Enacting market crisis: The social construction of a speculative bubble. *Administrative Science Quarterly* 33: 117–193.

Abolafia, M. 1996. *Making Markets: Opportunism and Restraint on Wall Street.* Harvard University Press.

Bacon, F. 1960. *Novum Organum (The New Organon)* [1620]. Bobbs-Merrill.

Baker, W. 1984. The social structure of a national securities market. *American Journal of Sociology* 89: 775–811.

Beunza, D., and D. Stark. 2003. The organization of responsiveness: Innovation and recovery in the trading rooms of wall street. *Socio-Economic Review* 1: 135–164.

Beunza, D., and D. Stark. 2005. A desk on the 20th floor: Survival and sense-making in a trading room. In *Wounded City*, ed. N. Foner. Russell Sage Foundation.

Boltanski, L., and L. Thevenot. 1991. *De la justification: Les économies de la grandeur.* Gallimard.

Boltanski, L., and L. Thévenot. 1999. The sociology of critical capacity. *European Journal of Social Theory* 2: 359–377.

Brown, J., and D. Paul. 2000. *The Social Life of Information.* Harvard Business School Press.

Callon, M. 1998. Introduction: The embeddedness of economic markets in economics. In *The Laws of the Markets*, ed. M. Callon. Blackwell.

Callon, M., and F. Muniesa. 2002. Economic Markets as Calculative and Calculated Collective Devices. Manuscript, Centre de Sociologie de l'Innovation, Ecole des Mines de Paris.

Callon, M., C. Méandel, and V. Rabeharisoa. 2002. The Economy of Qualities. Manuscript, Centre de Sociologie de l'Innovation, Ecole des Mines de Paris.

Castells, M. 1996. *The Information Age: Economy, Society and Culture*, second edition. Blackwell.

Clippinger, J. 1999. Tags: The power of labels in shaping markets and organizations. In *The Biology of Business*, ed. J. Clippinger. Jossey-Bass.

Dunbar, N. 2000. *Inventing Money: The Story of Long-Term Capital Management and the Legends Behind It*. Wiley.

Eymard-Duvernay, F. 1994. Coordination des exchanges par l'entreprise et qualité des biens.' In *Analyse Economique des Conventions*, ed. A. Orléan. Presses Universitaires de France.

Favereau, O., O. Biencourt, and F. Eymard-Duvernay. 2001. Where Do Markets Come From? From (Quality) Conventions. Manuscript, INSEAD, Paris.

Fligstein, N. 1990. *The Transformation of Corporate Control*. Harvard University Press.

Galison, P. 1997. *Image and Logic: A Material Culture of Microphysics*. University of Chicago Press.

Galison, P., and E. Thompson, eds. 1999. *The Architecture of Science*. MIT Press.

Garud, R., and P. Karnøe. 2001. Path creation as a process of mindful deviation. In *Path Dependence and Creation*, ed. R. Garud and P. Karnøe. Erlbaum.

Girard, M. n.d. Francis Bacon and the New Empire of Knowledge. Manuscript, Department of Anthropology, Harvard University.

Girard, M., and D. Stark. 2002. Distributing intelligence and organizing diversity in new media projects. *Environment and Planning A* 34: 1927–1949.

Gladwell, M. 2000. Designs for working. *The New Yorker*, December 11, 60–70.

Godechot, O. 2000. Le bazar de la rationalité. *Politix* 13: 17–56.

Graham, B., and D. Dodd. 1934. *Security Analysis: Principles and Techniques*. McGraw-Hill.

Grabher, G. 2002a. Cool projects, boring institutions temporary collaboration in social context. *Regional Studies* 36: 205–214.

Grabher, G. 2002b. The project ecology of advertising: Tasks, talents and teams. *Regional Studies* 36: 245–262.

Granovetter, M. 1985. Economic action and social structure: The problem of embeddedness. *American Journal of Sociology* 19: 481–510.

Heath, C., M. Jirotka, P. Luff, and J. Hindmarsh. 1995. Unpacking collaboration: The interactional organization of trading in a city dealing room. *Computer Supported Cooperative Work* 3: 147–165.

Hull, J. 1996. *Options, Futures, and Other Derivative Securities*. Prentice-Hall.

Hutchins, E. 1995. *Cognition in the Wild*. MIT Press.

Hutchins, E., and T. Klausen. 1991. Distributed cognition in an airline cockpit. In *Distributed Cognition and Communication at Work*, ed. Y. Engestrom and D. Middleton. Cambridge University Press.

Knight, F. 1921. *Risk, Uncertainty and Profit*. Houghton Mifflin.

Knorr Cetina, K. 2002. The Market as an Epistemic Institution. Paper presented at New York Conference on Social Studies of Finance, Columbia University.

Knorr Cetina, K., and A. Preda. 2001. The epistemization of economic transactions. *Current Sociology* 49: 27–44.

Knorr Cetina, K., and U. Bruegger. 2002. Global microstructures: The virtual societies of financial markets. *American Journal of Sociology* 107: 905–950.

Lane, D., and R. Maxfield. 1996. Strategy under complexity: Fostering generative relationships. *Long Range Planning* 29: 215–231.

Latour, B. 1986. Powers of association. In *Power, Action, and Belief*, ed. J. Law. Routledge.

Latour, B. 1987. *Science in Action: How to Follow Scientists and Engineers through Society*. Harvard University Press.

Latour, B. 1988. *The Pasteurization of France*. Harvard University Press.

Latour, B. 1991. Technology is society made durable. In *A Sociology of Monsters*, ed. J. Law. Routledge.

Lepinay, V., and F. Rousseau. 2000. Les trolls sont-ils incompétents? Enquête sur les financiers amateurs. *Politix* 13: 73–97.

Lewis, M. 1999. How the eggheads cracked. *New York Times*, January 24.

MacKenzie, D. 2002. Risk, Financial Crises, and Globalization: Long-Term Capital Management and the Sociology of Arbitrage. Manuscript, University of Edinburgh.

MacKenzie, D., and M. Yuval. 2003. Constructing a market, performing theory: The historical sociology of a financial derivatives exchange. *American Journal of Sociology* 109: 107–145.

Malkiel, B. 1973. *A Random Walk down Wall Street*. Norton.

March, J. 1991. Exploration and exploitation in organizational learning. *Organization Science* 2: 71–87.

Muniesa, F. 2000. Un robot walrasien. Cotation électronique et justesse de la découverte des prix. *Politix* 13: 121–154.

Muniesa, F. 2002. Reserved Anonymity: On the Use of Telephones in the Trading Room. Paper presented at New York Conference on Social Studies of Finance, Columbia University.

Podolny, J., and T. Stuart. 1995. A role-based ecology of technological change. *American Journal of Sociology* 100: 1224–1260.

Preda, A. 2002. On Ticks and Tapes: Financial Knowledge, Communicative Practices, and Information Technologies on 19th Century Financial Markets. Paper presented at New York Conference on Social Studies of Finance, Columbia University.

Reverre, S. 2001. *The Complete Arbitrage Deskbook*. McGraw-Hill.

Smith, C. 2001. *Success and Survival on Wall Street: Understanding the Mind of the Market*. Rowman and Littlefield.

Suchman, L. 1987. *Plans and Situated Actions: The Problem of Human-Machine Communication*. Cambridge University Press.

Stark, D. 1999. Heterarchy: Distributing intelligence and organizing diversity. In *The Biology of Business*, ed. J. Clippinger. Jossey-Bass.

Stark, D. 2000. For a Sociology of Worth. Paper presented at the Workshop on Heterarchy: Distributed Intelligence and the Organization of Diversity, Santa Fe Institute.

Thévenot, L. 1993. Essais sur les objects usuels. Propriétés, fonctions, usages. Les objets dan l'action. De la maison au laboratoire. *Raisons Pratique* 4: 85–111.

Thévenot, L. 1996. Pragmatic Regimes Governing the Engagement with the World: From Familiarity to Public "Qualifications." Manuscript, Ecole des Etudes en Sciences Sociales.

Thévenot, L. 2001. Organized complexity: Conventions of coordination and the composition of economic arrangements. *European Journal of Social Theory* 4: 405–425.

Thrift, N. 1994. On the social and cultural determinants of international financial centres: The case of the City of London. In *Money, Power and Space*, ed. S. Corbridge et al. Blackwell.

Thrift, N. 2000. Pandora's box? Cultural geographies of economies. In *The Oxford Handbook of Economic Geography*, ed. G. Clark et al. Oxford University Press.

Uzzi, B. 1997. Social structure and competition in interfirm networks: The paradox of embeddedness. *Administrative Science Quarterly* 42: 35–67.

Uzzi, B. 1999. Embeddedness in the making of financial capital: How social relations and networks benefit firms seeking financing. *American Sociological Review* 64: 481–505.

Weick, K. 1979. *The Social Psychology of Organizing*, second edition. Addison-Wesley.

White, H. 1981. Where do markets come from? *American Journal of Sociology* 87: 983–938.

White, H. 2001. *Markets from Networks: Socioeconomic Models of Production*. Princeton University Press.

Wolfe, T. 1987. *The Bonfire of the Vanities*. Farrar, Straus and Giroux.

Zaloom, C. 2002. Ambiguous Numbers: Trading and Technologies in Global Financial Markets. Paper presented at New York Conference on Social Studies of Finance, Columbia University.

Zaloom, C. 2004. The discipline of the speculator. In *Global Assemblages*, ed. A. Ong and S. Collier. Blackwell.

9   Trading-Room Telephones and the Identification of Counterparts

Fabian Muniesa

One important contribution from Science and Technology Studies (STS) to the understanding of markets is perhaps the introduction of market devices as a legitimate topic for economic sociology. But what is a "market device"? An STS approach tends to emphasize a material definition of the latter: objects, instruments, tools, and techniques (i.e., technologies in the largest sense) that enable market activities. Market devices include metering systems, communication technologies, calculating tools, allocation protocols, display techniques, payment facilities, and feedback methods. These are not just fancy gadgets that are there to support, garnish, or buoy up economic arrangements—economic arrangements whose rationale and operations would unfold plainly before the social scientist's eyes precisely if she were to avoid being distracted by such devices and concentrate instead on economy and society "alone." Much to the contrary, market devices play a crucial role in the formation (and the deformation) of economic configurations.

Market devices play a role, but "to play a role" is a rather ambiguous thing. Do market devices play a role in a purely instrumental sense, i.e., as an aid to an otherwise fully purposeful, plainly human action? Or do they imprint action with a technologically deterministic direction, i.e., as a driving force of markets? Students of technology have dealt with these sorts of agency conundrums since, at least, Karl Marx (MacKenzie 1984). One possible solution, which is sometimes referred to as the "performativity program" and which often conveys recognizable traits from actor-network theory (Hardie and MacKenzie 2007; Callon 2007; Callon and Muniesa 2005; Muniesa, Millo, and Callon 2007), is to consider an "economic actor" as a compound arrangement of people and devices and, more precisely, "agency" itself as the very product of this arrangement (or *agencement*, to use a particularly useful notion inspired by Gilles Deleuze's pragmatist philosophy).

In financial markets, for instance, there is serious evidence of "actors" being compound arrangements, artificial beings mixing persons and things, but mixing them in a way that allows them (the resulting *agencement*) to act in a particular economic way. Legal devices that organize liability and imputation are critical ingredients of such arrangements. Technologies that sustain transactions are crucial too, as abundantly studied in social studies of finance (e.g. Beunza and Stark, this volume; Knorr Cetina and Bruegger 2002; MacKenzie and Millo 2003; MacKenzie 2004; Godechot, Hassoun, and Muniesa 2000; Muniesa 2007; Lépinay and Hertz 2004; Preda 2003, this volume; Zaloom 2003). Note even how furniture and spatial layout can be at the center of the definition of what an actor is in finance: "desks" are commonly referred to as units of agency in the trading rooms of investment banks. The desk is a team (of traders and their assistants), but a team equipped with workstations, telephones, and data networks, and, of course, with a particularly important piece of "financial furniture": the open-space work table.

In this chapter, I focus on one particular device that is found on these financial desks: the telephone. Research in conversation analysis and workplace studies (e.g., Boden 1994; Luff, Hindmarsh and Heath 2000; Hutchby 2001) has pointed out the relevance of this technology in work settings. Studies on emerging repertoires of connectedness and availability in everyday life (Licoppe 2004) and on the organization of economic exchange (Mallard 2004) also emphasize the fundamental role of telephonic uses and strategies. Telephones are particularly pervasive in financial markets. The fact that computer screens have become a leitmotif of financial material culture does not preclude market actors' making extensive use of telephonic devices. The phone call is an essential feature of many market interactions, and telephony constitutes a non-negligible part of the financial technology business.

Instead of considering the telephone as a passive medium, a mere instrument serving human interaction as a vehicle for voice, the research I focus on in this chapter pays attention to the way in which this device's technical features shape action in particular ways and achieve a number of effects that are relevant to the functioning of markets. The telephone—and dedicated trading-room voice technologies in particular—configures interlocutory relations in several ways, with variable degrees of privacy and variable ways of making the market audible and actionable. This translates into diverse ways of making trades and of making prices. The way the telephone is handled in financial markets can be thought of as a process of "co-construction" of both users and technology (Oudshoorn and Pinch 2003). The device imposes its features and induces a "script" (Akrich 1992), but traders also engage in creative uses and combine the telephone with other market technologies, such as electronic trading systems. The telephone enters particular arrangements and shapes the way financial counterparts address one another and the way trade interactions are enacted.

Such considerations can also be developed in relation to sociological discussion of how economic transactions are or are not embedded in social relations (Granovetter 1985). It is obvious that "over-the-counter" markets—especially in the case of complex, custom-made financial products designed for specific corporate clients—are potentially more "socially embedded," in Granovetter's sense, than markets such as centralized, order-driven equity exchanges in which counterparts cannot choose one another bilaterally—especially in the case of automated, anonymous exchange protocols. Although financial markets have provided good occasions for sociological analyses in terms of social networks (Baker 1984), they have also prompted a renewal of the sociological repertoire, precisely because technical mediations challenge classical notions of market networks (Knorr Cetina 2003; Knorr Cetina and Grimpe, this volume). Telephonic practices themselves are crucial to the understanding of the extent to which the identification of a counterpart relies on bilateral recognition or not in one particular market configuration.

In this chapter I explore in this line of analysis telephonic practices in a number of relevant financial settings. My empirical material comes from a set of fieldwork interviews carried out between 2000 and 2002 in the trading rooms of several investment banks and brokerage houses in France.[1] In the first section I briefly describe professional telephony systems for financial trading rooms. In the next three sections I illustrate communicational usages involving telephones in three different financial trading environments: a market-making environment, a stockbroker environment, and a sales environment. For the market-making environment, I use the example of a bond trader trying to get a better price. For the stockbroker environment, I describe brokers' telephone strategies previous to the introduction of anonymity in an automated stock exchange. For the sales environment, I analyze new technological combinations aiming at enhancing the identification of clients. In the concluding section I address some sociological implications of the use of telephones in financial trading rooms.

Features of Trading-Room Telephony

Trading-room telephones are highly sophisticated devices. Several specialized vendors, including IPC Information Systems, Syntegra (British Telecom), and Etrali (France Télécom), provide dedicated voice trading technology for financial trading rooms. These services include aspects such as specialized voice and data networks, recording technologies, and desk turrets.

Unlike more conventional telephony systems, trading-room telephony systems allow handling several simultaneous calls. All calls can be treated frontally, without queuing. Such systems allow users to manage several open lines with several interlocutors—up to 28 simultaneous calls, for instance, in the case of Etrali's most widespread system (Etradeal). In daily activity, a line may still be open even if interlocutors are not currently engaging into a conversation, so they can be immediately available to one another if necessary.

Most trading-room telephone turrets include a board, often in the form of a touch-screen interface, where the user can select several pre-defined contacts. Several color codes indicate incoming calls and call status. Lines can be opened or closed with one touch—i.e., they do not have to be dialed, as communication is handled through specific digital voice and data networks. Some systems allow saving turret configurations in a personal card, so that a user can keep his private phone book and recover it in another desk. Incoming sound may be displayed through phone handsets, earplugs, or speakers. The user can speak up through a phone handset or through a microphone. (See figure 9.1.)

Permanently open lines can be handled through a special interphone module, often alluded to as "the box" (in French "la boîte") that allows traders to keep several brokers or market makers on the line permanently.

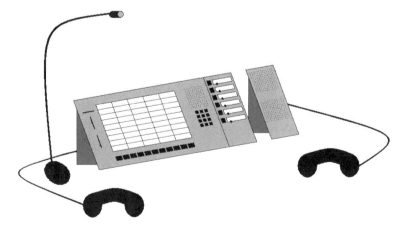

Figure 9.1
A typical trading room telephone turret.

The "box" is particularly relevant for market life. One trader put it this way: "If you don't have your brokers in the box, you are not in the market!" (Options trader and market maker at an investment bank, July 2001.) This communicational environment allows for complex arrangements. It is not unusual to witness customized uses of telephones in trading rooms. A trader, for instance, can physically bend his desk microphone toward the interphone module (the "box") in order to share with a client the morning analysis delivered on the phone, through the "box," by a broker. The extent to which a call can be fully private is adjustable. A user's availability is never straightforward, as the user can deploy many strategies for allocating call handling priorities. The absence of automatic call-queuing protocols gives the user the possibility of recognizing incoming calls, of deciding which call to take first, of discriminating among calls on the basis of heterogeneous strategies, and of managing calls through a wide variety of choices—pass the call to a colleague, handle several calls simultaneously, hold the line with or without sound, etc. It takes some time before a neophyte gets used to this technology. And mistakes—such as accidentally letting a client listen to one's conversation with a broker—are not rare.

Real-time identification (and authentication) of the interlocutor is an essential feature of this kind of communication devices. All calls can also be tracked through specific back-office technologies. In present-day trading rooms, telephone activity is always accompanied by recording technologies. All telephone conversations are recorded, time-stamped, and saved on a secure server. In the case of a trade dispute with a counterpart or some other controversial event, this memory allows for rapid identification and auditing of telephone interactions. Although it is not unusual to observe uses of mobile phones in trading rooms, the use of dedicated telephone turrets is commonly acknowledged—if not enforced—for most trade-sensitive, professional conversations. Recent technological developments, including Computer Telephony Interface (CTI), are often oriented toward an enrichment of identification. For instance, some vendors allow their telephone technology to be combined to Customer Relationship Management (CRM) software: a telephone event triggers a computer action such as the real-time display in a PC screen of relevant information about the interlocutor that is on the line.

Trading-room telephone technology is flexible and permits diverse communicational arrangements. Overall, one important feature characterizes it: the recognition of the counterpart. Although not always strictly private and bilateral, telephone interactions are primarily based on the identification of interlocutors and the deliberate engagement into conversations. In this respect, this particular exchange technology is quite different from other technologies (such as centralized electronic trading systems) now operating in many stock exchanges. As opposed to telephones, such exchange technologies are often compatible with—if not purposefully based upon—the anonymity of counterparts. Of course, as will be emphasized below, different sorts of exchange technology are not necessarily exclusive. Telephones and screens may cohabit and combine.

Getting a Better Bid-Ask Spread

Our first empirical illustration of telephone practices in financial trading rooms deals with market-making environments. "Market making" is a particular, quite widespread way of organizing trade in financial markets. Generally speaking, in such environments exchange is handled through intermediaries called "market makers" or "dealers." In order to purchase or to sell a specific product—for her own account or on behalf of a client—a trader in an investment bank must get in touch with a market maker who is able to propose a selling price or a buying price to her. Market makers publicize their bid-ask spread, i.e., the prices at which they are ready to trade, for the products they are specialized in. They trade for their own account and make economic profit out of the difference between the prices at which they sell and the prices at which they buy. Several market makers can compete to capture trades for the same product. Markets for which there are no such intermediaries and where trades are directly handled between bilateral counterparts are commonly referred to as "over-the-counter" (OTC) markets.

Bond markets often work on a market-making basis, especially when the scale of the issuer, the liquidity perspectives, and the stability of the product are considerable. In present-day bond markets, market makers can publish their indicative quotes or indicative "bid-ask spread" (i.e., the price at which they would be ready to buy or to sell a specific government or corporate bond) in specific dealing screen interfaces, developed by companies such as Bloomberg or Reuters for this particularly complex informational environment (Brière 2005). Typically, a trader can consult these publicly available indicative quotes and then make a phone call to the relevant market maker in order to close a transaction or to refine the price offer. But present-day screen interfaces include also specific features that make it possible not only to consult these publicly available indicative quotes but also to engage in electronic interaction with market makers (through an electronic messaging system) and to "take" a specific price from the screen. However, and despite of the usefulness of these screen-based tools, exploratory and trading activity is still frequently handled through the telephone.

While exploring the market for a counterpart, this corporate bond trader will use the indicative quotes displayed on the Bloomberg interface, but not exclusively nor exhaustively:

It's not centralized. You have to go and compare yourself. Comparing between several electronic systems is not so used, because it's quite new. But comparing on the telephone is a normal activity. (corporate bond trader at investment bank, November 2000)

This informant emphasizes the fact that comparison is an essential part of her activity. In this market-making environment, different competing prices might be available at the same time for the same financial product. It is quite remarkable that, despite the fact that screen interfaces can provide a sound comparative space, telephone interaction is often preferred (at least this was the case at the time of my fieldwork). The telephone, of course, is not particularly helpful to aggregate information in a single metric space. The use of the telephone lies more on the issue of "price improvement," i.e., on the capacity to mobilize close networks in order to obtain favorable conditions.

Telephone contact gives this trader the feeling that she is negotiating, as opposed to trading in an order-driven platform. She compares these trading practices in corporate bond markets to French listed future contracts which are traded, in this case, on an electronic order-driven platform:

For the future contracts [traded on an electronic platform] you don't have any negotiation power. You can't ask them for a better price. With the market-making system, you can say "Listen, you're joking with such a price, try to get me a better one, I can find a better one elsewhere." You can get in touch with the salesperson or directly with the market maker if he's at our own bank. (corporate bond trader at an investment bank, November 2000)

This trader works in a large investment bank. For some products, she can use the services of a market maker working at the same bank, but not always. For many products, she must get in touch with a salesperson at another bank or brokerage house in order to obtain a suitable price.

This circumstance can lead to interesting interactions where the recognition of the counterpart is used to improve the quote for a particular transaction. Telephone practices are central to these interactions—and little flaws in the adjustment of the telephone sound settings like the following give good evidence of this:

If you call a salesperson, you ask for a price, and you say "Listen, could you get a bet-ter price for me?" sometimes you can hear him talking to the market maker, if he keeps his box opened, "Hey! Could you give me a better price? It's for [name of the bank]." (corporate bond trader at an investment bank, November 2000)

Overall, these communicational practices correspond to a market ar-rangement in which social networks play an important role in trade. Find-ing a trade counterpart is based more on the exploration of local networks (in the sense of close professional networks) than on exhaustive com-parison for quote improvement. Within these networks, the identifica-tion of a counterpart cannot be easily separated from a series of business entanglements—starting with contractual duties and brokerage commis-sions (Ortiz 2005). A trader will also feel somewhat compelled to trade with the counterpart that provided analysis for that particular transaction:

If an idea is coming from Morgan Stanley, it is quite fair to go and deal with Morgan Stanley. You see? It's fair play. Because they pay for their research, in order to bring such ideas. They give you the idea, so you should deal with them. Well, they won't try to find out if you dealt with JP Morgan instead, but … it's a kind of a moral obli-gation. Of course, if their price is really bad, you go elsewhere. But it is not usual. (corporate bond trader at an investment bank, November 2000)

These particular market arrangements are concomitant with the use of the telephone. To put it briefly, in a networked market the personal address book becomes a crucial trading tool. A trader will complain, for instance, if she is not allowed to customize the settings of her telephone turret touch screen, or if she has lost the configuration of her turret. Cursory observa-tion of telephone configurations in a trading-room denotes extensive use of nicknames or first names (instead of full names, professional functions or bank names) in the identification of pre-defined contacts in turrets' touch screens. A convenient counterpoint to this kind of communicational market configuration is electronic trading or, more specifically, fully auto-mated order-driven equity exchanges in where the identification of the counterpart is meant to be irrelevant.

Identifying a Counterpart below the Order Book

Order-driven exchanges are often defined in opposition to market-making systems, the latter also being referred to as "quote-driven" markets. The basic functioning principle of order-driven exchanges is the auction mech-anism: the matching of buy and sell orders is not handled through a net-work of decentralized intermediaries but through direct confrontation in the typical form of a double auction—electronically or in an open outcry

(Cohen et al. 1986; Lee 1998). A central feature of order-driven markets is the "order book"—the file in which standing buy and sell orders for an equity are queued and publicly displayed. A "market order" is executed against the best available counterpart as soon as it reaches the market, and a "limit order" (an order with a price limit) is stored until a compatible counterpart is available. Other, more complex order types can also come into play. In automated order-driven stock exchanges, such as the Paris Bourse (now Euronext), the counterpart to a trade is defined, in principle, through an automated allocation protocol.

Many equity exchanges are combinations of order-driven and quote-driven procedures. For instance, the New York Stock Exchange is often defined as a centralized order-driven market, but some intermediaries (the "specialists") play a market-making role: they buy and sell for their own account in order to guarantee stock liquidity (Abolafia 1996). Even fully automated order-driven exchanges can rely on intermediaries whose task is to guarantee liquidity for specific stocks. But when they do exist, these intermediaries must compete with a "public," centralized order book.

The Paris Bourse is often presented as an extreme example of order-driven mechanisms. The CAC (Cotation Assistée en Continu) system, implemented in the late 1980s, allowed for a full computerization of price determination, order matching, and shares allocation.[2] The Nouveau Système de Cotation (NSC), introduced in the late 1990s, is based on similar principles and is the technological core of the Euronext market platform. Quotation is handled through a centralized double auction mechanism. Each stock is traded on a public, electronic order book: a single price is determined by an order-matching algorithm. Market participants do not "trade" against one another directly but against this "electronic auctioneer." A clearinghouse facility allows for an aggregate settlement of all trades. However, bilateral contacts between counterparts are also possible at the Paris Bourse. Block orders (i.e., large orders that could disrupt the liquidity of the order book) can be traded on a parallel OTC platform called ACT, where trades are concluded on the basis of bilateral negotiation.[3] Moreover, in some market segments, such as the Nouveau Marché (a section of the Paris Bourse for technology stocks), some participants are authorized to develop market-making functions in order to activate market liquidity—but in compliance with the existence of a single, public order book for each stock (Revest 2001).

Let us focus on one particularly interesting combination of bilateral telephone communication and screen trading at the Paris Bourse. Trade-oriented telephone activity did not disappear from Parisian stockbrokers' trading rooms with automation. In the 1980s, stockbrokers fought to embed into the electronic system some of the prerogatives they had in the open-outcry market regime (Muniesa 2005). One was the ability to recognize one another's identity in the screen display of the order book (as they did on the floor of the Paris Bourse before its dismantlement). The CAC system maintained the anonymity of investors, but it allowed stockbrokers (although only stockbrokers) to identify one another in the screen display of the electronic order book with an "agent code" (i.e., an identification number) for each order hitting the electronic order book. One of the architects of the Paris Bourse early automation put it this way:

We did not succeed in imposing anonymity. Why? Because during the CAC negotiations we did not succeed in saying "You stockbrokers will not be able to know who is there anymore, you will not know that this is this particular stockbroker and this is that other one." Stockbrokers were clearly saying "We want to keep this advantage, we know that this stockbroker works for this investor and we want to keep that information." . . . For and against anonymity: this is a question of power. Of power left to stockbrokers or to investors. In the end, we decided to favor stockbrokers. (engineer at the Paris Bourse during the automation process, July 2000)

Full anonymity of the Paris Bourse's electronic order book was introduced much later, in 2001. Until then, identification of counterparts (at stockbrokers' level) was possible. Through the "agent code" displayed on trading screens, traders could know the identity of the potential counterparts that were posting limit orders to the order book. This allowed for interesting practices involving telephone interaction. Traders could identify an interesting move in their screens, and then give a call to the corresponding agent in order to arrange a transaction "below" the order book—for instance, through the block-orders channel:

You're in front of me, you are agent number 512. I am 521. I know you are willing to sell. I see it in the screen. You sell systematically. So I call you. And I tell you "OK, I'm a buyer for 100 000, what's your volume?" You're going to tell me "Listen, I sell 100 000." "So, mine." We make a trade. It's done. We are not in the order book anymore. We do this on the ACT. (trader at a brokerage house, October 2000)

The electronic order book can thus be used to make signals about potential trades. Traders can interpret its content in search of an opportunity for a block order. An agent posting recurrent small orders, hiding her global size while waiting for an interesting counterpart to come up, will be contacted by an interested stockbroker. They can match their overall volume in the order book itself. If they want to avoid any risk of mismatching (due to the presence of third parties or "intruders" in the order book), they can use the parallel OTC system for block orders (ACT).

The functional link between the electronic trading interface and the telephone is made explicit through cases like the following, in which a broker developed some proprietary software that facilitated phone access (automatically matching the agent code on the screen with the identity and telephone number of the counterpart):

Here, you see [showing the order book of a stock on the screen], you have the market maker's prices for this stock [at the Nouveau Marché, where some stockbrokers are officially entitled to perform market-making functions]. If you want to give him a call, you just have to click, and his name and telephone number are displayed on the screen. (trader at a brokerage house, February 2000)

This mixture between public activity on the order book and private telephone communication was quite usual in Parisian stockbrokers' trading rooms until full anonymity was implemented in April 2001. Other informants explain that, even when the purpose of the phone call is not to trade immediately, it is good to be able to call:

It's good to see who does what. It's important to be able to call the person that pops up into the order book. Even if the trade is already gone. It's useful to call, just to check if he has some business left. Or just to make contacts, to show that you are there. (trader at a brokerage house, July 2001)

But since the implementation of the Paris Bourse's new Euronext market model (which involved a merger with several European exchanges), the disclosure of agent codes has been removed from the order book. On April 23, 2001, the market became fully anonymous. The introduction of anonymity responded, in part, to the demand of big investors and big Anglo-American market actors of the brokerage and investment banking sector. In a context of increasing international competition between stock exchanges it seemed appropriate to favor the interest of important actors that were potentially interested in market configurations where their actions would be less visible. As was mentioned above, the identification of counterparts at the stockbrokers' level could give modest brokers some clues about who was doing what in the market, which was a significant strategic advantage. Big actors that could originate important market movements had thus an interest in reducing this source of strategic information. As some informants commented unofficially, the introduction of anonymity at the Paris Bourse corresponded, in part, to efforts to retain these actors in a context of fierce competition.

How did Parisian stockbrokers react? The French financial media reported some disagreement:

The removal of the code identifying the intermediary placing orders on the order book, introduced yesterday with the new Euronext market model, has provoked some protests from several operators in the Parisian marketplace. It will be more difficult indeed for small brokers to follow market trends. According to some market professionals, this will lower market liquidity and will represent a disadvantage for arbitrageurs. As [a risk arbitrage fund manager] puts it, "It was quite useful to know which brokers were buying or selling. It was possible to contact interested counterparts directly." (translated excerpt from "L'anonymat gène les professionnels" ["Anonymity disturbs professionals"], *La Tribune*, April 24, 2001, p. 1)

A number of financial actors expressed their concern and emphasized that many of them were seeking at the Paris Bourse precisely what was relatively rare in other stock exchanges: the identification of counterparts.[4] However, the market did not suffer a considerable disruption of liquidity, and this technical reform did not have a significant impact on trading levels. But it consistently transformed telephonic practices in Parisian stockbrokers' trading rooms.

The transformation of telephone practices in turn probably translated into reconfigurations of market networks. By suppressing resources for counterpart identification on the trading screen, the introduction of anonymity can potentially disrupt the development of a network-shaped market in the sense that it reduces the possibility of profitable telephone contact. How can social networks develop and play a role in trading activity if the identity of potential counterparts in the market is not public anymore? But anonymity can also trigger a somewhat opposite phenomenon: the development of an important network of counterparts among "big" players that co-exists with a "public" order book that no longer works as a device for signaling trade availability. In other words, anonymity would hinder the openness of market networks but would nonetheless protect established contacts between a few important brokers who are able to retain and direct an important part of order flow. Our data do not allow us to explore this hypothesis further, but it is interesting to observe that, in the months previous to anonymity, actors themselves were developing similar hypotheses:

Anonymity will definitely help developing the electronic order book. But it will be very hard to get a counterpart, especially for market makers here at the Nouveau Marché. They won't completely disappear, but they will need to have extended networks. They'll have to be able to call here and there, without relying on the screen as a point of reference. We'll not be able to say "Look, here comes 512 selling." If you are 512 you will get a call because you are in a network, not because you are seen. (trader at a brokerage house, October 2000)

Interestingly enough, some informants unofficially commented that some market participants were developing techniques for signaling counterpart identity through the trading screen anyway. In principle, the only information that one can post to an anonymous trading system is price and quantity. But it is possible to "disguise" an agent code into the quantity. The Parisian system allows traders to trade "odd lots" (i.e., order lots that are not multiple of a determined "board lot" of, say, 10, 50, or 100 shares). A trader can thus announce an order to buy 100,512 shares instead of just 100,000 in order to signify the presence of agent 512 and his willingness to engage into bilateral interaction (on the phone).[5] Again, we observe the evolution of strategies coupling the telephone with the screen in order to adjust a suitable market arrangement.

Monitoring the Client

Relevant uses of the telephone, often combined with the screen, can be observed in trading activity in several market configurations. Our two preceding examples focus on the trading side of financial operations: traders handling buy and sell orders for specific products need to match them against a counterpart (through a network of brokers and market makers, as in the first case, or through an organized order-driven market, as in the second). Other relevant uses of the telephone can be found, however, in the commercial side of financial operations, i.e., at the level of the salesperson (or sales trader). In a trading room, these uses are located at the sales desks, as opposed to the trading desks. A salesperson at a sales desk enters into interaction with corporate clients (corporate treasury departments, asset managers, etc.) in order to propose financial operations, which eventually translate into the origination of buy and sell orders on behalf of these clients. The salesperson then typically passes these orders for execution to a colleague at a trading desk or to a distant broker. In such activities, the telephone is a major tool. At sales desks, voice technology can coexist with computer developments. But, unlike at trading desks, those developments are often explicitly oriented toward an enhancement of the bilateral recognition of counterparts, and rarely informed by principles such as anonymity or publicity. As was mentioned above, CTI (Computer Telephony Interface) and CRM (Customer Relationship Management) technologies are coming to the forefront of telephony innovation in trading rooms: this is particularly true for sales desks. Many developments aim at identifying the client on the line in real time, so to automatically trigger the display of relevant information on the workstation's screen, in order to enrich the resources for commercial conversation.

The marketing arguments upon which these new customer relationship technologies rely are often based on knowledge relevance and real time: to know the customer, to focus on her precise needs, to optimize data availability within time constraints, to enhance proactive contact. In other words, it is a matter of refining the salesperson's response in a context where there is a strong tension between relevance and urgency in the course of action.

A salesperson in a trading-room usually handles a reduced number of clients:

Each salesperson here will handle no more than twenty clients. If we only consider the good clients, the ones that are contacted on daily bases, we can talk about six or height clients per salesperson. (sales desk manager at a Foreign Exchange dealing room, January 2002)

I have fifteen clients in my box. This is too much. The best would be to have three or four big clients, plus four or six less important. (fixed income salesperson at an investment bank, December 2001)

The relation with a client is far from distant. It is in this commercial side of trading-room activity that more "socially embedded" market relations are to be found. A salesperson typically meets with her client regularly. Close sociability is a generalized practice. Professional telephone conversations include all the sociological leitmotifs of personal closeness: asking about holiday trips, scheduling time for going out together, knowing first names of the client's family members, and so on.

However, closeness to commercial counterparts does not always translate into an easy client monitoring. In order to capture business opportunities, a salesperson must refine her knowledge of her client's interests and strategies—which is far from straightforward, as the following testimony shows:

[A client] wanted to know what we were doing with [name of equity]. We know that there is a broker trying to sell at 3.24. My client just asked me to sell a big amount at 3.25. He asks me to sell, so I guess he has bought it somewhere else. Not with us, anyway. We cannot grasp his full strategy. (sales trader at a brokerage house, November 2001)

Corporate clients—and especially large scale corporate clients (other investment banks, treasury departments of important companies, fund managers, etc.)—will usually distribute their strategy among several intermediaries. In many cases, it is virtually impossible for one salesperson to monitor the overall client's strategy. In order to propose relevant operations in a proactive manner (instead of just hoping that the client will call with a particular request), the salesperson will try to gain knowledge of the client from heterogeneous sources:

A month ago, my client [treasury department of a large corporation] told me that he had an exposure in Argentina. Now, there are big disorders with the Argentinean currency. I can try to anticipate my client's needs and give him a few ideas about how to cover this currency risk.... I can also guess about his exposure in Argentina by other means. For instance, I can take a look at his firm's balance sheet, and look for subsidiaries in Argentina. (sales desk manager at a Foreign Exchange dealing room, January 2002)

Proactive sequences often take place in the early morning: the salesperson will call her clients and report some relevant news, some information from her trading room's morning meeting, and some comments about possible trends and strategies. The possibility of customizing this kind of information will vary on the basis of several elements, including the salesperson knowledge about the client's needs and strategies. During the rest of the day, such proactive sequences become rarer. The salesperson will be waiting for her client's requests, and managing the subsequent deals with the trading team.

Many informants reported interest in any technical device that could help salespersons to keep a memory of the interests expressed by each particular client, to trigger an alert message when relevant market circumstances are met, or to get instant display of relevant information with incoming calls. The idea underpinning such CRM developments is not just to recognize a client but to couple her identity with some heterogeneous information: market events associated to her interests, information about dealing activity, about the characteristics of the product sold, information about the client's account. Of course, this kind of information is available to the professional salesperson through her desk's various computer terminals. The issue at stake is its rapid connection to telephone events, i.e., to the commercial conversational situation. What recent developments in CTI confirm is the tendency to render explicit the functional connection, especially in the salesperson's environment, between the phone call and the screen.

The particularity of telephone-based interaction in a sales environment stems from the fact that corporate counterparts (i.e., the salesperson's interlocutors) are considered to be more demanding in terms of interlocutory attention than inter-bank counterparts (i.e., the trader's or the broker's interlocutors). The former are less likely to easily fit a streamlined treatment—often alluded to as STP (Straight Through Processing)—in part because of the prevalence of the telephone as a crucial medium for sales interaction. Generally speaking, commercial interaction with corporate counterparts requires a more qualitative approach, thus more conversational. This is not only due to the idiosyncrasies of corporate clients' demands. In a sales environment, sound description of products is crucial, especially in the case of complex products whose properties and behavior must be carefully explained to the client. The more "mature" a product is—i.e., the more standardized becomes its description—the less crucial its conversational account will be.

Concluding Discussion

There is sociological and historical evidence that conversations are a constitutive part of financial life. This applies to sociability practices of "conversations-about-the-world" but also to the financial matter of "conversations-qua-transactions" (Preda 2001). It is true that, with the development of modern market technologies (especially electric and electronic technologies), price display and commercial transactions partially abandon their conversational nature. The stock ticker and the price chart, for instance, consistently transform the way in which things are rendered public and discussed about in financial markets (Beunza and Muniesa 2005; Preda, this volume). In present-day markets, however, the financial conversation—as both "conversation-about-the-world" and "conversation-qua-transaction"—is far from being a pure remnant of more traditional practices. The pervasiveness and innovativeness of voice technologies in financial trading rooms give good evidence of this, despite of the "all-computerized" syndrome that characterized financial mythology in the early 2000s. As a financial journalist put it in 2003:

A way of life whose days seemed numbered three years ago, because of the advance of dealing platforms that match buyers and sellers electronically, has not only survived, but flourished. Brokers who for a while put their faith in pure electronics have had to dust off their handsets. (excerpt from "Voice squad," *The Economist*, 11 January 2003, p. 69)

We also share the following analysis from this same journalist:

[B]rokers these days use a mixture of voice and electronics. That said, they believe that more of the less complex trading will be carried out electronically; and the more liquid the instrument, the likelier it is that it will be traded on screens. For example, a few big spot Foreign Exchange trades are still done by telephone, but most, perhaps 70%, are done automatically on an electronic brokerage system, known as EBS [Electronic Brokerage System], built by big banks nine years ago. That leaves plenty of complex deals to be haggled over by brokers—say a credit-derivative transaction combined with an interest-rate swap and the purchase of bonds, involving several buyers and sellers. (ibid.)

"Mature" markets (listed equities, spot markets), with a tendency to mass trading and with decreasing arbitrage opportunities, conform more straightforwardly to liquidity-enhancement protocols such as public, anonymous auction mechanisms (or "screen markets" at large). Conversely, trading activity accompanying complex derivative contracts—especially when large amounts of money are involved—is likely to rely consistently on telephone interaction. The stability of the description of a product and the calibration of the engagement of commercial counterparts are clearly at stake in such transactions.

The issue of identification of the counterpart is central to this shifting aspect of market configurations. By "identification" I mean the detection and recognition of a specific and singular counterpart, with such specific characteristics as corporate identity, professional qualification, and/or possible personal acquaintance. Although often corresponding to a physical person, a "counterpart," a "client," a "trader," or a "broker" is always engaged in trade as instances of moral persons such as a bank, a corporation, or an investment fund. "Identification" applies here to any such elements of personality (physical or moral). In this context, the identification of the counterpart diverges from market arrangements characterized by anonymity, in which the counterpart is not identified bilaterally and trades are executed against an aggregate (and thus abstracted) counterpart.[6]

As the examples presented in this chapter show, the use of the telephone is a good indicator of the intensity of the identification of counterparts in a particular market arrangement. Because of its technical qualities, the telephone is a critical tool for counterpart identification. Of course, the telephone also typically serves the conversational features of a particular market arrangement—although financial conversations can also be supported by open-outcry architectures or by electronic messaging systems.[7] My point, however, is restricted to the problem of identification and to its relation to telephones. (I do not elaborate here on the conversational nature of the interaction.) Also, I do not wish to claim that the telephone is the pivotal technology of counterpart identification. As was observed above, electronic trading terminals can be also used for that purpose. My point is that telephone practices provide good indications of the concrete empirical arrangements that market actors may deploy for (or against) counterpart identification, and also, more generally, that it is not possible to fully understand the structure and functioning of a market without a sociological analysis of the technologies that underpin it.

The widely accepted fact that social networks matter in market formation (Granovetter 2005) often translates into a rather straightforward claim: the fact that markets are "socially embedded" means that (to variable degrees and depending on the circumstances) there is a point in trading with "known" counterparts, especially with "personally known" counterparts. But this is not always the case. An account of the concrete technologies that allows for an activation of such personal knowledge of the counterpart can significantly enrich this sort of sociological analysis. In a certain sense, this research direction is to explore the correspondences between what are usually called "social networks" (in the sense of the economic sociology tradition) and the material networks of communication or "technical networks" that allows for a "tie" to be expressed and articulated in a particular code and manner. One way of putting this is saying that market arrangements are made of "social networks" and of "technical networks" as well, the latter being made explicit to sociological analysis when the material enactment of the former is under consideration. But starting with a clear-cut distinction between what is "social" and what is "technical" about market networks might also be somewhat misleading, at least if one considers issues (such as the identification of counterparts) that explicitly mix the "social" and the "technical."

In this chapter I have provided a brief contribution to this perspective by focusing on the use of trading-room telephony in three market circumstances: a situation involving interaction with market makers in bond markets, the strategic use of the telephone in an automated stock exchange before full anonymity, and client monitoring practices in sales desks.[8] In all cases, telephony combines with other electronic media. In the first case, the telephone combines with screen display of prices in a network-shaped market configuration and gives traders the opportunity to activate resources for price improvement. In the second case, the telephone offers decisive resources for liquidity search, especially to small players that try to compete in an electronic market platform that is about to censor the public identification of counterparts. In the third case, conversational attention to corporate clients is accompanied by attempts at electronically increasing the relevance of the identification.

Although in the first and the third cases there is a point in trying to trade with "known" counterparts (as a way to maintain a certain play of trust and fairness, for instance), the general concern in all three cases is rather about knowing which counterpart might be available and ready for a profitable trade, and getting in touch rapidly. In the three cases I have studied, telephone interaction is typically performed with eyes staring at a screen, often inspecting on the computer terminal (price display services, trading systems, or customer database facilities, respectively) an aspect of the counterpart who is on the line—a configuration that sharpens the sense of counterpart identification, in a kind of "I-hear-you-here-and-I-see-you-there" pattern.

Different market arrangements are characterized by different material configurations. The presence or absence (and the features and combinations) of such technological ingredients as telephones or computers are fundamental to understanding the behavior of the market as a collective arrangement. A trader with a telephone differs consistently from a trader without a telephone. More than that, traders with telephones and traders without telephones constitute "economic actors" that are different and that act differently. Differences sharpen when we consider all the functionalities of different telephony systems and all their possible combinations with other surrounding market technologies. When I talk about "differences" I do not refer exclusively to a matter of trading-room local material culture. This difference affects how transactions are shaped, how prices are set, how strategic actions are performed, and how connections between market counterparts are enacted. In particular—and this corresponds to a "performativist" concern (Fourcade 2007)—close attention to how telephones populate markets can help us to understand differences in how these markets fit different social-scientific programs. The telephone is likely to be a vital ingredient in constructing the kind of markets preferred by social network analysis, whereas this technology could somewhat hamper an attempt at configuring markets the way game theory likes them. Are markets composed of actors who know one another personally? Or are they rather composed of agents who communicate only through prices? Markets might indeed be performed either way (and these are only two possibilities among many)—but only if a set of suitable devices make the particular market arrangement hold together.

Notes

1. I conducted this research when I was working at the social sciences laboratory of France Télécom R&D, in Issy-les-Moulineaux. Most of the empirical material corresponds to a research partnership with Etrali, a France Telecom company specialized

in voice technology for financial markets. I had the opportunity to follow Etrali's innovation processes in the area of trading room telephony for one year. I visited ten different trading rooms (corresponding to five investment banks, four brokerage houses and one corporate treasury department) located in Paris and Ile-de-France and interviewed several operators at their desks. These included both traders and salespersons working in a variety of markets (listed stocks and derivatives, bonds and currency markets). I also use material from the work carried out for my doctoral dissertation on the automation of the Paris Bourse (Muniesa 2003). All interviews were carried out in French between 2000 and 2002 (excerpts provided here are translated by myself). I would like to thank France Telecom R&D and Etrali for their support for this research. I would also like to thank Valérie Beaudouin, Daniel Beunza, Marie Brière, Michel Callon, Eric Cassimatis, Karin Knorr Cetina, Vincent-Antonin Lépinay, Olivier Godechot, Christian Licoppe, Alexandre Mallard, Anne-Sophie Marie, Alex Preda, Valérie Revest, David Stark and Pascal Zératès for helpful comments and suggestions. A preliminary version of this paper was presented at the New York Conference on the Social Studies of Finance (Columbia University, 3–4 May, 2002).

2. The CAC system was developed on the basis of Toronto Stock Exchange's CATS (Computer Assisted Trading System). For a sociological analysis of the transition from open outcry to CAC, see Muniesa 2005.

3. In 1994, a block trading facility was introduced at the Paris Bourse, based on Nasdaq's ACT system (Automated Confirmation Transaction), later replaced by a new technology called TCS (Trade Confirmation System).

4. For an analysis of how this issue connects with the ambiguous notion of transparency, see Grossman, Luque, and Muniesa 2008. The identification of counterparts is compatible with a 'literal' notion of transparency. But the enforcement of anonymity can also be justified in terms of a more "abstract" notion of transparency.

5. The literature on collusive bidding analyzes similar practices, for instance in the case of the FCC spectrum auctions (see Guala 2001).

6. For a study of the implications of central counterpart methods in financial markets, see Millo, Muniesa, Panourgias, and Scott 2005.

7. For an analysis of screen-based conversational interactions in the Foreign Exchange spot market, see Knorr Cetina and Bruegger 2002.

8. I do not provide in this paper proper social network analysis of these situations, mainly because of the impossibility of gaining access to exhaustive, quantitative data on telephone activity in one or several trading rooms. In different research conditions, interesting outcomes could be obtained using network analysis methodologies on telephone activity.

References

Abolafia, M. 1996 *Making Markets: Opportunism and Restraint on Wall Street*. Harvard University Press.

Akrich, M. 1992. The de-scription of technological objects. In *Shaping Technology/ Building Society*, ed. W. Bijker and J. Law. MIT Press.

Baker, W. 1984. The social structure of a national securities market. *American Journal of Sociology* 89, no. 4: 775–811.

Beunza, D., and F. Muniesa. 2005. Listening to the spread-plot. In *Making Things Public*, ed. B. Latour and P. Weibel. MIT Press.

Beunza, D., and D. Stark. 2004. Tools of the trade: The socio-technology of arbitrage in a Wall Street trading room. *Industrial and Corporate Change* 13, no. 2: 369–400.

Boden, D. 1994. *The Business of Talk: Organizations in Action*. Polity.

Brière, M. 2005. *Formation des taux d'intérêt: Anomalies et croyances collectives*. Economica.

Callon, M. 2007. What does it mean to say that economics is performative? In *Do Economists Make Markets?* ed. D. MacKenzie et al. Princeton University Press.

Callon, M., and F. Muniesa. 2005. Economic markets as calculative collective devices. *Organization Studies* 26, no. 8: 1229–1250.

Cohen, K., S. Maier, R. Schwartz, and D. Whitcomb. 1986. *The Microstructure of Securities Markets*. Prentice-Hall.

Fourcade, M. 2007. Theories of markets and theories of society. *American Behavioral Scientist* 50, no. 8: 1015–1034.

Godechot, O., J.-P. Hassoun, and F. Muniesa. 2000. La volatilité des postes: Professionnels des marchés financiers et informatisation. *Actes de la Recherche en Sciences Sociales*, no. 134: 45–55.

Granovetter, M. 1985. Economic action and social structure: The problem of embeddedness. *American Journal of Sociology* 91, no. 3: 481–510.

Granovetter, M. 2005. The impact of social structure on economic outcomes. *Journal of Economic Perspectives* 19, no. 1: 33–50.

Grossman, E., E. Luque, and F. Muniesa. 2008. Economies through transparency. In *Transparency in a New Global Order*, ed. C. Garsten and M. Lindh de Montoya. Elgar.

Guala, F. 2001. Building economic machines: The FCC auctions. *Studies in History and Philosophy of Science* 32, no. 3: 453–477.

Hardie, I., and D. MacKenzie. 2007. Assembling an economic actor: The *agencement* of a hedge fund. *Sociological Review* 55, no. 1: 576–580.

Hutchby, I. 2001. *Conversation and Technology: From the Telephone to the Internet.* Polity.

Knorr Cetina, K. 2003. From pipes to scopes: The flow architecture of financial markets. *Distinktion*, no. 7: 7–23.

Knorr Cetina, K., and U. Bruegger. 2002. Global microstructures: The virtual societies of financial markets. *American Journal of Sociology* 107, no. 4: 905–950.

Lee, R. 1998. *What is an Exchange? The Automation, Management, and Regulation of Financial Markets.* Oxford University Press.

Lépinay, V.-A., and E. Hertz. 2004. Deception and its preconditions: Issues raised by financial markets. In *Deception in Markets*, ed. C. Gerschlager. Palgrave Macmillan.

Licoppe, C. 2004. "Connected" presence: The emergence of a new repertoire for managing social relationships in a changing communication technoscape. *Environment and Planning D: Society and Space* 22, no. 1: 135–156.

Luff, P., J. Hindmarsh, and C. Heath., eds. 2000. *Workplace Studies: Recovering Work Practice and Informing System Design.* Cambridge University Press.

MacKenzie, D., and Y. Millo. 2003. Constructing a market, performing theory: The historical sociology of a financial derivatives exchange. *American Journal of Sociology* 109, no. 1: 107–145.

MacKenzie, D. 1984. Marx and the machine. *Technology and Culture* 25, no. 3: 473–502.

MacKenzie, D. 2004. Social connectivities in global financial markets. *Environment and Planning D: Society and Space* 22, no. 1: 83–101.

Mallard, A. 2004. From the telephone to the economic exchange: How small businesses use the telephone in their market relations. *Environment and Planning D: Society and Space* 22, no. 1: 117–134.

Millo, Y., F. Muniesa, N. S. Panourgias, and S. V. Scott. 2005. Organised detachment: Clearinghouse mechanisms in financial markets. *Information and Organization* 15, no. 3: 229–246.

Muniesa, F. 2003. Des marchés comme algorithmes: Sociologie de la cotation électronique à la Bourse de Paris. Ph.D. dissertation, Ecole des Mines de Paris.

Muniesa, F. 2005. Contenir le marché: La transition de la criée à la cotation électronique à la Bourse de Paris. *Sociologie du Travail* 47, no. 4: 485–501.

Muniesa, F. 2007. Market technologies and the pragmatics of prices. *Economy and Society* 36, no. 3: 377–395.

Muniesa, F., Y. Millo, and M. Callon. 2007. Introduction to market devices. In *Market Devices*, ed. M. Callon et al. Blackwell.

Ortiz, H. 2005. Evaluer, apprécier: Les relations entre brokers et gérants de fonds d'investissement. *Economie Rurale*, no. 286–287: 56–70.

Oudshoorn, N., and T. Pinch, eds. 2003. *How Users Matter: The Co-Construction of Users and Technologies*. MIT Press.

Preda, A. 2001. In the enchanted grove: Financial conversations and the marketplace in England and France in the 18th century. *Journal of Historical Sociology* 14, no. 3: 276–307.

Preda, A. 2003. Les hommes de la Bourse et leurs instruments merveilleux: Technologies de transmission des cours et origines de l'organisation des marchés modernes. *Réseaux* 21, no. 122: 137–164.

Preda, A. 2005. The stock ticker. In *Making Things Public*, ed. B. Latour and P. Weibel. MIT Press.

Preda, A. 2006. Socio-technical agency in financial markets: The case of the stock ticker. *Social Studies of Science* 36, no. 5: 753–782.

Revest, V. 2001. Le Nouveau Marché: La construction d'une identité. *Revue d'Economie Financière*, no. 61: 193–202.

Zaloom, C. 2003. Ambiguous numbers: Trading technologies and interpretation in financial markets. *American Ethnologist* 30, no. 2: 258–272.

IV Technology, Economy, Use

10 Understanding and Reframing the Electronic Consumption Experience: The Interactional Ambiguities of Mediated Coordination

Christian Licoppe

Economic sociology has shown how the stabilization of markets, which was considered a given by many economists, required explanation (Smelser and Swedberg 1995). Markets may be considered as institutions whose "architecture" is made of rules that organize and stabilize the coordination of the varied parties involved in the economic exchange (Fligstein 2001). Among such stabilizing rules, the systematic organization of elementary transactions (such as buying and selling) as ordered interaction sequences has been rarely studied from a technology-sensitive perspective, though anthropologists have investigated the temporal organization of bargaining (Geertz et al. 1979) and conversation analysts the situated ordering of the "hard sell" (Pinch and Clark 1988). At that level of analysis, transactions must be described as practical accomplishments, embedded in and mediated by technological artefacts. Their comparison in different mediated contexts reveals the interplay of sociality, economy, and technology underlying any consumption-oriented action sequence, and its transformations.

E-commerce is an interesting case, for many of the questions that have been raised about it have focused on precisely the kind of commercial relationship that this medium created. Concern about its slow development has been related to issues of trust in transactions performed at a distance. Conversely, the potential growth of e-commerce has been thought to depend on how it might allow the transformation of occasional isolated commercial contacts into a personalized, continuous relationship, thus giving substance to an old dream of mass distribution.[1] The electronic tools that support the transaction have held the promise of making it possible to track digitally all contacts with a particular consumer and to use this information to adjust commercial proposals more closely to consumers' needs.

The economic paradigm that prompts us to conceive of commercial transactions as occasional, with no memory, often causes us to overlook the fact that these transactions are not instantaneous and the fact that they require the accomplishment of a series of reciprocal actions by both buyer and seller.[2] In a supermarket, for example, the actual purchasing of goods involves collaboration between the consumer and the cashier in performing a set of reciprocal generic actions in a standardized environment:

- The consumer displays his or her wish to purchase particular goods by taking to the checkout products selected from the shelves.
- The cashier validates the order and announces the total price.
- The consumer pays.
- The cashier hand the consumer a receipt (and, in some cases, change).
- The consumer collects the goods.

This typical sequence is so conventional that it can be carried out without a word being exchanged between the actors concerned, or even without their eyes meeting (Rafaeli 1989). According to Goffman (1981), it may or may not be accompanied by greetings or bits of conversation without that affecting the instrumental, targeted, and sequential nature of the encounter.

Mirowski (2002) has described markets as "automats" and market transactions as "algorithms." He and others highlight the likelihood of the commercial transaction being computerized, for a process that can thus be formalized is in a sense preformatted and can be transformed into a computer algorithm operating through a formal syntax (Agre 1995). Conversely, such an algorithmization of economic transactions reinforces their sequential and functional character. Information and communication technologies (ICT) have contributed toward translating economic interactions into algorithms, especially in the context of financial markets and e-commerce. The economic consequences of the computerization of the microstructure of markets is an active field of research (Mirowski and Somefun 1998; Callon and Muniesa 2003).

However, commercial interactions themselves cannot be entirely equated to a finalized algorithmic process and reduced to a sequence of discrete steps leading toward a predefined goal. An exchange constitutes a situated and sequential accomplishment, as ethnomethodology-inspired studies so clearly show. An action performed here and now redefines the meaning of the turns that preceded it and projects a distribution of relevance and normative expectations over possible next turns (Schegloff 1988). From such a situated perspective, the accomplishment of a sequential procedure such as the purchase of a consumer good appears to be a collaborative and public performance by the participants, the meaning of which emerges as the action unfolds. The functional purchasing sequence broken down into the five stages described above must be treated as a typification, an ideal mapping of the purchasing procedure. It should be considered to be a resource available to the participants when they try to make sense of the transaction situation as it develops. This suggests that a purchase can be read in two different ways when problems arise: (1) a utilitarian reading, in which the sequence is evaluated in terms of its goal and its correspondence to a standard sequence; (2) a moral and interactional reading, in which it will be evaluated in relation to the moral, practical, ritual, and syntactic conventions that are at the heart of ordinary sociability. In the case of e-commerce, an order that stays unanswered for too long may be treated either as a sign of commercial inefficiency or as a consequence of asymmetry between the buyer's engagement (completely engaged by his or her action) and that of the seller (visible only when he or she has reacted). This asymmetry might threaten the interactional order underlying the ongoing transaction. This duality stems from the fact that the transaction is simultaneously oriented toward a goal and toward others and is therefore subject to moral standards that regulate the way in which the participants pursue their objectives. The actors apply an interactional competence to pursue goals without alienating themselves from others, and this principle guides their own interpretations of their actions.[3]

The aim of this chapter is to show how the most controversial aspects of e-commerce, such as Internet users' alleged lack of trust and loyalty, are related to the protocols proposed by the e-commerce site and to the actors' interpretations of transactions collaboratively and sequentially accomplished through the use of such technologically embedded protocols.

First, I will consider how the respective relevance of various typified accounts of the consumer experience (the consumer with a plan, the consumer who engages open-mindedly with the commercial environment) is transformed according to the characteristics of the algorithm through which the purchasing sequence is enacted. In the case of e-commerce, the use of search engines is prominent. It introduces a time gap between revealing and displaying preferences and access to goods. I will show how this relates to the salience of the model of the consumer as an intentional actor—an "entrepreneur" of his or her own consumption—in Web users' descriptions of their e-commerce experiences.

I will then analyze the emergence and resolution of problems concerning the sequentiality of the e-commerce purchasing process. The way in which Internet purchases are sequentially performed leads to equivocal interactional moments. The lapse between the sending of the order by the consumer and the sending of goods is often read by e-consumers as a silence that marks potential interactional trouble and that renders the participants' commitments and mutual obligations questionable. Consumers act consequentially enough on this interactional interpretation of economic transactions to simultaneously require the development of call centers and the rapid emergence of a conventional regulation embedded in the software architecture of e-commerce (namely the automatic sending of an e-mail acknowledging the order within 24 hours), which specifically addresses the interactional problem thus revealed.

The case studies analyzed here can then be understood as a step in a more general research program that aims to bring together economy, sociality, and technology at a level of analysis where transactions must be considered as interactions. This helps us to reconsider the way in which some of the problems of e-commerce are currently addressed. E-merchants try to develop consumers' loyalty by enlivening the sales relationship with multiple electronic contacts and by building up a continuous relationship through them. Yet in various spaces of electronic discussion (e.g., forums and chats) messages often are ignored without the participants' getting particularly troubled (Herring 1996; Velkovska 2004). The electronic sociability deployed on the Internet thus seem to be characterized by a high level of tolerance with respect to the participants' actual or potential defection. Is it not therefore paradoxical to want to repair the supposedly volatile nature of Internet consumption by "augmenting" it with devices supporting various forms of electronic sociability?

The empirical data on which this work draws were obtained through two field studies on e-commerce in the cultural goods mass distribution sector (books, records, video and DVD, etc.). One was a study based on extensive interviews with electronic consumers (Licoppe et al. 2002); the other was an ethnographic study of the service work at the call center of one of the leading French websites for the sale of cultural goods (Licoppe 2002).

Mass Distribution and E-Commerce: Wandering Around or Searching for Product

This section focuses on interviews with consumers in Paris and Toulouse who have repeatedly bought books, CD-ROMs, and multimedia goods on the World Wide Web. As is now well known, such consumers almost always buy both online and in physical stores, and combine both experiences into a variety of heterogeneous practices (Lunt 1999). Although the consumers investigated here are no exception, my discussion will focus on the way their accounts describe very differently the isolated experiences of buying cultural goods online or in "brick-and-mortar" stores, within actual practices that combine these experiences or alternate between them.

Impulsive Buying and Planned Buying in "Brick-and-Mortar" Stores

Mass distribution is organized around a promise (making accessible a maximum of goods in a single place) involving the construction of places of consumption. These places are occupied by rows of similar products, which are distinguished by the way in which they are "positioned" in relation to one another by market professionals, thus orienting and shaping consumers' choices (Barrey et al. 2001). The consumer may be described as an individual who approaches the market with his or her own resources (for example, the conversations held with his or her own social network, or the advice or tests seen in the consumer press) (Mallard 2000). If there is an asymmetry between buyer and seller, it stems not only from the intrinsic properties of each but also from their cognitive and material resources, and the way in which their activity and interactions are distributed (Callon and Muniesa 2003).

The work of positioning products (and, more generally, of configuring the encounter between goods and consumers) has become the subject of analytical and normative practical knowledge, the construction of which was professionalized in the twentieth century as the subject of marketing (Cochoy 1999). This applied discipline has defined two ideal types for the representation of consumers' engagement in the places of mass distribution (Bowlby 2001). On the one hand there is planned buying: the consumer faces the goods with a precise intention that guides all his or her cognitive processes. This intention may also be materialized in external repositories, such as the shopping list. On the other hand there is impulsive buying: the potential consumer is a mobile body whose attention wanders and whose mind is vacant. Via his or her mobility, the potential consumer encounters the goods. His or her attention is captured by the articles displayed on shelves, for instance through the brand names or cues embedded in the packaging. Once focused, attention turns into desire, then into an intention to buy, and finally into the actual purchase. As will be seen, these two typologies run throughout consumers' discourses. They constitute core interpretive resources to account for consumption experiences.

The places and especially the practices of consumption are multiple as regards the ways they enhance or play down the relevance of one or the other of these two typical accounts. In the case of planned buying, for example, research undertaken in the perspective of situated action has shown that action could not be undertaken in the form of a plan. The plan is just one resource among many to which participants reflexively orient to the moment-by-moment unfolding of the transaction as a practical accomplishment (Suchman 1987). In supermarkets, even with an explicit intention and a list, consumers face both orders and contingencies in the way they encounter the spatial organization of the shelves and products. This leads to specific forms of anticipation (as when a shopping list is drawn up taking into account the order in which products appear in the shop), and to possibilities of unloading part of the cognitive burden on the environment, of adjusting to the contingencies of an embodied encounter with spatially ordered products, in a kind of improvisation resembling impulsiveness (Lave 1988).

Commercial spaces are overloaded with devices that allow widely diverse forms of encounter between consumers and products. The shelves overflowing with records or books lend themselves to strolling and surprises. As Michel Callon would say, this type of arrangement of products "calculates" an impulsive buyer. By contrast, the seller equipped with a computer and catalogs caters more to a consumer who is looking for something. A configuration thus equipped "calculates" a consumer who is engaged with a relatively clear intention. As a whole, the place of consumption constitutes a plural space equipped with multiple material devices that invite one to partake in varied forms of consumption. At a given point in the specialized supermarket, depending on the nature and visibility of the equipment presented, various actors and practices of consumption will be preferable and preferred.[4]

During sociological interviews, when the researcher asks consumers to describe their consumption practices in a few words, they often apply a principle of narrative economy. They recount their consumption experiences by relating them to generic forms of narrative, thus rendering them immediately intelligible. Impulsive buying and planned buying are two narrative modalities that prevail in Web users' accounts. These two descriptions of consumption constantly circulate from scientific discourse (produced by marketing professionals) to accounts of ordinary experiences or vice versa. Respondents often reduce the complexity of their consumption practices and the plurality of the ecologies to which they are attuned by framing their practice in one of these two shared and authorized ideal types: planned buying or impulsive buying.

The specialized department store is seen as aiming toward making visible and available a carefully arranged profusion of records, books, and multimedia products, and becoming a place of temptation where objects are approached with a disposition for exploration. The customer is not the dummy of merchants and marketers, for he or she often plays the game knowingly, displaying a disposition to be caught in the particular mood of consumption that the large store promotes the most, and to account accordingly for his or her experience:

At the FNAC there are too many people and it stresses me out, so when I'm looking and I'm pressured I search for someone to inform me, they're all too busy.... [She prefers another bookshop.] ... I like variety, mind you at the FNAC they've also got a wide variety, but what I like at Ombre Blanche is the view, the room is huge so you see lots of things, I can see all the sections, at the FNAC they've made little corners all over like this and when you're in a space, in a section, you don't see the others, and what I like is to see everything a bit, because it gives me ideas, and when it gives me ideas I change ... when I'm at Ombre Blanche I go up, down, left, right, I do everything, foreign literature, religion, everything, but actually it's crazy because I'm interested in everything so, from one idea to the next my ideas spring up, so I need to look at everything.

Although not all consumers go so far, the majority see large retail stores specialized in cultural goods as places in which the commercial experience—built around the mass distribution model and its promise of virtually unlimited availability of a type of good—revolves around a kind of situated temptation. Such stores are arranged and organized to frame the encounter of goods and consumers as an opportunity for the emergence of impulse. Consumers are assumed to be sensitive to such an invitation to let go, to readily allow themselves to be captured, and they account for their own experiences in such terms:

I went just to buy two or three books but once I get there, there's such a temptation that when you like books you can't resist.

By contrast, such stores seem less inviting to a planned consumer experience, where the consumer looks for a specific book or record in the shop:

For example, if I ask for Debussy's Arabesques ... it's something very precise, if I was in the store I wouldn't know how to find it. Often, in the store, I have to ask because I don't know where to look for such specific things in the store.

The way in which Web users account for their experience reduces the plurality of the department store. Despite the department store being highly equipped to allow a wide diversity of commercial mediations, it appears most prominently in such accounts as a place for the display of a

profusion of goods, inviting open-minded consumers to situated encounters likely to produce impulsive attachment of consumers and goods. The
department store seems to them to be a place in which the commercial environment is entrusted with the task of providing the resources required to
reveal their preferences to them. Very different accounts are used to make
sense of their online experience.

E-Commerce, Another Understanding of the Consumption Experience

Even more than specialized supermarkets, the websites mass-selling cultural
goods (books, disks, films, etc.) promise to make a huge, exhaustive catalog
available so that visitors can expect all their consumption desires to be
satisfied there. Most Web users say that they go onto these sites with a specific intention ("It's always when I'm looking for something specially, that
I've already got something in mind") and are unlikely to be distracted ("I
buy what I feel like buying and that's it!"). Thus, a shop and a website
lend themselves to consumer behaviors that are intelligible through two
dominant, diametrically opposed accounts: impulsive buying and planned
buying. During an interview, this contrast emerges from most attempts at
comparison:

In a physical shop you can move around randomly.
Is that what you do when you're in a shop?
No, I don't do it but what I mean is that on Internet you have to go
somewhere . . . whereas in a shop you can go nowhere.

Yet the equipment of the website is also varied enough to allow a wide
range of commercial mediations. A customer entering the book or record
section on the site is immediately faced with at least two salient modes
of access to the product: showcases in which promotions and the latest
releases are highlighted, and a search engine that provides lists of answers
to searches based on a few key descriptive categories (author, title or words
in title, musicians, etc.). Use of the engine requires the Web user to have
certain skills, and frequent use is the sign of a basic expertise (Assadi and
Beaudouin 2002). Many novices go no further than the first showcase
screens and ignore the search engine:

Well, I've got to admit that I'm pretty stupid, but I didn't understand that
on Alapage you could buy other records, apart from the ones that you see
on the first screen, like the month's specials. So, er, when it comes to
records, I bought only things I saw on the home page.

If one takes into account only the products highlighted on the screen, the
size of the accessible catalog remains imperceptible:

I didn't take note of the size of the choice ... I mean my first purchase, the one I was looking for was part of the new releases, so it was already on the page, I didn't need to do a search.

Only when the search engine is used is a world of goods suddenly revealed, and the initial promise of the website actualized. The sites are therefore primarily organized around the exploration of the offer via the search engine (since their promise is to make an exhaustive catalog available to the Web users, accessible in its entirety only through the search engine). The search engine is the most salient mediation between the Web user and the goods. Whereas in the shop the goods are displayed on countless shelves, most of them remain invisible if consumers only browse through the pages as they would the aisles of a store. The search engine itself makes only a tiny fraction visible at a time, in limited explorations. The exhaustiveness of the offer is perceptible only in the fact that almost no request, no matter how particular, is left unanswered. This feature is considered important enough to be shared, once discovered:

So, to test, just for fun actually, I even had fun about it afterwards with friends, I said: I'm going to show you a site, you type in anything, you click and you find it.

Descriptions of use of the search engine on the largest e-commerce sites highlight intention and planning:

And now you first go onto the Internet and then maybe somewhere else? Yes, when I'm looking for something specific, when it's for the enjoyment I tell myself I've got no more books to read so I want to stock up on detective stories and others then I won't go onto Internet ... the disadvantage of Internet and it's advantage is that its funnel function is good when you've got a precise idea of what you want, you don't get the showcase effect.... Because the things you didn't think about, you'll buy them in town. A thing you didn't think of, well, for me it's like that.

One feels that one is required to feed fairly precise data into the search engine, and explicitly to reveal a well-defined preference for a certain good, which must be enacted through its typing and entering it into the space reserved for that purpose.

Those customers who arrive with a vague idea have the impression of being constrained by the search engine's technological affordances, in the sense of having to specify their request ever more clearly. They experience a sequential convergence, which they interpret as a characteristic of the search engine and ascribe metaphorically to its "funnel function":

I typed in 'gift'. It was really too broad. I had 'specialized gifts', I really didn't know what I was looking for. I found 'engravings'. I typed in 'gift engraving' and that's more or less where I found what I wanted.

In this case, intentions are revealed and revised as the search proceeds. However, at each step in this sequential process the list of results provides unanticipated and contingent information. For instance, a search may reveal unexpected results, such as other titles or other versions of a work:

I did let myself diverge a little from the list of records to buy, even if in general, when they are on the list it means that they're not available. But then it gives me ... it makes me think of certain musicians that I don't always think of right away. So I tried....

The elicitation of more precise and explicit category-based searches therefore also leaves room for improvisation, for branching into new searches toward objectives that were impossible to anticipate in advance. Such contingency does not feature much in e-consumers' experiential accounts, though.

The search engine enacts a particular ordering of the purchase sequence. First, the demand is expressed through successive, more narrowly focused searches that allow an exploration of the otherwise invisible and inaccessible catalog. Each search requires a written clarification of its goal, which is projected in the linguistic expression on which the search is based. Second, the final choice that materializes the demand is separated from the steps through which the consumer negotiates access to the product. The use of the engine therefore creates a sequential gap between the formulation of the intention and access to the good. In the case of e-commerce this gap, similar to the one observable in old-fashioned libraries where borrowers have to fill in a form and wait for a librarian to fetch the required book, "calculates" a consumer whose preferences are revealed before the concrete encounter with the desired goods. Accounts based on the representation of consumption as intentional are more salient than those based on impulsive behavior, because of the particular kind of sequentiality that search-engine-based procedures "afford." Even if the reference to impulsive behavior is minimized in standard accounts, they still retain some relevance: they may become useful interpretive resources again to account for those particular cases in which the search proceeded through a contingent path.

Sites such as Amazon.com have tried to remedy this pragmatic and interpretive bias, which they perceive as a shortcoming of websites with respect to large stores. They have developed innovative functionalities to reintroduce new forms of contingency and unexpectedness into search-engine-based exploration of their catalog. Their aim is thus to prompt the Web user to engage in purchases without having to enter a linguistic explanation of his or her intentions. One particularly well-known feature adjusts the choice of products proposed on the page to the details of the Web user's past electronic path profiles (composition of an individualized homepage based on previous navigation, suggestions based on past purchases, etc.). Another uses collaborative screening techniques to highlight the preferences of consumers interested in the same goods as the current user. This marks a deliberate strategy on the part of Web merchants to cue impulsive consumption behavior as much as possible. The aim is to correct the intentional bias that stems from the sequential properties of search-engine-based explorations of available goods, and particularly the gap it introduces between the linguistic expression of demand and mediated access to the relevant goods.

The Evolution of E-Commerce and the Dialectic between Screen Interactivity and Commercial Interaction

Any economic exchange is simultaneously a targeted procedure and an interaction in which the participants cooperate to accomplish a satisfactory transaction. Any disturbance in the transaction sequence is likely to be evaluated both in relation to the goals of the economic exchange and as an interactional problem that brings into play the moral and practical normativities of ordinary sociability. This duality inherent in concrete economic interaction also conditions the resources available to participants to solve these problems and repair commercial interactions. More specifically, the development of new types of technology-embedded mediation for consumption, such as e-commerce, creates many opportunities for problems to occur in the enactment of a transaction. In this section I will use the example of the reception of orders to show how such an oscillation between treating electronic transaction as procedural interactivity or as collaborative interaction is relevant to describe and explain the regulation and stabilization of (now conventional) transactional sequences for e-consumption.

The Ambiguity of the Time Lapse That Occurs after an Internet Order Has Been Placed

One of the nagging problems raised by e-commerce concerns the low level of consumers' commitment. Web users are considered to be inconsistent and unfaithful, hopping from site to site and seldom appearing to be inclined to finalize their transactions. This problem is often related to a lack of trust. Invited to participate in "virtual" interactions and a screen-formatted dialogue, consumers lack all the information they need on the entities that are supposed to respond to their actions, especially to evaluate whether they are credible, authorized to sell, and capable of doing so. The question of trust affects the rules and conventions governing interaction and its step-by-step organization. To be able to treat the subsequent events displayed on their screen as responses to their actions, consumers must be able to readily imagine that they come from authorized and legitimate authorities, committed to the orderly accomplishment of the transaction. Moreover, these reactions have to comply with certain syntactic rules and ritual conventions governing the cooperative accomplishment of an acceptable transaction. The participants' engagement is continually renegotiated through compliance with these rules and conventions. E-commerce transforms not only interactional formats and media (by proposing screen forms to send off at a click, rather than a salesperson or a paper catalog) but also the pace at which the successive actions required to accomplish the transaction are carried out and made visible to the other party.

In contrast with a supermarket, where products available on the shelves are immediately available for sale, in traditional, paper-based mail-order selling there is necessarily a time lapse between when a consumer engages in the transaction (by mailing an order, and often the payment) and when the trader actively commits himself by sending the purchased goods and a receipt. This latent period is the sum of the time required for the seller to receive the order and of the time taken to process the order and send the goods. These two processes both usually take days. With electronic mediations—both Minitel and Internet—that same interval is now split into highly asymmetrical parts. On the one hand, just a few seconds are needed to send an order, by clicking on an icon or an active box that signals and highlights that this simple click involves a strong commitment on the part of the consumer[5] (for example, "confirm your order") and for the electronic order thus "sent" to reach its destination on the seller's information system. On the other hand, the time required for the order to be processed and the goods sent can be counted in days or weeks, for it depend on the time scale of logistic processes, as in the case of mail-order selling (with the notable exception of dematerialized goods). This asymmetry between the immediacy of the order (which constitutes a very strong commitment by the consumer) and the material response of the e-merchant aggravates the potentially worrisome character of this moment of suspension of the exchange between ordering and receiving the goods. Such a pause in the transaction may then be treated as a problem in two respects: it is perceived either as an excessive delay (when the economic transaction is interpreted as an algorithmic procedure) or as a silence that lasts too long (when the economic transaction is interpreted as an interaction).

Perceived as a delay, the same time interval is a reminder of the vulnerability of the transaction to various risks of inefficiency: unavailability of goods, inadequacy of their description in catalogs or on websites, various dysfunctions in the delivery process, dishonesty of the virtual trader,[6] etc. These risks are framed by forms of legal regulations that aim to stabilize the commercial procedure, for example the obligation to refund a dissatisfied customer, as well as institutional guarantees, such as the quality labels issued by professional organizations (e.g., the Fédération de la Vente à Distance[7]).

Perceived as a silence, this time interval may also be seen as the consequence of mistakes in the syntax that govern the proper accomplishment of the relevant turns of interaction. Using the Internet requires skills in manipulating hypertextual screens. When a procedure does not take place as expected and nothing happens after a particular action, certain users question their own competencies. The longer the silence after the order, the more they doubt their ability to have accomplished and transmitted the corresponding turn of action in a proper way. Only the other party is able to dispel this uncertainty through his subsequent reaction:

It's always a bit unnerving when you've got a problem, it happened to me once with the games site, when there's a crash in the middle of the connection, I mean, of the transaction, it blocks, and you don't really know what state you were in, should you order again, should you not order again, it happened to me once to wait two three days before sending the order again to see if I'd get the e-mail confirming or not.

This way of treating the absence of a system's reaction to person's actions on it is very general. Researchers in cognitive science—e.g., Norman (1991), who relies heavily on interactional metaphors to describe human-computer interaction and to orient design practices—have posited that "good" design is design that ensures that any user action is accompanied by a "rapid" reaction from the technical system, which makes visible to the user the alteration of its state accomplished by his or her action. Such a reaction may then be treated by the user as the "response" of the system.

An e-commerce site is a kind of technical system maintained by humans to mediate between consumers and the goods proposed by a seller. It therefore represents a salesperson. The silence that may follow an order may then take on an ethical connotation and be interpreted as a problem relative to the ritual constraints of the interaction. The longer it lasts, the more it will highlight the vulnerability of the consumer who, by clicking on Enter, has clearly indicated a commitment to the purchase of a particular item, in a way assumed to be immediately visible to the seller. For consumers, long silences followed by the seller's refusal to sell (a situation that occurs most often in the case of unavailability of the product) are treated as a breach of the site's promise and of the moral contract underlying it. They are particularly worrisome because of the high level of asymmetry between the instantaneousness and force of the consumer's engagement (the order) and the seller's apparently nonchalant acceptance of it (due to the far longer interval required for a response in the form of ensuring that the consumer receives the goods). It is an even more sensitive issue in e-commerce than in mail-order selling, where the temporal asymmetry is reduced.

This ethical dimension of e-commerce is evidenced by the strong emotion and sometimes indignation expressed by many Web users in case of difficulty. Consumers interviewed say they were "irritated" or "hurt" to discover the unavailability of the book that they thought they had bought and paid for online a few days earlier. Being refused what mass distribution promises to everyone—access at the same price to all listed products—is interpreted as being treated as a "non-person" in Goffman's sense, all the more so when the seller is slow to decline the order. In e-mails received by call centers, some customers are indignant about the fraudulent promise represented by public visibility, on the site, of products that are actually unavailable. They demand compensation in the name of human dignity. They expect all trace of this broken promise to be removed and for the reparation to be public and visible to all:

Further to my telephone call today, I confirm that I refuse this cancellation, unless you delete this reference from your on-line catalog. I ordered this article on 22/12/2000, no. 431 723 and you were unable to obtain the article, despite several reminders. So, either you obtain this game, or you remove it from your site. I expect to hear from you very soon.

Since users consider themselves to be completely committed, as customers, when they enter and send their order on their computer, the unavailability of products displayed on the site pages is treated as breach of a contract assumed to be known and applicable to everyone. This is particularly true insofar as, following Amazon, sites selling books online claim that their online catalogs are far more exhaustive than the inventory of any "brick-and-mortar" store.

The instrumental framing of consumption as an algorithmic sequence of targeted actions may be transcended[8] by its treatment as an interaction when problems arise, for the ordered accomplishment of the transaction requires the observance of ritual and systemic constraints. The ethical and instrumental dimensions of the economic interaction are inseparable, because consumption is part of the social order in public places.[9] In the case of e-commerce, the procedural logic of the transaction is embodied in human-machine interactivity, which may always be seen as an interaction between the consumer and the "merchant-in-the-machine." By taking this duality between algorithmic interactivity and dialogical interaction into account, we are able to understand certain aspects of the way in which e-commerce has evolved and grown since 1995.

The Dialectic of Process and Interaction: Framings and Overflowings of the Social Order Characterizing E-Consumption

In the event of a problem, consumers often react by asking to speak to a human representative of the firm. They see conversation as a powerful and flexible resource for solving everything that appears to be a hitch in the course of an online transaction. The possibility of having such contact by telephone or e-mail at a reasonable cost is also interpreted as proof of the distributor's commitment to provide a service of quality and treated as a sign of consideration for the consumer.

Online consumers explicitly ask for interaction with a person (and not a machine), to deal with problems that seem to be easy to solve with the resources of ordinary conversation:

When I ask a question I would like it not to be a robot telling me that my order has been taken into account. There's a problem, a hitch, it should be recognized and an answer given to the precise question.

If a robot answers by e-mail (which makes the interactional dimension of the exchange underway less salient and highlights its algorithmic dimension), this consumer becomes angry or flees.

The greater a consumer's uncertainty about what could happen between the order and the delivery, the more problematic the ethical frame in which the transaction is set will seem to that consumer, and the stronger his or her demand to speak to someone or to interact in writing with someone in case of a delay will be. The following e-mail received at a call center attests to the intensity of the claims that may be triggered by excessive delays:

I'm going to try to be clearer, I bought a Strasbourg-Munich ticket for the 19/12, to be posted to me, I didn't know that a week was required for this postal service, the order was nevertheless taken into account and thus accepted. I didn't receive the ticket and the address indicated for posting it is no longer valid from Monday. I'll repeat my question VERY CLEARLY: What must I do to receive my ticket in time? Can it be issued at the station?

The actors of e-commerce have had to organize themselves to take into account the overflowing of the electronic transaction frame, in relation to this request for oral or written interactions. This overflowing is more pronounced when the algorithm of the transaction is faulty in terms of procedure, or deemed inappropriate by users. In the journalistic chronicle of the origins of Amazon, the emblematic electronic mass distribution site, its founders are said to have intuitively perceived that e-mail was as important as the e-commerce site itself (Spector 2000). Most e-commerce sites now have multimedia call centers that process mainly telephone calls and/or e-mails from consumers. In some sectors where complex goods are sold, such as the travel industry, virtual agencies have had to build call centers with dozens of advisers to answer Web users' questions. These call centers often had to be set up in a very short time, sometimes from scratch (Licoppe 2002). Even today, despite the fantasy that entrepreneurs and company managers have entertained since the start of the industrial era, of professional environments where technology would remove the need for human actors,[10] the head of an e-travel agency undergoing reorganization cannot imagine a future without organized human resources to support the need for interpersonal dialogue: ("And afterwards of course, a human being will be needed, in case the person has a question, he or she will always need someone"). He adds that this way of solving problems through conversation has a heavy financial cost and therefore has to be closely supervised ("but we can't grow unless there are more and more customers who reserve online, because it consumes little or no manpower"). On several sites studied, the ratio of transactions accomplished in a purely electronic mode (that is, without any dialogue with a call center adviser) to the total number of transactions was considered to be one of the main indicators of the organization's overall performance. Defective procedures or improper site design (often defined as such retrospectively) translated directly into an increase in the volume of requests by telephone or e-mail. The entire organization was aligned on the movements of this indicator, thus attesting to the inevitability of the entanglements between interactive man-machine procedures and human dialogue in the accomplishment of the transaction.

Another resource for "market professionals"[11] consists in exploiting the possibilities afforded by the technology to alter the algorithmic ordering of the transaction sequence, so that its enactment is less likely to be perceived as problematic with respect to the systemic and ritual conventions of ordinary interactions. In recent years many sites have introduced an automatic procedure in which a confirmatory e-mail is sent within 24 hours as an acknowledgement of the initial order. This also confirms the site user' identity as a consumer. These automatic e-mails often provide an order number, which marks the seller's commitment to complete the transaction. Finally, this procedure sets an upper limit to the lapse of time between the order and the subsequent actions that may legitimately be treated as a response to it. The time gap between order and response, even if the goods are not there yet, is then less likely to be treated as a silence marking the existence of a problem from an interactional perspective.

On the supply side, this convention was introduced and implemented in technical protocols for transactions at the end of the Internet bubble. It cannot be said to have been invented by anybody. Its production and stabilization were based on material, cognitive and social resources collectively afforded by the "new economy" world: fairs and other professional gatherings, electronic media, various kinds of evaluation performed and disseminated by consultants, etc. In 2001, for example, consultants compared firms' quality of e-mail service to consumers. The results of their study were disseminated by *Le Monde* Interactif (a subsidiary of the French newspaper *Le Monde*), which commented that there was nothing more annoying than an unanswered e-mail.[12] Other consultants empirically evaluated the causes of Web users' deception in relation to the way in which their e-mails were treated, and identified delay (absence of an answer or answer too late) as the main factor.[13] Three years later the consensus became a norm that was sufficiently naturalized to be able to be stated in the form of a rule that could be cited without any further justification. The e-magazine *01.net* suggested the following basis for an acceptable management of e-mails: "To give the sender the feeling of being taken into consideration, it is important to return an acknowledgement of receipt, if an immediate detailed answer can't be given."[14] This axiom circulated from one professional magazine to the next, in almost exactly the same words.[15] In this way, a link was established between taking the customer into consideration, and respecting the customer, and the seller's commitment to responding rapidly. The most concrete and sure form of this commitment is found in the automation of the acknowledgement of receipt procedure and of the time it takes.

From the users' point of view, insertion of this particular interactional move within in the algorithmic sequence has become a strong expectation and even a criterion for assessing sites. For instance, a young woman interviewed uses only sites that send confirmatory emails within a maximum of 24 hours. In her opinion this email is reassuring in two respects. First, it confirms the purchase and its author's position as a customer ("While I haven't received my order at least I have proof that I've ordered something"). Second, it confirms that on the other side of the screen there is a seller ("You've just got a confirmation of the purchase that serves as a kind of invoice because after all this is all very virtual") who is committed and is proving to be worthy of trust, as evidenced by the order number on most of these confirmatory emails. Almost all online consumers archive these e-mails on their computer. A large proportion of them print them, especially when they think there is a chance of the seller not meeting its commitment. It is as if putting the e-mail answer into a material, paper format added a "legal" dimension to the confirmation and to the commitment that it represented.

The 21 June 2004 law to promote confidence in the digital economy in France[16] was intended to provide a legal frame for the sequential ordering and meaning of the turns comprising interaction in e-commerce. Article 1369-2 requires site designers to comply with two prescriptions: (1) the "double-click rule"[17] (according to which the consumer must confirm his order by clicking on it a second time) and (2) that an order must be confirmed by a recapitulative e-mail from the cyber-trader "without any delay." The order and the acknowledgement of receipt are considered valid only when received at their addressee's e-mail address. The law aimed explicitly at a closer adjustment of the sequential order of transactional algorithms and the social order of interaction, to minimize the consequences of possible misalignments by making some of them subject to legal action. Technology is society (it incorporates a sequential order institutionalized into a legal norm) and society is technology (meaningful interaction sequences are reified within the e-commerce software).

Conclusion

E-commerce sites and software have developed through the to-and-fro between their role as tools to implement a transactional sequence and their role as a mediation between consumers and merchants. The designers of the electronic offer have configured the procedures of e-commerce so that they incorporate and reproduce as much as possible the algorithmic logics and interactional conventions governing those older and more stable forms of consumption that occur in supermarkets and large stores. But electronic procedures have to rely on the resources, properties, and uses peculiar to the Internet, such as use of the search engine (to employ search data to explore an offer that cannot be represented satisfactorily on a small screen) and the supposedly instantaneous character of the execution and transmission of actions performed on screen, online. These properties and expectations provide for a particular form of temporal ordering and sequencing for online purchases, different from what would be observed in a store.

When they use a search engine, consumers reveal their preferences through the series of requests they perform to get the search engine to function "properly." Each request is presented both as a move in a procedural sequence (punctuated and formatted as such by a final click on the Enter or the Send button) and as a query projecting a response in the forms of a list of relevant items. The use of the search engine introduces a sharp separation between the moments at which the elaboration, disclosure, and display of preferences take place and that at which the qualities of the desired product become accessible. In a store, it is often the availability and accessibility of goods to manipulation that triggers the emergence of a preference. With e-commerce, on the other hand, the linguistic elaboration of the preference has to come first. We have seen how this sequencing of the purchasing procedure, characteristic of e-commerce and its reliance on the use of search engines, reconfigures the way in which online buyers account for their consumption experience. That experience is often typified as an action motivated by an intention or a plan prior to the consumer's involvement with the site. By contrast, accounts based on the impulsive purchase ideal type lose much of the relevance they had with respect to the "brick-and-mortar" consumption experience, with its situated coproduction of the intention to buy an item and of the desirability of the latter.

E-commerce alters both standard purchasing sequences and the way in which the unfolding of actual sequences may be interpreted. The time lapse between order and reception of goods is, for instance, treated as a sign of inefficiency but also, simultaneously, as a deviation from what constitutes a proper interactional sequence. It is interpreted as a silence, signaling the occurrence of a problem that needs to be solved, with consumers moving from the website to their phone and e-mail to dialogue directly with a representative of the seller. To limit the high costs of such direct negotiations, the designers of e-commerce sites have amended the original transaction algorithms, exploiting software resources to minimize the interactional problems that the purchasing sequence might cause. In the case of the time lapse between order and reception of goods, their solution has been to introduce a new step in the purchasing sequence: the automatic sending by the seller of an acknowledgement of receipt of orders within 24 hours. This became standard practice at the time of the Internet "bubble," to fill the "gap" between the order and the dispatching of the goods. During this gap, Web users proved to be extremely sensitive to any event that could be interpreted as a sign of disturbance in the interaction since they saw the time lapse as a silence signifying a problem. The generalization of this practice was particularly swift because it was relayed by consultants and the specialized press, which systematically prescribed it and erected it as a tenet of customer relations management. Sending an acknowledgement of receipt has symmetrically become the object of a normative expectation on the consumers' side. This rule has recently turned into a legal norm.

The evolution in the e-commerce transaction procedure, toward the reification of a new, built-in interactive sequence, is driven by the need to minimize the risks of overflow with respect to the favored (by the e-merchants) standard e-consumption on-site procedure. The problems and their solution simultaneously play on the fact that it is always possible to treat e-consumption as an instrumental step-by-step procedure (which may be assessed with regard to its efficiency, in relation to an economic target) and as a collaborative and situated accomplishment (which may be assessed with regard to the normative expectations governing the management of interaction in the public sphere). Such an embedding of the economic exchange in the social order, characteristic of ordinary sociabilities, is a general fact. The e-commerce case is interesting in the way it reveals their continuous interplay through the transformations of the algorithms and artifacts that provide a material substrate and a pragmatic environment for such transactions. The continuous redesign of the technologies of consumption strives to stitch the economic exchange and the social interaction into a seamless web.

In the case of Web-based practices, our analysis suggests that some tensions may still endure in the resulting fabric. The least costly alternative for sites is to multiply and automate electronic contacts with consumers, particularly by e-mail. It is just as easy to systematize the sending of an acknowledgement of receipt of orders as to automate the sending of messages prior and subsequent to the transaction. More and more sites inundate their customers with newsletters, notifications of promotions and events, etc. This communication flow aims toward the build-up of a continuous and personalized commercial relationship, instead of the (apparently) one-shot and anonymous one characterizing mass retail. Such a possibility has appeared as a specific and fairly attractive feature of e-commerce. It has also provided some hope that such a construction and individualization of the commercial relationship will solve problems of trust and customer loyalty. From the consumer's point of view, this type of relationship is likely to be understood in relation to the norms governing forms of electronic sociability (since it is conveyed essentially by e-mail and the media of electronic chatting). But, as many studies of computer mediated communication have shown, electronic links are weak, ephemeral, and revisable.[18] It is known that a majority of the visits to forums or blogs does not lead to any kind of visible action (according to the sites, considered, 60–80 percent of visits are by "lurkers"), and those Web users who do get involved often do so under multiple identities and only temporarily. Messages can go unanswered without its necessarily upsetting their authors. To construct a relationship in which consumers are strongly committed, e-consumption is somewhat paradoxically made accountable with respect to electronic socialities in which exit is a low-cost, well-established strategy. This tension may prove decisive for the future of e-commerce.

More generally, I have tried to study in detail the way economic transactions, as actual actions, are founded on (too often) taken-for-granted elementary procedural sequences, such as buying and selling. I have argued that, at that level of analysis, transactions must be described as practical accomplishments mediated by different technological artifacts. They appear equivocal, in the sense that they can always be treated as a goal-oriented procedural action sequence or as ongoing interactions, which are not vulnerable in the same way to trouble and contingencies. Their meaning therefore emerges from the situated interplay of sociality, economy, and technology. This kind of analysis constitutes one of the steps needed for the development of a technology-sensitive, interaction-oriented economic sociology research program.

Notes

1. Apart from the numerous efforts to create databases from sales receipts, it is for this purpose that firms use customer loyalty cards, for example, whose name evokes this relational commitment.

2. Or the seller's representative, who may be a supermarket cashier or a computer procedure incorporated into a website.

3. Primatologists and anthropologists who defend the hypothesis of social cognition situate it in an evolutionary interpretation: social cognition is said to be an acquired feature, reinforced by a process of evolution that bridges the gap between primates and man (Goody 1995).

4. E-commerce sites often select several products and highlight them on their home pages (they describe such web-page formatting as metaphorically analogous to "aisle-end display" in stores). Symmetrically, large stores specialized in the distribution of cultural goods usually make available for their customers, in certain strategic places, paper catalogues (to check or complete a reference) and salespeople equipped with computers and software (to check the availability and locality of a product). But in both cases such resources are less salient and accessible than the goods available via the search engine in the case of websites, or browsing in the aisles in large stores.

5. In some experiments the engineers thought of exploiting the performative nature of this validating action by asking users to repeat it, to distinguish acceptance from legal commitment (Akrich 1993).

6. The disclosure of confidential information to a dishonest trader is a risk common to most forms of distance trade. Some risks are specific to e-commerce. Generalized connection makes the "connected" user's terminal vulnerable to strangers that he or she can neither qualify nor locate, for instance hackers. Web users thus always remain vaguely pre-occupied with the possibility of certain actions by a third party seriously harming him or her, whether intentionally or not.

7. Formerly the Fédération de la Vente par Correspondance, it was renamed in 2001 to include e-commerce sites in its perimeter.

8. On this notion of an economic transaction that appears as such only through a process of framing, constantly subject to overflowing by different forms of external-ity, see Callon 2000.

9. Goffman (1961) considered that the social order is defined as the consequence of any set of moral standards that regulates the way in which people pursue objectives.

10. On the social history of the managerial fantasy of the "factory without workers," see Noble 1986.

11. This term was introduced to describe the occupations and activities of the actors on the supply side working on the positioning and merchandizing of products in the mass distribution context (where the challenges of singularizing equivalent products are strong) (Barrey, Cochoy and Dubuisson-Quellier 2000). It is interesting to extend this notion to the actors of supply who, like here, manipulate the algorithmic order of the transaction.

12. Le mail en mal de réponse, *Le Monde* Interactif, 21/02/2001.

13. http://www.digiway.fr/html/email.htm

14. 01.net of 29/09/04, http://www.01net.com/article/214579.html

15. "Yet, to give senders a feeling of being taken into consideration, it is important to plan an initial instant reply informing them that their request has been received and will be treated rapidly." (Traitement des e-mails. Outils et Techniques, *En Contact* no. 15, summer 2003, p. 17)

16. *Journal Officiel* no. 143, 22 June 2004, p. 11168.

17. Although the article does not explicitly mention the double click, the document presenting the bill in parliament on 15 January 2003 describes the article as designed to ensure that acceptance of the offer by the consumer is in the form of a "double click" or an equivalent protocol, accompanied by recapitulative information from the contract' (http://www.assemblee-nat.fr/12/p10528.asp). This notion of a double click was then relayed by the general press (see, for example, Règles d'or du cyberacheteur, Multimédia section, *Le Figaro*, 23 December 2004.

18. See, e.g., Herring 1996 or Velkovska 2004.

References

Agre, P. 1997. Surveillance et saisie. Deux modèles de l'information personnelle. *Raisons Pratiques (Cognition et information en société)* 8: 243–265.

Akrich, M. 1993. Les objets techniques et leurs utilisateurs. *Raisons Pratiques (Les objets dans l'action)* 4: 35–57.

Assadi, H., and V. Beaudouin. 2002. Comment utilise-t-on les moteurs de recherche sur internet. *Réseaux* 116: 171–198.

Barrey, S., F. Cochoy, and S. Dubuisson-Quellier. 2000. Designer, packager et merchandiser: trois professionnels pour une même scène marchande. *Sociologie du travail* 42: 457–482.

Bowlby, R., 2001. *Carried Away: The Invention of Modern Shopping*. Columbia University Press.

Callon, M., and F. Muniesa. 2003. Les marchés économiques comme dispositifs collectifs de calcul. *Réseaux* 122: 189–233.

Cochoy, F. 1999. *Une histoire du marketing, discipliner l'économie de marché*. La découverte.

Fligstein, N. 2001. *The Architecture of Markets*. Princeton University Press.

Geertz, C. 1979. Suq: The bazaar economy in Sefrou. In *Meaning and Order in Moroccan Society*, ed. C. Geertz and L. Rosen. Cambridge University Press.

Goffman, E. 1963. *Behavior in Public Places*. Free Press.

Goffman, E. 1981. *Forms of Talk*. University of Pennsylvania Press.

Goody, E. 1995. *Some Implications of a Social Origin of Intelligence: Social Intelligence and Interaction*. Cambridge University Press.

Herring, S., ed. 1996. *Computer Mediated Communication: Linguistics, Social and Cross Cultural Perspectives*. John Benjamins.

Lave, J. 1988. *Cognition in Practice*. Cambridge University Press.

Licoppe, C. 2002. Le traitement des courriers électroniques dans les centres d'appel. *Sociologie du Travail* 44, no. 2: 381–400.

Licoppe, C., A.-S. Pharabod, and H. Assadi. 2002. Contribution à une sociologie des échanges marchands sur internet. *Réseaux* 116: 97–140.

Lunt, P. 1999. E-Commerce and the Restructuring of Consumption: Social, Psychological and Economic Perspectives. http://www.ergohci.ucl.ac.uk.

Mirowski, P. 2002. *Machine Dreams: Economics Becomes a Cyborg Science*. Cambridge University Press.

Mirowski, P., and K. Somefun. 1998. Markets as evolving computational entities. *Evolutionary Economics* 8: 329–356.

Noble, D. 1986. *Forces of Production: A Social History of Industrial Automation*. Oxford University Press.

Norman, D. 1988. *The Psychology of Everyday Things*. Basic Books.

Pinch, T., and C. Clark. 1988. *Hard Sell*. HarperCollins.

Schegloff, E. 1988. Goffman and the analysis of conversation. In *Erving Goffman*, ed. P. Drew and A. Wooton. Polity.

Spector, R. 2000. *Amazon.com: Get Big Fast*. HarperCollins.

Smelser, N., and R. Swedberg, eds. 1995. *The Handbook of Economic Sociology*. Princeton University Press.

Suchman, L. 1987. *Plans and Situated Action: The Problem of Human-Machine Communication*. Cambridge University Press.

Velkovska, J. 2004. Les formes de la sociabilité électronique. Une sociologie des activités d'écriture sur internet. Ph.D. thesis, EHESS, Paris.

11　Six Degrees of Reputation: The Use and Abuse of Online Review and Recommendation Systems

Shay David and Trevor Pinch

We are living in the midst of one of the biggest infrastructural changes of our time: the establishment of the Internet and its increasing penetration into more and more areas of life. In this chapter we are particularly concerned with changes brought about by the new sorts of reputation systems which accompany online shopping and e-commerce. We examine in particular the world of books and CDs.

When we move from an offline world to an online world, some things change and some things stay the same. If you walk into a physical bookstore or CD store hoping to buy a new book or CD, even before speaking to anyone you will have already gotten significant feedback from the environment. The music playing in the background, the items featured on the "best-sellers" shelf, the number of copies of a particular book or CD that are on the shelf, the chatter of the people standing in line in front of you praising one item over another, the items they are holding and are about to buy—all these are inputs into a multi-dimensional, material, and symbolic means of assessing and establishing reputations. When we shift settings to online shopping and e-commerce, where such direct feedback from the material environment in which retail commerce occurs is absent, new mechanisms have to compensate for the changes in materiality that the commerce environment offers.

This chapter is concerned with understanding the promises and perils of reputational systems that are based on user-generated product reviews, which in the last few years have become the primary mechanism fulfilling this reputational gap.

A Cultural Lake Wobegon?

Charles McGrath, a former editor of the *New York Times Book Review*, recently posed the rhetorical question "Has there ever been a book that wasn't acclaimed?" (Safire 2005) What McGrath was lamenting, of course, was the inflation of accolades in the universe of book promotion, which, much as in the case of CDs and other cultural products, is influenced by commercial interests more than by standards of accuracy in the representation of a product's quality.

Traditionally, the critics employed by respected institutions (for example, the *New York Times* and other leading newspapers and trade magazines) served as cultural gatekeepers and proprietors of quality. However, the growing abundance of books (more than 100,000 new titles were published in the United States in 2006), CDs, and similar "information goods" precludes any wide coverage or quality assessment. Recently, the small group of paid experts who are hired by these select institutions have been aided by a wide variety of trade publications and websites, which cover ever more specialized sub-fields of the culture industry and which are employing systems that harness the power of user communities.

Evidently, user reviews are mushrooming as an alternative to traditional expert reviews in many areas of cultural production. It was established long ago that reviews and recommendation systems play a determining role in consumer purchasing,[1] and recent qualitative research adds weight to the claim that reviews have causal and positive effects on sales. To nobody's surprise, books that have garnered more and better reviews sell better (Chevalier and Mayzlin 2003). With people in the culture industries increasingly realizing this, many of the reviews are positively biased, and it is hard to distinguish the "objective" quality of reviews. In addition, owing to the large variance in the quality of the reviews and the varied agendas of the reviewers, user input is often untrustworthy, leaving consumers with little ability to gauge an item's actual quality. Do we live in a "cultural Lake Wobegon" where (to paraphrase Garrison Keillor) all the books are above average? Is there a way to review the reviewers, to guard the guards?

As will be discussed in detail below, new reputational systems such as those employed by Amazon.com (2005) and eBay.com have tried to take advantage of the direct feedback which purchasers can supply concerning their purchases in an online environment (and, in the case of eBay, their transactions.) But these new options, which we describe as *affordances*, also bring with them new problems as the participants adjust to what is at stake in the new economy of reputation.

The new user input systems that are burgeoning on the Internet employ various types of user input to assess the quality of books and CDs (Amazon .com, BN.com), news (Slashdot.org, Kuro5hin.org), consumer electronics (Shopping.com), home-recorded music (ACIDplanet.com), teaching quality

(RateMyProfessor.com), drugs (DrugRatingZ.com), and many more types of information, information goods, and information-embedded goods (i.e., goods whose value derives from the information embedded in them). These systems exemplify peer-production systems (Benkler 2002) in which communities of users pool their resources in order to produce higher-quality information goods and information-embedded goods, in some cases replacing the traditional mechanisms of firms and markets altogether.

There is disagreement among scholars regarding the novelty and the potential long-lasting effects of these systems. Proponents claim that peer-production systems will revolutionize the production, the consumption, and the use of information, primarily through the cost reduction they offer and the enhancements they enable in assessing and allocating human creativity (Benkler 2006; Raymond 1999; Himanen 2000; Lessig 2004; Coleman 2005). Skeptics ponder the ways in which such systems are being expropriated and appropriated by existing actors and refer us to a lengthy tradition of user involvement.[2] In and of itself, we should remember, harnessing the power of community members for the purpose of evaluating the quality of products is not a new idea. The *Whole Earth Catalog*, during its heyday in the early 1970s, was distributed in more than a million copies, introduced members of the back-to-the-land movement to products from fertilizers to computer displays, and offered for each product a summary of users' experiences and recommendations. This community-based product can be viewed as an early model that directly influenced later systems, including online communities (Turner 2005). But the proponents of novel systems would argue that the Internet, by virtue of its sheer scale and immediacy, offers something genuinely new.

Regardless of the future prospects of these systems, however, both sides of the debate would agree that practices as well as norms are not yet stabilized in this domain, and that the mechanisms that control the "reputation economy" are not yet well understood. Clearly, the nascent systems introduce new variations of old problems concerning authority and expertise as we experience the shift in domains brought on by the internet. The primary objective of this chapter is to explore the reputational underpinning of these systems in the face of alarming discoveries concerning the authenticity of some of their content.

We start with two episodes that drew us to this research. These episodes attest to the role of user reviews in our culture and introduce some of the concerns we wish to address. We later discuss how reputation is manifest in six different degrees, which span the gap between existing offline reputation and the meta-reputation that is based on reviewing of the reviewers and their reviews. We conclude by enhancing Larry Lessig's analytic framework that shows how norms, laws, markets, and code interact to regulate human activity in this changed material world.

Why Look at the Copying of Book Reviews?

One of us, while sitting in a café in Ithaca, New York, witnessed a meeting between a local author, Barry Strauss, and one of his fans. The fan, an enthusiast for naval histories, was very excited by the fact that Mr. Strauss's latest book, *The Battle of Salamis* (2004), had been well received. As proof of this he mentioned a review he had read on Amazon.com that compared Strauss's style to that of Tom Clancy. The review read as follows:

"A Good Story Well Told" / December 1, 2004 / Reviewer: John Matlock "Gunny" (Winnemucca, NV)
... This extensively researched book is centered on the naval battle, but it is set in its place with descriptions of other parts of the war.... It also includes an amazing amount of detail on the two countries, their cultures and the times in general.... I have to say that the author's writing style makes this read like a Tom Clancy novel....[3]

From a sociological point of view, this encounter is interesting in several respects. First, users' book reviews have become conversation pieces in the "offline" world, where what is otherwise a "secondary" information good becomes the primary topic of a conversation. Second, the fan does not appear to know the reviewer, but this does not impede him from invoking the reviewer as a legitimate authority; the materialization of the review, in writing, on a reputable website is enough to warrant quotation. In this specific case, the reviewer is a "Top 50" reviewer, with more than 1,500 reviews to his credit, but does this make a difference? Third, the content of the review makes use of tiered reputation: the reviewer credits Strauss by comparing him to a reputable thriller writer (Clancy). Taken together, these three points suggest that users' book reviews are an interesting topic for analysis.

In a second case, we came across something more alarming: review plagiarism. This instance concerned the book *Analog Days* (Pinch and Trocco 2002). *Analog Days*, which chronicled the invention and the early days of the electronic music synthesizer, was well received by reviewers both offline and online. Shortly after its publication in fall 2002, the Amazon.com editors quoted a review from the *Library Journal*:

... in this well-researched, entertaining, and immensely readable book, Pinch (science & technology, Cornell Univ.) and Trocco (Lesley Univ., U.K.) chronicle the synthesizer's early, heady years, from the mid 1960s through the mid 1970s.... Throughout, their prose is engagingly anecdotal and accessible, and readers are never asked to wade through dense, technological jargon. Yet there are enough details to enlighten those trying to understand this multidisciplinary field of music, acoustics, physics, and electronics. Highly recommended.[4]

A similar but distinctly different book that had appeared earlier, *Electronic Music Pioneers*, by Ben Kettlewell (2001), received the following user review on Amazon.com on April 15, 2003:

This book is a must. Highly recommended. April 15, 2003 / Alex Tremain (Hollywood, CA USA)
... In this well-researched, entertaining, and immensely readable book, Kettlewell chronicles the synthesizer's early, years, from the turn of the 20th century—through the mid 1990s.... Throughout, his prose is engagingly anecdotal and accessible, and readers are never asked to wade through dense, technological jargon. Yet there are enough details to enlighten those trying to understand this multidisciplinary field of music, acoustics, physics, and electronics. Highly recommended.[5]

The "similarity," of course, is striking. The second review is simply a verbatim copy of the first one, with the names of the authors and the period book covers replaced. The word 'heady' has been removed, and the period Mr. Kettlewell covers is lengthier than the 1960s (the focus of the Pinch-Trocco volume), but the comma after 'early' has been left in and thus the review is now grammatically incorrect. Other reviews of Mr. Kettlewell's book posted in subsequent weeks contain other sentences lifted from Pinch and Trocco's accolades (in this case, from readers' reviews). Furthermore, an inspection of the entire set of readers' reviews of Mr. Kettlewell's book suggests that the copying from Pinch and Trocco's reviews might have had an ulterior motivation. Just before the copied reviews, the following reader review of Mr. Kettlewell's book appears:

Dissapointing, for a $21 book ... / March 12, 2003 Reviewer "djminiwjeats" (Chicago, Il USA)
This book, although comprehensive to be sure, often paints in extremely broad or disconnected brush strokes, leaving me wishing there was more detail at times. This was especially evident in the first section, a seemingly endless series of brief bio's of various figures who are presented as key players in the development of electronic music, with very little indication of how they might actually fit into the historical continuum, or how they might relate to each other.... I also have read Frank Trocco's book on the Moog synthesizer (which also covers the Buchla, ARP, and others), and found it to be far superior. I'd recommend anyone just getting into this subject to start there instead.[6]

What does this copying strategy suggest? Is this simply an attempt to influence sales of a less well-received book, stealing the credit from a better-established publication, or is there a deeper undercurrent provoked by the above reader review comparing this book unfavorably to the Pinch-Trocco book? When Pinch contacted Amazon in 2003 and alerted them to this plagiarism (which Pinch had discovered by accident), Amazon's response was to recite Amazon's policy, which states that Amazon gives users complete freedom in posting reviews and does not intervene in the process (a policy that has since changed). Clearly issues concerning the freedom of expression clash here with notions of integrity and the imputed genuineness of reader reviews.

Pinch felt disappointed by Amazon's response. He was proud of his positive reviews. (It is rather unusual for an academic book to get any reader reviews at all.) Pinch felt that his own positive reviews, which he had taken to come from genuine enthusiastic readers, were now compromised by a system that allowed blatant copying by possibly non-existent readers.

With these two cases in mind, we decided to further explore the extent to which reviews are being used and abused in Amazon's system and in similar systems. Clearly the electronic media allow perfect copies of both primary and secondary information goods. Usually we are concerned only with the authenticity and the quality of the primary artifacts, but reviews play a determining role in assessing those characteristics. To what degree can we count on those reviews? A preliminary literature search revealed evidence suggesting that many reviews are not authentic, that users are using various techniques to game the system, and that this phenomenon may be widespread. For example, in 2004 both the *New York Times* (Harmon 2004) and the *Washington Post* (Marcus 2004) reported that a technical fault in the Canadian version of Amazon.com exposed the identities of several thousand of its "anonymous" reviewers, and alarming discoveries were made. It was established that a large number of authors had "gotten glowing testimonials from friends, husbands, wives, colleagues or paid professionals." A few had even "reviewed" their own books, and, not surprisingly, some had unfairly slurred the competition.

In view of these early observations, and in view of the enormous volume of books sold through Amazon and similar systems, the task of understanding the mechanisms that control these reputation and expertise tools is of prime importance. Cases in which reviews are plagiarized or otherwise abused seem to be good starting points for understanding the issues.

Preliminary results from our investigation follow.

Research Methods

Our study involved three kinds of activities: participant observation, interviews, and downloading and analysis of data from Amazon.com using custom-built software.

Participant Observation

Both of us are regular browsers of Amazon.com and other product-review systems. One of us is an author with many books listed and sold on Amazon.com, and we both use Amazon.com to buy new and used books and CDs. We both read and write reviews, rank reviews, review the profiles of expert users, and more. In addition, we have encouraged our students to write book and CD reviews and post them, and we have observed the dynamics of that interaction.

Interviews

We conducted a small number of interviews with prominent authors (novelists and nonfiction writers). The main question we were trying to answer was "In what ways have relationships between authors and readers changed as a result of the introduction of these tiered reputation systems?" We also investigated the continuities and discontinuities that authors perceived between older models of quality control and reputation management and the new models. We encouraged our interviewees to recall cases in which book reviews had been abused in the offline world, and we asked them to reflect on how the review system has changed with the introduction of new systems.

Custom-Built Software for Downloading Data and Detecting Plagiarism

The primary quantitative tool for our research was a set of software programs that one of us wrote specifically for the purpose of evaluating the prevalence of review copying, plagiarism, and abuse. The software included communication modules for downloading data from Amazon.com and copy-detection algorithms for detecting re-used text in downloaded data.

The first task for the software was to identify which books or CDs might have copied text contained in their reviews. A brute-force comparison of all the reviews ever published is technically feasible; however, we did not have access to the full database, and we saw little point in comparing books from different categories. As a working solution, we decided to compare only those items that are somewhat similar to one another. For a "similarity"

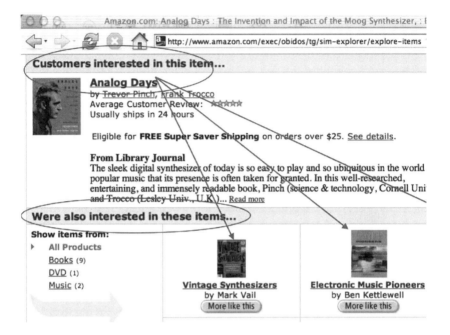

Figure 11.1
Using similarity data.

criterion we used data available from Amazon's public application programming interfaces (APIs). Using eXtensible Markup Language (XML),
these APIs programmatically expose various types of data including similarity matching which is based on collaborative filtering algorithms that use
customers' past purchasing behavior to deduce similarity among items and
project customers' interest. Evidently, such similarity algorithms can be
very powerful. For example, the similarity algorithm for Pinch and Trocco's
Analog Days finds Kettlewell's *Electronic Music Pioneers* to be similar (figure
11.1). Our data-download algorithm, then, was seeded with any book or
CD title as its source and was able, using recursive calls to the similarity
API, to build a virtual graph representing other items which are "similar"
to the original book. Such items included both books and CDs.

Having built a similarity graph, the algorithm proceeded to make calls to
further APIs that exposed the content of user reviews (figure 11.2). Trying
not to exceed limitations set by Amazon on the amount of data that can
be accessed using the APIs, for each book we accessed only the five most
recent reviews.

Figure 11.2
Detecting copies.

The next step of the algorithm was to compare the reviews to one another. (See the appendix for a more detailed technical description of the algorithm.) Simply put, the algorithm looks for text re-use at the sentence level, and produces lists of re-used texts ranked by the amount of similar text and the probability of the copying (figure 11.3). Importantly, the algorithm is able to detect re-use of text even if the re-use order within the paragraph is different. For the examples described here, we selected cases in which more than one sentence was re-used.

The system of reviews we are examining and how they are presented on Amazon.com can be thought of as a reputational system. That is, Amazon presents people who visit the website with discursive information from which they can infer the reputation of a particular reviewer. Amazon's system is largely typical of other electronic retailers' systems; thus, the

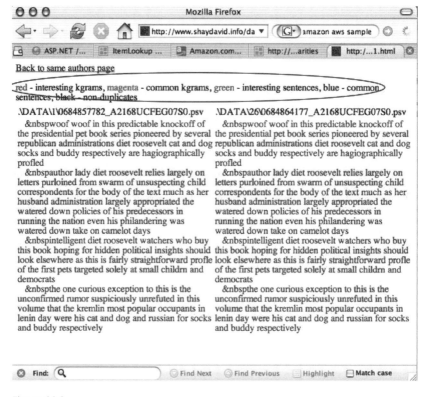

Figure 11.3
Comparing similar reviews.

description below is intended to be illustrative of a whole class of similar systems that might vary in details of implementation but which are based on the same principles. Importantly, the system features six discrete levels of reputation management, which in the case of Amazon.com are layered in a structure that we call "six degrees of reputation":

1. At the first level, authors' reputations and credentials accrue to their benefit directly from the association of the items with their names. This level of reputation is influenced by activities that take place outside the on-line recommendation system. It might include official credentials, past performance, reviews offline, and sales history.

2. At the second level, paid editors write reviews and try to influence buyers to buy a specific book. Sometimes these are quoted from other media sources; sometimes they are produced for Amazon by its own

employees; mostly, they are commercial in nature, resembling the promotional material found on the back-cover blurb which McGrath decries. The primary objective of these reviews is to offer readers a professional, well-written, mostly positive review of the item for sale.

3. At a third level, expert users (reviewers) write free-form reviews and compile best-of lists which are ostensibly non-commercial and unbiased as much as they are opinionated. In the reviews, the expert users also rank the books on a numeric scale, assigning from one to five stars.

4. At a fourth level, lay users (readers) rate expert user reviews on a binary usefulness scale (useful or not useful). A summary of past ratings is available (for example, "5 out of 10 people found this review useful"). In addition, a "report this" feature (introduced by Amazon.com recently) allows lay users to report inappropriate content, which is then evaluated by Amazon's staff. The primary objective of these mechanisms is to offset the reviewers' power. Ostensibly, poor reviews will receive an inferior usefulness ranking, and thus future readers will be alerted and will pay less heed to such negatively rated reviews.

5. At a fifth level, some reviews are highlighted and given more visibility based on the usefulness scale and the ranking of the expert users who wrote them. Reviews that are found to be useful by more readers, or reviews that are written by credible reviewers (i.e., those whose reviews consistently get high rankings) are displayed first.

6. At the sixth level, the expert users (reviewers) are credentialed on the basis of the number of reviews they post and the usefulness of their reviews as evaluated by lay users. Reviewers who reach the top of the list receive visual endorsement of their status in the form of an icon next to their name that states "Top X Reviewer" (where X is 1,000, 500, 100, or 20), and reviewers who hit the top 20 get to write a profile page that includes a biography and a picture.

In this system, the notions of quality, reputation, and expertise are tightly bound and (as will be explained in detail below) are often conflated. There is a direct transition from evaluating the quality of an artifact (both quantitatively, using a numerical scale, and qualitatively, using words) to meta-evaluating the usefulness of that evaluation (on a binary scale) to evaluating the expertise of the expert user (based on the level of participation and the evaluation of that participation by other lay users). Higher levels of expertise and authority are directly tied to participation; formal credentials and reputation enter the system only at the first two of the six levels; expertise is measured on a continuous scale and is affected by levels of participation and expert-lay interaction and by lay-user review of expert-user activity.

To understand what expertise means in this context, we can compare this system with traditional book-review venues where only certified experts get to express their opinion (other than, perhaps, in a minuscule "letters to the editor" section where all readers can participate). We can think of this in terms of Collins and Evans's (2007) discussion of levels of expertise in regard to science. Collins and Evans distinguish "contributory expertise" (possessed by experts who contribute to some technical specialty—typically scientists in a field writing papers in that field) from "interactional expertise" (held by people who do not have enough expertise to contribute but who have enough to meaningful understand a field—typically peer reviewers of a field who do not publish in the field, or skilled sociologists of science who are able to understand the technical terms of a field but who are not themselves able to contribute). Last, Collins and Evans mention individuals who lack any expertise. If we think of book reviewing in these terms, we can see that the standard (offline) system of book reviewing uses reviewers who have contributory expertise. That is to say, editors typically farm out reviewing to people who have expertise in the topic being written about—typically to individuals who are experienced in the topic or who have written a book on the same subject. This applies even to novels; leading novelists are asked to review the works of other novelists. But even in the offline world this ideal of contributory expertise breaks down. Many reviewers for newspapers appear to be "professional reviewers" whose expertise lies precisely in writing book reviews and who may have no contributory expertise in the field being written about. Such reviewers may signal their lack of knowledge of the field, but can still write an influential review by, for instance, revealing how much they learned and recommending the book on that basis (and perhaps also for its literary quality). Such reviewers may have what Collins and Evans call "interactional expertise" in the topic of the book, and they will certainly have contributory expertise as expert writers of book reviews. It is extremely unlikely in the world of offline reviewing that a reviewer has no expertise either in the topic under consideration or as a writer of reviews.

The radical shift with online reviewing is precisely over the matter of expertise. It is possible in a system like Amazon's for a reviewer with no expertise at all in terms of Collins and Evans's discussion to write a review. The only thing required, other than basic language skills, is participation. Users demonstrate their expertise by writing reviews and compiling lists.

Whether the reviews are sufficiently expert is determined not by editorial fiat or by credentialed qualifications but by the community. The reviews are subject to the community's scrutiny: the community evaluates their contribution on a usefulness scale, which in turn subjects the review to increased or decreased levels of visibility. At the top of the expertise scale, Amazon tries to mimic offline mechanisms of accreditation by ranking the top reviewers and by giving visual indication of their status as such. We read this as an attempt to overcome the problem of information overload, to which this system, of course, is not immune. By giving the readers (lay users) a clear indication of the expert's status, the site saves the users the effort of trying to evaluate the experts—a saving that can be significant when one is browsing items that have hundreds or even thousands of reviews. There are two significant differences between this accreditation system and the traditional educational accreditation system: (1) The only accreditation of the "experts" is tied to participation, attests to "real-world" performance within the system, and does not rely on external factors such as credentialed markers of expertise or reputations garnered by subject area. This is very different from, say, a training certificate that attests to the completion of some educational activity but says little about the expert's ability to partake in hands-on activities. (2) The accreditation is relative. A reviewer's only way to reach the top is by topping other reviewers. As we have already seen, though, the actual working of this system is more nuanced. This system makes "expertise" substantially performative. Unlike in Collins and Evans's discussion, there is no way to determine whether reviewers are non-experts, whether they possess interactional expertise, or whether they possess contributory expertise other than by their performance.

In summary, the six degrees of reputation are underwritten by various mechanisms for displaying and expressing expertise through writing, reading, and evaluating reviews. Admittedly, the form of expertise being performed and displayed here is complicated—almost certainly a mix of reading and writing skills, real-world knowledge of the subject matter, and (occasionally) a degree of strategic acumen in selecting a book to attempt to review. But the radical nature of Amazon's system is precisely the attempt to reduce this complexity by substituting simple numerical rankings. Just as the complex success of a book is reduced to its sale ranking on Amazon, so the complex matter of reputation and expertise is reduced to simple numerical valuations—for the author, how many readers' reviews and how many stars the book gets, how many people found the review useful, and, if the reviewer is highly successful, his or her ranking among Amazon's "top reviewers."

At this point one may raise several objections. First, reviews of books are particularly powerful because they help establish the meaning of the artifacts in question more than, say, reviews of electronic gadgets. In Collins and Evans's terms, it is easier to gain contributory expertise in evaluating a printer or a food mixer than in evaluating a book on high-energy physics. While this might be true, it only serves to make our point that meaning-making, quality assessment, expertise, and reputation are conflated in such systems. A second objection will alert us to the point that people participating in this system are probably a self-selected group who, to begin with, ascribe weight to tiered systems of evaluation—in short, they seek meta-information. Are these observations, then, generalizable to larger audiences? Can we see parallels in other contexts? Anecdotal evidence suggests that we can. For example, we have seen "user reviews" scrawled on billboards in New York subway stations. It seems as if user input as a form of review is becoming an accepted form of cultural expression. A third objection will note that levels 4, 5, and 6 of the reputation system concern only a small group of lay experts, and that their activities do not necessarily reflect the activities of reviewers as a whole. This, of course, is true, but it does not limit our claim that the system offers built-in incentives for such activities, and that at the outset such incentives might be a part of the solution to the free-rider problem. All these claims are explored in detail below.

Empirical Findings: Strategies and Practices of Reputation Management

In this section we report conclusions from evaluations of more than 50,000 user reviews of more than 10,000 pseudo-randomly selected books and CDs. Our findings allow us to estimate that about 1 percent of all review data is duplicated, either verbatim or with some variations. The similar patterns observed across different genres of books and CDs suggests that our findings will be corroborated with larger data sets, but further research is necessary.[7] In most of these cases the copying is done by a person who writes an original review and copies it into reviews of different items with or without an attempt to modify his or her reviewer identity (i.e., name and email address). Importantly, we have not found further cases of reviews that were copied from a book or a product to a competitive book or product, as was done in the case that triggered our investigation.

The numerous cases of review re-use can be grouped into several categories:

• reviews copied from one item to another in order to promote sales of a specific item

- reviews posted by the same author on multiple items (or multiple editions of the same item) trying to promote a specific product, agenda, or opinion
- reviews posted by the same author using multiple reviewer identities to bolster support for an item
- reviews (or parts thereof) posted by the same reviewer to increase his own credibility and/or to build his identity.

We will now describe examples of how these strategies are employed.

Product, Opinion, or Agenda Promotion

The most common use we identified for duplicated reviews was the promotion of an agenda, a product, or an opinion. In several cases, review space was used simply for free advertisements or spamming. For example, in the review space for drummer Bill Bruford's CD *Earthworks*, which was sold on Amazon.com for $19.98, we found the following "review":

AMAZON IS THE WRONG PLACE TO BUY THIS / As with all the BB remastered Winterfold/Summerfold titles, this CD can be found @ $14.98 direct at billbruford .com.[8]

The same reviewer posted the same ad on other items and also participated persistently in attempts to boycott products and promote alternative products. Surprisingly the ad stayed in the review space; Amazon took no action against it. Not surprisingly, perhaps, users did not find it necessary to report this "review" as inappropriate content using the "report this" feature. Many of them probably found this information useful. Users we have talked to also reported more extreme cases found on other CD product pages, where sometimes the review space is used to post links that point directly to digital copies of the music itself (often as torrents, a file type supported by the popular peer-to-peer software BitTorrent, which is hosted on users' computers and not on a central server and is thus more resistant to the standard threats from copyright owners). Under this scheme, the Amazon system is used simply as an easy-to-use index of copyrighted music. (Ironically, providing such an index was the original goal of Napster before it was sued out of existence by the music industry.)

Another strategy is agenda promotion. Two examples from the political memoir genre follow. In one case, review space for many of the books critiquing President George W. Bush was used to promote a conspiracy-theory video. Variations of the following excerpt were copied multiple times:

… check out the movie 911 in Plane Site. www.911inplanesite.com. This movie shows suppressed news footage and video evidence from 9/11 that proves the government's "official story" is ludicrous. For example, the Pentagon was hit by something, but it wasn't a passenger airliner. That's why we haven't seen Pentagon footage. Please, check out this movie and show as many people as possible.[9]

This reviewer's strategy is sophisticated. By inserting this morsel of information into review space of books that are critical of the Bush presidency, the writer is sure to be speaking to a specific audience that is likely to be receptive to his agenda. A more complicated use of the same strategy was employed against Henry Kissinger. In this case, text was entered in the review spaces for all of Kissinger's books and for many books about him. The reviewer, of course, proposes an alternative in the form of another book:

... if you want the evil truth about Dr K and how he undermined the 1968 peace talks, read "No Peace, No Honor: Nixon, Kissinger, and Betrayal in Vietnam" by Larry Berman. This book explains how Nixon and Kissinger illegally colluded with SVN and Nguyen Van Thieu—he was told by Nixon via Anna Chenault to "hold on, we are going to win" and "you will get a better deal with us." So Thieu says he won't talk peace, Nixon wins, Kissinger openly changes sides after working with the Democrats, and together they crank up the war. The point is: The War could have ended in 1968 if it were not for this man—Dr Death himself, Henry Adolf Kissinger![10]

A subtler strategy in this category calls for posting the same review multiple times for the same item under different reviewer names. Our data include many cases of the use of "sock puppets"—virtual identities which are used to back up other users' opinions. With this strategy, the same review will be repeated under different reviewer names. For instance, the reviewer "bookcritic.com" shows up with the same reviews under the name "faithful _reader." This strategy is especially useful for popular items that receive dozens of reviews. In these cases the reader usually is not aware that the reviews are duplicated; the reviews are separated into multiple pages, and a limited attention span usually keeps a reader from browsing through more than a handful of reviews. It is important to note, however, that in several cases there seemed to be little commercial intent in this strategy. In some cases it was obvious that a user had, for some reason, lost access to his older identity, had built a new user profile, and had manually copied all his earlier reviews and posted them anew under the new identity. This suggests that reviewing is highly connected to identity building.

Identity Building

Several cases demonstrate how online reviewing is becoming an activity that, in addition to being aimed at assessing the quality of information goods for sale, provides a way for reviewers to construct their own identities. In one case, a reviewer took the opportunity to use review space for the purpose of communicating with the rock musician Bruce Springsteen. In a review titled "YO BOSS! UP OFF YOUR BUTT & GET BUSY REMASTER-ING!" the user writes:

Hey, BRRRRRUUUUUCCCE! What's the deal? Just about every major Top 40 artist has had their catalogs sonically updated EXCEPT YOURS. Why can we buy the "Tracks" editions, and get glorious HDCD-encoded sound, but "Touch" sounds like it's coming out an AM radio? OK, you did your best, but Dubya's back in the White House for the next four years, and Kerry's home in his underwear watching the Weather Channel. You should have plenty of time on your hands now … GET BUSY! Let's see some remastering![11]

Such direct dialog speaks both to the artist and to the audience. The reviewer tries to establish himself as an expert who is in a privileged position to tell the artist what to do. This is a reversal of the usual relationship between artist and audience.

Our interviews with authors corroborate the premise that artists often read user reviews, looking for feedback. Interestingly, one of our interviewees, the novelist J. R. Lennon, reported that he stopped doing so when he understood that some reviewers were not seriously assessing the literary merits of the book but rather were pursuing their own agendas. Lennon had initially liked the access to readers' comments, and indeed on receiving his first negative comment he actually posted a reaction to it. He told us:

I wanted to defend myself to people who were being really nasty in their customer comments. . . . I was shocked that people were not discussing the book in any analytical or rational way. They were just sort of blurting out their gut reaction to it under the cloak of anonymity. The reaction to it was shocking. I think that's why I posted a comment. Of course that's not shocking anymore, because that's what the internet is full of now, and customer comments are sort of accepted as a reasonable form of book review.

Soon, however, Lennon asked Amazon.com to withdraw his comment:

I called Amazon and asked them to remove my own comment on my book, [I felt that] the whole concept of commenting on your own book on the internet was really stupid.
Q: Why is it stupid?
Because if somebody wants to ask me about it they can, I feel the book is self explanatory. If someone wants to ask me about the book I don't mind talking about it. So when I talked to the Amazon guy about removing it, he said "yeah everyone's doing it."

Here an author can be seen adjusting his own practices to life in the new and unaccustomed digital medium. There is no doubt that this author's experience of the new reviewing system has, over time, become largely negative:

Basically I hate Amazon customer comments. I despise them, because any anonymous prick can wreck your day without even looking at your book. If your book is coming out at the same time as someone else's they can just send their friends in to sabotage your rating.

Lennon found that early on he himself was drawn into the game of writing reviews simply to defend a book against what he perceived as unfair comments:

The only review I ever remember doing was for a guy named Stewart O'Nan, who was in the writing program in Cornell. He lives in Connecticut now. He wrote this terrific book [*A Prayer for the Dying*], a very short novel about a priest in a small village in Wisconsin when plague and fire hit at the same time. It's very dark and in second person. And someone had written in an Amazon customer comment who hated Stewart O'Nan, saying it was plagiarized. . . . It was really nasty and had nothing to do with the book, and I actually wrote a review defending the book.
Q: Did O'Nan ask you to do that?
No, I barely know him. I was furious.

We are finding many users have felt (and occasionally succumbed to) the temptation to attack others by posting negative reviews, the temptation to write in support of oneself, or the temptation to fend off attacks on one's friends and colleagues. It is easy to do, and—with the possibility of posting anonymously—it is almost free of cost. In the course of our research, we have encountered numerous stories of publishers, friends of authors, and others engaging in these practices. It seems likely that a significant percentage of Amazon book reviews (and probably of online product reviews in general) are generated in such a way. As a result, the worth of even positive reviews becomes more and more difficult to gauge.

One interviewee, an academic author, was frustrated when she found out that a person who had reviewed her book favorably was less genuine than she had once thought. She commented to us in an e-mail:

[This reviewer] wrote a very lengthy and kindly review of my own book on Amazon, but upon surfing semi-randomly over the years for books by others that interest me, I see he has been everywhere, and every damned book he reviews gets 5 stars. It turns out he is "reviewer #47" [and] has written over 600 (five-star) book reviews. Is he really so easy to please? Has he also read hundreds of other books, which don't merit five stars and which he generously declines to review so as not to post criticism? Does he perhaps pilfer his prose from other places? It might bear looking into. I can say that I never paid a dime for his flattery, but maybe he sells it to others.... I'm curious as to what he gets out of all these five-star reviews. Does he hope to ingratiate himself with authors, is he a frustrated ABD, or is he just a history whore?

Our initial data suggest that indeed many of the reviews generated by top-ranked reviewers are positively biased, and that inadvertently these reviewers might continue to generate this bias in order to maintain their high rankings. Further research is needed, however.

An individual who reviewed several Tom Hanks–Meg Ryan movies posted the same review for *Sleepless in Seattle* as for *You've Got Mail*. Each was "a film about human relations, hope and second chances, but most importantly about trust, love, and inner strength."[12] As we know, especially with the demands for producing one blockbuster after another, Hollywood movies are sometimes strikingly similar, yet posting the same review for two different films suggests that the reviewer is interested less in accurate representation of a movie's content or qualities and more in the sort of reputation and identity that he or she can build as someone who posts numerous reviews.

One might think that there is little harm in posting duplicated reviews like those cited above. After all, a reviewer's statements could independently be true of more than one movie, and there are no serious consequences if they are not. A reviewer who regularly posts music reviews on ACIDplanet.com, an online music site where musicians can post their own music and can review and download the music of others, told us that he posted only positive reviews, knowing that the chart rankings on this site were based mainly on such customer reviews. He had found that if he reviewed other artists favorably he would get positive reviews in return (a phenomenon that users call "R = R"). But such positive reviewing stretched his capacity to find enough words to post the same plaudits without repeating himself:

If I feel that I've written a lot, the same thing, I'll get the dictionary out and change a few words. And I just, say, "I'll change 'creative' to 'inventive.' Oh, that's good! I'll change, you know, 'cool' to 'groovy.' Oh, that's good, you know!

This artist said that he found it unethical to copy a review, but that he encountered copied reviews often:

And there's people who post the same review over and over again.

Q: You've seen that?

Oh yeah, yeah. I've gotten a review, and it's just all positive stuff, and then I've gone and I've listened to someone else's music, "Hey, you know, this guy reviewed—oh it's the same review!" And he just cuts and pastes, you know, a positive review. Or a negative review, for that matter. Uh, I see that a lot.

The problem takes on even more salience when the product being reviewed is not a book or music but a drug. We found such a case on DrugRatingZ.com, a site used by people to post reviews of prescription drugs. In this example, the review concerned the drug known by the generic name Lorazepam or by the brand name Ativan:

Works well if you have never used benzodiazepines regularly. It is very habit-forming and stops being effective if used too often. Makes you very drowsy, but can be extremely useful for panic attacks.[13]

At a first glance this seems to be a simple, authoritative statement, but further browsing on the site reveals a review for the drug known by the generic name Clonazapam or the brand name Klonopin that reads as follows:

Works well if you have never used benzodiazepines regularly. It is very habit-forming and stops being effective if used too often. Makes you very drowsy, but can be extremely useful for panic attacks.[14]

A short consultation with a medical doctor reveals that these drugs are based on the same molecule. Arguably, under those conditions, it makes sense to have the review copied, but does the fact that in the latter case we're dealing with reviews of drugs that might have significant effect on people's lives (or deaths) suggest that the norms for such reviews should be stricter than those for book and music reviews? We came across this one case by accident, and further research of drug product sites is clearly called for.

Another common form of copying we found was when some of the review space was used to write summaries of a topic that were then used over and over again. For example, a top-20 reviewer with more than 1,350 reviews has made a point of reviewing books dealing with black history, and in several of his many reviews has used the same paragraphs as excerpts within specific reviews. Is there anything wrong with doing so? Should such self-plagiarizing be condemned? The answers to these questions are not clear and depend on the framework of analysis. Samuelson (1994) proposes to consider such cases within the framework of fair use.

What is clear, however, is that the aforementioned individual's well-written reviews (which, to be sure, demand a lot of time and attention) have been highly ranked, allowing him to sustain his status as a top reviewer and to continue to spread his ideas. In cases like this we see how strongly a reviewer's strategies of identity building can be tied to agenda promotion.

Self-plagiarism appears to be common in other settings too. The journal *Nature* recently published a special report, titled "Taking on the cheats," that discusses self-plagiarizing in academic settings and makes the following assertion:

Self-plagiarism, in which authors attempt to pass off already published material as new, is a particular problem. In an increasingly competitive environment where appointments, promotions and grant applications are strongly influenced by publication record, researchers are under intense pressure to publish, and a growing minority are seeking to bump up their CVs through dishonest means.... And although most cases are never discovered, almost all of the editors and publishers contacted by Nature agreed that self-plagiarism is on the rise. (Taking On the Cheats 2005)[15]

The question that remains open is the question concerning norms. Unlike in academic settings, where the motivation for such practices is clear and there is good moral and practical basis for scorn, here it is not clear what the standard should be. We will return to this point later.

Why Write Reviews?

All book reviewing takes time. Obviously, professionals who write reviews for, say, the *New York Times* get paid. But more importantly, reviewing for such a prominent newspaper adds to their identity as authors. A well-written review by a well-known author in a well-regarded place can itself become a topic for further discussions, and on rare occasions a literary event in its own right. The *Times* will often review a book only if a well-known person with special relevance for the topic can be located and is willing to do the review. One novelist told us that there is also a possibility of "payback": if you review for the *Times*, you are more likely to get your own work reviewed there. Academic book reviewing pays nothing (other than a free copy of the book) and counts for little on academics' résumés. Again it would seem that reviewers use such a medium to build their academic identities so that they can act as gatekeepers evaluating new work in their fields.

Why do people engage in online reviewing? What is in it for them? One possible answer is that it seems that some reviewers on Amazon.com hope to break into the offline world of paid reviewing. For instance, in an example of what we might call "market signaling," a top-20 reviewer below the age of 30 declares in his profile: "My objective is to do what I do here for a living." With hundreds of high-ranked reviews under his belt, this reviewer is signaling to the market that he is worthy of a job as a professional book reviewer. In other cases, people (many of them adolescents) write reviews as a social practice and as conversation pieces with their significant others. It seems to be empowering to them to see their name and review on a website attached to a famous movie. We have witnessed cases in which children wrote DVD reviews for movies they had seen in a theater. They would later send links to their reviews to their friends, and take pride in their ability to "publish." In other cases, our observations of students who participated in online book and CD reviewing as a form of writing practice suggest that such activities are empowering to them, especially when those reviews received high rankings on the usefulness scale. In those cases, young students who have never "published" before were very excited to see that their reviews were read by many people. Reportedly, that moment of interaction demonstrated to many of these students how effective their writing could be.

Users as Active Agents

Taken together, the examples cited above suggest that the new (and still evolving) systems of online product reviews do not function straightfor-wardly as a democratic pooling of expertise. As we have shown, people who write product reviews are engaged in a variety of activities: promoting agendas, making personal attacks, boosting their own and others reputa-tions, building their own identities as reviewers, experiencing for the first time the empowerment of publication, and so on. Of course the majority of reviews can probably be taken at face value and are reviewers' attempts to give their own honest appraisals of products. But the multiple uses of the new digital technology that we have encountered demonstrate a theme that is becoming increasingly prominent in the history and sociology of technology: the power and agency of users (Kline and Pinch 1996; Oud-shoorn and Pinch 2003). Not only are users a source of new innovations (von Hippel 2005); they can radically reinterpret the meaning of existing technologies—a process known in the sociology of technology as "inter-pretative flexibility" (Pinch and Bijker 1987). For example, early rural users of Ford's Model T found that the automobile could be used to power vari-ous appliances (Kline and Pinch 1996). In online reviewing, the "user" is a multi-faceted category. It can be argued that the user is really a complex of different enactments (reading, writing, interpretation, protest) that may be embedded in the same person or in the collective. Some users find loopholes in the system's design that help them game the system to their own ends, but at the same time the same user may give what he or she considers an honest appraisal of a reader review. Just as identities are complicated in the online world, so too are users.

Continuities and Discontinuities with Earlier Models and the Offline World

New Affordances?

To what degree do the copying behaviors we have documented hinge on specific digital system designs, and to what degree do they represent departures from existing offline models? Clearly the changed materiality of the online encounter is mediated by technology. As a way of helping to think about the role of technology here, we would like to start with the concept of technological affordances. 'Affordance' was a term originally coined by perceptual psychologists and then used in the fields of cognitive psychology, environmental psychology, industrial design, and human-computer interaction. Introduced by the psychologist James J. Gibson in 1966, it was explored more fully in Gibson's 1979 book *The Ecological Approach to Visual Perception*, which investigates affordances for action (i.e., the empty space in a door-path affords walking through the door). Donald Norman further developed the concept in *The Psychology of Everyday Things* (1988). His definition of an affordance is the design aspect of an object or a system that suggests how the object can and should be used: "... the term affordance refers to the perceived and actual properties of the thing, primarily those fundamental properties that determine just how the thing could possibly be used. A chair affords ("is for") support and, therefore, affords sitting." (ibid., p. 9) However, this definition of affordance is too tied to the psychological literature and assumes too essentialist a use of a product for our purposes. We want to make the notion of 'affordance' consistent with the sociology of technology. It is clear that users come up with new and unexpected uses of technology, and that it is problematic to read off a definitive or "best" use from the design of an artifact. How a chair will be used depends on the context in which it is used—for example, some chairs are always used as footrests and never sat in. Furthermore, what a technology is good for, or what can be done with it, is in itself a process of social construction. Affordances cannot be analyzed by looking at the technology alone. The social, historical, economic, and legal contexts are decisive in shaping how technologies are interpreted.

An instructive example is that of Minitel, an early French videotex system (Schneider et al. 1991). Often seen as a precursor to the Internet,

Minitel enabled users to use their home terminals to exchange and post information about restaurants and other products and services (including, famously, sexual services). The rapid take-up of the interactivity dimension of Minitel surprised the engineers who had designed the system. They had included interactivity because it was technically feasible to do so, but they did not expect it to become the defining feature of Minitel. Other videotex services (e.g., British Telecom's Prestell) had no such interactivity, and their users were limited to broadcast announcements about television programs and services, posted from only a few central locations. If we think of this in terms of affordances, we can say that use is constrained by the affordances but that users determine the actual use of the technology.[16] For example, the affordance of a chair allows it to become a footrest but does not allow it to become a Minitel terminal.[17] We should be careful to distinguish physical limitations on affordances from social (including political, cultural, and economic) limitations. For example, the physical limitation on the affordance of a book review does not constrain its length (one could write book reviews as long as novels), but the social limitation does constrain length.[18] In summary, we can say that affordances are socially constructed, and we can ask what new affordances the tiered reputation systems offer. How do authors and readers construct these affordances, and how are they reflected in their practices? What affordances make it easier to abuse the review space? Are there parallels in the offline world?

One of the main affordances the digital systems allows is that the barrier to entry into participation is dramatically lowered. All one needs to take part is access to a computer connected to the Internet and some rudimentary keyboarding skills. No one cares if you have ever written a book review, what your age is, what your qualifications are, or whether you can write. Even children can participate. The Internet, of course, does not permit limitless participation (there are "digital divides" both in access and in skill), but it does give greater affordance to participation than any offline medium. Perhaps this is best demonstrated by the removal of geographical barriers. People around the world can participate in one conversation, regardless of their physical location.

A second difference in affordances concerns the length of the reviews. In the physical world, book reviewing is constrained by length, especially in important publications such as the *New York Times*. One novelist told us:

This is not to say that longer does not always equal better because short reviews can force the reviewer to be more meaty and succinct and lead to a more quotable quote for blurbing, but there is no doubt in the *Times* longer and more towards the front part of the weekly review is best.

In online reviewing, a pre-specified word-length limitation is removed. A review can consist of a few sentences or can be an extended essay—it is the reviewer's choice. The review system gives affordance to very lengthy reviews. In practice, we find, very few online reviewers actually take up the offer of extended space. Most reviews are rather short, and yet the lack of length limitation allows a multitude of review lengths to co-exist.

A third and more important physical limitation that is removed in the online world is that imposed by time. One can take as long as one likes to produce a review, and can post it at any time. Some books are best read at a rate of only a few pages a day, but this is not how a reviewer with a deadline to meet reads them. Online review systems, with their lack of deadlines, give affordance to a greater variety of reading practices among reviewers.

A fourth affordance facilitated by online review systems is the ability to refer to earlier reviews so a dialogue of sorts can ensue. Several instances show how reviewers disagree with earlier reviews—something that is rare in the physical world of reviewing. Interestingly, when Amazon changed the ordering of the reviews from chronological to "most useful" it messed up this affordance. This affordance can be abused too. One interviewee reported a case in which "anarcholiterists" systematically discredited earlier positive reviews of a book in an attempt to get revenge on the author, who had criticized their group in a newspaper article.

As we saw earlier, ostensibly a significant affordance of the new class of online review systems is the ability to copy or to cut and paste reviews from one item to another at a negligible cost. This affordance is the basis of many of the practices we described in detail above. But is this really a new affordance? One of our interviewees, the novelist Brian Hall, recounted a story of offline review plagiarism in English newspapers stemming from a blurb for one of his books:

I had noticed six months or so after the publication of the hard cover, when the clippings of the really little papers are coming in, that there was two or three [of these duplicated], and it may be of course some of these little papers in England, like here, are just syndicated with each other. So it may have been one person who did this. That then got farmed out to like three or four of these different little papers. It was basically just publishers blurb. "This in enthralling journey into. . . ." That kind of publisher talk. And so they just copied that, and at the end they added one sentence of like "so all in all a good read," and then signed it and got paid for it. I chuckled when I saw it and stuck it in a box. "Oh look some lazy guy scamming his editor, and the editors to lazy to even notice." . . . And then six months later when [my publisher] came out with the paperback, and they were putting together the back and trying to quote things, sure enough they quoted one of these things. Of course it sounded great, it was the publishers copy! And there too the editors of the paperback hadn't noticed that it was exactly the same language as the flat copy on the hardcover. So I called them up and said, "Yes this is a wonderful review, it sounds great. I don't think we should have it because if you'll just turn to your hard cover you'll notice that we wrote that." So they took it off.

In summary, we are not claiming that plagiarism and copying do not take place in the offline world (clearly they do); it is just that in the online world copying is easier. Speaking in the language of affordances, we can say that the new online recommendation systems offer opportunities for greater participation and for larger variance in review style and length, but also for easier copying and abuse. More people participate in the review process, but more foul play occurs.

Cross-Influences

An often overlooked aspect of the new systems is their influence on older systems. Such cross-influence can work in any of the six degrees of reputation. For example, one author encouraged all his friends to buy his book at a special "Amazon hour." The resulting spike in sales brought the book to the fourth position among the best-sellers in its category. The author then received "Amazon best-selling author" status, which has no time limit. This author can now enjoy that status for the rest of his career, online or offline, despite the fact that his best-selling tenure was only a few hours long.

We have been in offline bookstores that allow users to "publish" book reviews written on "sticky notes" (figure 11.4). This cross-influence shows the profound ways in which online participation changes offline practices. Once user reviews have been legitimized online, there is little resistance offline to users' intervening in a process that formerly was a simple business transaction between a shop and a customer. Evidently, shops that allow this practice do so in the belief that, overall, the shopping environment they foster will allow their business to flourish. At the same time, they are engaging in "me-too" behavior, offering their customers an online-like experience.

As we noted earlier, we have seen product reviews written on billboards in New York subway stations. In those cases, random users, some anonymously and some by name, chose to share their opinions of a well-known

Figure 11.4
Sticky notes as users' reviews.

manufacturer's stereo system. Started by a negative review, a dialog of sorts ensued. What these cases show is that the boundaries between on-line and offline practices are permeable if not non-existent, and that cross-influences work both ways.

Conclusion

Clearly, the world of online recommendation systems is a world in transition. New systems in the changed materiality produced by online purchasing introduce new technological affordances that make certain activities easier then others. Authors, artists, editors, and users find creative ways to interpret those technologies and often use the system in richer ways then the designers originally intended.

Laurence Lessig's (2000) model of norms, law, markets, and code and their interplay is useful as a way of making sense of this changing world. Lessig argues that the above four categories constitute regulatory regimes that influence the behavior and the freedom of individuals within a society, and particularly in cyberspace. Law, norms, market forces, and architecture (code) work together to set constraints and limitations on what we can or cannot do. All forms of regulation on human practices, they determine how individuals, groups, organizations, communities, or states regulate and are regulated. Lessig argues that we must consider these four forms of regulation as they pertain to one another because they interact and can compete, and that each of these concepts can reinforce or undermine another. Lessig further describes in detail how power nexuses (such as the movie industry) use their control to influence a combination of these categories in order to fortify their long-standing interests. In Lessig's view, we should understand those interactions, and we should intervene in them if we want to create a more equitable and just society.

We have seen that what is ostensibly an activity within a market (online book and CD sales) is actually influenced directly by code (the technological affordances that the system offers), by norms (which, as we have seen, are often not yet stabilized), and to a lesser degree by the law. (The legal aspects of this system are beyond the scope of the present work.) Indeed, we can expect all four components to come into play as the system matures. Lessig's account, however, does not place enough emphasis on the power of users to interpret technologies and their use in ever-new contexts. The user practices we have documented are not things that can be read formally into the system in advance. User practices in the form of local resistance, much before they stabilize into norms, can have a significant influence on the stabilization of a technology. Norms, laws, markets, and code—in essence the things that make up our material infrastructure—are often, as this case shows, in flux in the early days of a technology, and will further change as users evolve new practices and as designers and operators respond to users. And here our intervention as scholars produces another interesting potentiality in the system. The sorts of practices we have documented in this chapter could have been documented by Amazon (and for all we know may indeed have been documented). Furthermore, if we can write an algorithm to detect copying, then it is possible for Amazon to use such algorithms to alert users to copying and, if necessary, to remove material. If Amazon were to write such an algorithm and to remove copied material, that would not be the end of the story. Users will adapt to the new feature and will, no doubt, look for new ways to game the system. But we assume that at some point, as with offline reviewing, the system will become fairly stable.

In this chapter we were able, by studying a technology in transition, to focus on novel user practices. What we hoped to gain from studying these cases is a better understanding of the system as a whole—not only of where it "fails" but also of its potential when it functions properly. There is little doubt that online book reviewing is here to stay and that it is already changing many aspects of the book world. Whether these changes are for the good depends significantly on users. As we have shown, technology unfolds as user practices evolve. Users can and must play a part in shaping this technology, and can and must exercise their agency in living in a material world.

Acknowledgments

We wish to thank Daria Soronika of Cornell University's Computer Science department for her work in the area of plagiarism-detection algorithms that were essential to our project. We wish to thank Joseph (Jofish) Kaye and Phoebe Sengers for alerting us to several examples of review copying practices. We wish to thank our interviewees and students for participating in this study. A draft of this chapter was presented at the Annual Meetings for the Society for the Advancement of Social Economics in Budapest, July 2005; we wish to thank SASE participants for useful comments. Further drafts have been presented at Cornell University during the Cornell Conference on Economic Sociology, September 2005, and as part of New York University's Information Technology and Society colloquium series. We wish to thank the participants of these forums for insightful ongoing discussions, especially Mike Lynch, Bruce Lewenstein, Helen Nissenbaum, Gaia Bernstein, Anindya Ghose, Kim Taipale, Eddan Katz, and Jack Balkin. We also wish to thank Harry Collins and Robert Evans for a useful in-depth discussion concerning the role of reviews in meaning-setting and purchasing behavior, and John Lesko and his team for their detailed review of this work. We would also like to thank Knut Sorenson for helpful comments on an earlier draft.

Appendix: Technical Description of Copy-Detection Algorithm

The copy-detection algorithm we used is based on a modification of the winnowing algorithm (Schleimer et al. 2003). Originally the algorithm worked on the level of symbols. Daria Soronika's adaptation of this algorithm to text problems moves it to the level of words and sentences. Winnowing has two parameters: k and t. Sequences of k symbols are called k-grams. Each k-gram can be converted into a number by some hash function. A document is represented by the set of fingerprints—a subset of all possible k-grams extracted from this document. The main idea of winnowing is the following: For each window (sequence) of size $t > k$ there is at least one *k*-gram from this window that is chosen into the set of finger-prints. This *k*-gram depends only on the content of the window and does not depend on its place in the document. Therefore, if two documents share substrings of length at least *t*, then their sets of fingerprints will share some fingerprints.

In summary, in our version of the algorithm we are comparing texts based on sentence level. Therefore, we don't need *k*-grams crossing sentence borders. And *k*-grams crossing word borders don't make much sense, so we redefine *k*-gram as a sequence of *k* words, not *k* symbols, and we are considering only those *k*-grams that fit inside some sentence. Sentences are considered similar when they share at least one non-widespread *k*-gram. By our definition these are *k*-grams used by three or fewer authors.

Notes

1. For an early review, see Resnick and Varian 1997.

2. For a neutral account of larger trends and the rise of the network society, see Castells 1996.

3. Source: http://www.amazon.com/gp/product/customer-reviews/0743244508/ref=cm_cr_dp_2_1/102-6460714-8298569?%5Fencoding=UTF8&customer-reviews.sort%5Fby=-SubmissionDate&n=283155

4. http://www.amazon.com/exec/obidos/tg/detail/-/0674008898/qid=1124793312/sr=8-2/ref=pd_bbs_2/102-6460714-8298569?v=glance&s=books&n=507846

5. http://www.amazon.com/exec/obidos/tg/detail/-/1931140170/qid=1124793411/sr=1-1/ref=sr_1_1/102-6460714-8298569?v=glance&s=books

6. Source: http://www.amazon.com/gp/product/customer-reviews/1931140170/ref=cm_cr_dp_2_1/104-8520953-6108702?%5Fencoding=UTF8&customer-reviews.sort%5Fby=-SubmissionDate&n=283155

7. Further research is needed primarily to establish correlations between review type and copying practices. Further research is also necessary to fully evaluate editorial review practices and its uptake within user reviews. As the research progresses, it is our intention to stabilize the software and offer it for free download as an open source package so that interested parties can corroborate our data and use it for further research.

8. http://www.amazon.com/gp/cdp/member-reviews/AGEJE3WH26UBR/ref=cm_cr_auth/103-4232317-9497407

9. http://www.amazon.com/gp/cdp/member-reviews/AF4XMZXOYWR7L/ref=cm_cr_auth/103-4232317-9497407?%5Fencoding=UTF8

10. http://www.amazon.com/gp/cdp/member-reviews/AHSOTSV5VRTAH/ref=cm_cr
_auth/103-4232317-9497407?%5Fencoding=UTF8

11. http://www.amazon.com/exec/obidos/ASIN/B0000028SR/qid%3D1115574821/
sr%3D11-1/ref%3Dsr%5F11%5F1/103-4232317-9497407 and http://www.amazon
.com/exec/obidos/ASIN/B0000026E5/qid%3D1115575375/sr%3D11-1/ref%3Dsr
%5F11%5F1/103-4232317-9497407

12. http://www.amazon.com/exec/obidos/ASIN/B0000AOV4I/qid%3D1115573323/
sr%3D11-1/ref%3Dsr%5F11%5F1/103-4232317-9497407 and http://www.amazon
.com/exec/obidos/ASIN/6305368171/qid%3D1115573451/sr%3D11-1/ref%3Dsr
%5F11%5F1/103-4232317-9497407

13. http://www.drugratingz.com/ShowRatings.jsp?tcvid=618

14. http://www.drugratingz.com/ShowRatings.jsp?tcvid=754

15. A section of the report discusses anti-plagiarism measures taken in the e-print repository arXiv.org. For this project we have used an adopted version of the software developed and used by arXiv. See appendix for details.

16. A related notion here is Madeline Akrich's (1992) notion of a technological "script." As with a literary script, designers try and designate particular patterns of use which are "scripted" into the technology.

17. This issue is much debated within the field of Science and Technology Studies. See, e.g., Grint and Woolgar 1992.

18. For a nuanced study of how social and technical processes work together to enable new sorts of affordances in online newspapers, see Boczkowski 2004. Hutchby (2001) uses the notion of affordances in discussing how an interactionist approach to telephone conversations can be integrated with the sociology of technology.

References

Akrich, M. 1992. The de-scription of technological objects. In *Shaping Technology/ Building Society*, ed. W. Bijker and J. Law. MIT Press.

Amazon.com. http://www.amazon.com.

Benkler, Y. 2002–03. Coase's penguin, or Linux and the nature of the firm. *Yale Law Journal* 112, winter: 369–389.

Benkler, Y. 2006. *The Wealth of Networks*. Yale University Press.

Boczkowski, P. 2004. *Digitizing the News: Innovation in Online Newspapers*. MIT Pres.

Castells, M. 2000. *The Rise of the Network Society*, second edition. Blackwell.

Collins, H., and R. Evans. 2002. The third wave of science studies. *Social Studies of Science* 32: 235–296.

Collins, H., and R. Evans. 2007. *Rethinking Expertise*. University of Chicago Press.

Coleman, G. 2005. The Social Production of Freedom. Ph.D. dissertation, University of Chicago.

Chevalier, J., and D. Mayzlin. 2003. The Effect of Word of Mouth on Sales: Online book Reviews. NBER Working Paper 10148Document1. Available at http://www .nber.org.

Gibson, J. 1979. *The Ecological Approach to Visual Perception*. Erlbaum.

Grint, K., and S. Woolgar. 1992. Computers, guns and roses: What's social about being shot? *Science, Technology and Human Values* 17: 368–379.

Harmon, A. 2004. Amazon glitch unmasks war of reviewers. *New York Times*, February 14.

Himanen, P. 2000. *The Hacker Ethic and the Spirit of the Information Age*. Random House.

Hutchby, I. 2001. *Conversation and Technology*. Polity.

Keillor, G. 1985. *Lake Wobegon Days*. Penguin.

Kettlewell, B. 2003. *Electronic Music Pioneers*. ArtistPro.

Kline, R., and T. Pinch. 1996. Users as agents of technological change: The social construction of the automobile in the rural United States. *Technology and Culture* 37: 763–795.

Lessig, L. 2000. *Code and Other Laws of Cyberspace*. Basic Books.

Lessig, L. 2004. *Free Culture: How Big Media Uses Technology and the Law to Lock Down Culture and Control Creativity*. Penguin.

Marcus, J. 2004. The boisterous world of online literary commentary is many things. but is it criticism? *Washington Post*, April 11.

Norman, D. 1988. *The Psychology of Everyday Things*. Basic Books.

Oudshoorn, N., and T. Pinch. 2003. How users and non-users matter. In *How Users Matter*, ed. N. Oudshoorn and T. Pinch. MIT Press.

Pinch, T., and F. Trocco. 2002. *Analog Days: The Invention and Impact of the Electronic Synthesizer*. Harvard University Press.

Pinch, T., and W. Bijker, eds. 1987. *The Social Construction of Technological Systems: New Directions in the Sociology and History of Technology*. MIT Press.

Pink, D. 2005. The book stops here. *Wired* 13, no. 3: 99–103. Available at http:// www.wired.com.

Raymond, E. 1999. *The Cathedral and the Bazaar: Musings on Linux and Open Source by an Accidental Revolutionary*. O'Reilly.

Resnick, P., and H. Varian. 1997. Recommender systems. *Communications of the ACM* 40, no. 3: 56–58.

Safire, W. 2005. Blurbosphere. *New York Times*, May 1.

Samuelson, P. 1994. Self-plagiarism or fair use? *Communications of the ACM* 37, no. 8: 21–25.

Schleimer, S., D. Wilkerson, and A. Aiken. 2003. Winnowing: Local algorithms for Document Fingerprinting. In Proceedings of the ACM SIGMOD International Conference on Management of Data.

Schneider, V., J. Charon, L. Miles, G. Thomas, and T. Vedel. 1991. The dynamics of Videotex development in Britain, France and Germany: A cross-national comparison. *European Journal of Communication* 6: 187–121.

Strauss, B. 2004. *The Battle of Salamis: The Naval Encounter That Saved Greece—and Western Civilization*. Simon & Schuster.

Taking on the cheats. 2005. *Nature* 435, May: 258–259.

Turner, F. 2005. Where the counterculture met the new economy: Revisiting the WELL and the origins of virtual community. *Technology and Culture* 46.

von Hippel, E. 2005. Democratizing innovation: The evolving phenomenon of user innovation. *Journal für Betriebswirtschaft* 55, no. 1: 63–78.

12 Transfer Troubles: Outsourcing Information Technology in Higher Education

Nicholas J. Rowland and Thomas F. Gieryn

This chapter is about transfer—but not the kind of transfer one might expect from two specialists in science and technology studies (STS). We do not consider "technology transfer" the movement of scientific knowledge from laboratories into industry for development and commodification (Etzkowitz 1990). Nor do we consider the international transfer of scientific knowledge and technology from developed to less developed countries (Shrum and Shenhav 1995). Our focus is on the transfer of processes and practices from one organizational setting to another. Specifically, we consider a transfer that has, in recent years, thoroughly changed how universities and colleges conduct their organizational routines: outsourcing information technology (IT) to external vendors that sell diverse packages of software for data management. In general, outsourcing transfers some function or operation from within an organization to outside vendors who assume responsibility for handling the tasks—and who seek to profit from providing these services. Lately, colleges and universities have outsourced a wide range of auxiliary services in order to save money and enhance performance: college bookstores, residence halls, food services, student health care, campus security, real-estate management, laundry services, custodial labor, vending machines, parking lot management, printing, alumni relations management, bus services, golf courses, building design and maintenance, travel agencies, and even movie theaters (Biemiller 2005; Blom and Beckley 2005; Davies 2005; Gose 2005; Lund 1997). Perhaps the most visible marker of campus outsourcing is the "food court": golden arches and branded pizzas have migrated from shopping malls and airports to dormitories and student activities buildings, replacing cafeterias that once were fully operated by the university itself. As universities and colleges increasingly begin to outsource information management systems, many of them are discovering (often painfully) that IT is not exactly like hamburgers.

Our analysis centers on the recent experiences of Indiana University, which has traded in its long-standing "legacy" information systems for new integrated software packages marketed by PeopleSoft—one vendor of Enterprise Resource Planning systems (ERPs). Indiana University is a large, state-assisted research university. Its Bloomington campus has more than 35,000 students and an annual budget for campus operations of more than $1 billion. Like most public universities, Indiana University offers a huge array of undergraduate degree programs as well as graduate and professional schools. The Kelley School of Business is among the most prestigious and best-endowed professional schools on the Bloomington campus.

When the university switched over to PeopleSoft in 2004, transfer troubles began to appear as administrators and counselors at the Kelley School of Business sought to implement the crown jewel of its undergraduate degree program: "I-Core," an "integrated core" of courses in finance, marketing, operations and strategy that enables students to relate what they have learned in one course to problems raised in another. These courses are rigorous and demanding, and successful performance in I-Core is a student's only pathway to an undergraduate degree in business at Indiana University. Much of the prestige of the Kelley School stems from I-Core, and graduated students praise I-Core as good preparation for the real world of interconnected business challenges. However, I-Core is administratively organized in ways that set it apart from the curriculum of courses in all other teaching units at Indiana University. As we shall see, the distinctive process through which students register for courses in I-Core is not well aligned with the PeopleSoft system purchased by Indiana University, which was designed to handle the ordinary class registration procedures standardized throughout the rest of the Bloomington campus. Transfer troubles ensued, and officials at the Kelley School squawked about the time and money needed to protect their precious I-Core.

Our discussion of transfer troubles as a chronic feature of outsourcing will draw on two bodies of research. For new institutional economists, transfer of organizational practices creates economic costs that are calculated by organizations trying to decide whether or not to outsource. For STSers, transfer creates cognitive and communication problems stemming from the impossibility of articulating exactly and fully the practices to be outsourced.

Why Universities Outsource IT

When critics bemoan the corporatization of higher education, they often point to outsourcing as an instance where local traditions and collegiality have given way to the bottom line. Never mind that students now might actually prefer to eat franchised fast food instead of the "mystery meat" traditionally served up on ancient steam tables run by university-owned food services. Universities, in effect, have little choice: retrenchments in state support for higher education, coupled with increasing competition both for those qualified students with the ability to pay skyrocketing tuition and fees and for lucrative research grants, have made outsourcing a necessary tool of fiscal management (Barnett 2000; Breneman, Finney, and Roherty 1997; Goldstein, Kempner, and Rush 1993; Slaughter and Rhoades 2004). The decision to hire PeopleSoft to run a university's information system is easily justifiable these days in terms of good business practices—and it is no accident that ERPs were first developed to serve profit-seeking corporations (Klaus, Rosemann, and Gable 2000).

The use of ERPs is so widespread in business—and so profitable for its vendors—that the early 21st century is being called the "Enterprise Resource Planning Revolution" (Ross 1998). In the United States, a majority of Fortune 500 firms are running ERPs (Kumar and van Hillegersberg 2000; Langenwalter 1999; Norris et al. 2000), and globally more than 60 percent of multinational firms have adopted them. In 1998, ERP vendors made nearly $17.5 billion in profits (PricewaterhouseCoopers 1999). By 2000, approximately $40 billion in revenue was generated by ERP software producers (including their consulting services) (Willcocks and Sykes 2000). The use of ERPs is so widespread that even other software companies purchase ERP systems to manage their businesses, including such industry giants as IBM and Microsoft (O'Leary 2000). Organizations in higher education may have jumped on the ERP bandwagon a little late, but they are unhesitatingly going in now with both feet. By 2002, 54 percent of the 480 universities and colleges included in Kvavick et al.'s (2002) extensive study had implemented an integrated IT system such as PeopleSoft, Oracle, SCT Banner or SAP. Overall, American universities and colleges have invested more than $5 billion in ERPs, making higher education a small but growing source of revenue for ERP vendors.

Before ERPs arrived on campus, universities and colleges relied on "legacy" computer systems to handle their organizational information (student records, admissions, budgets, payroll, human resources, and course scheduling). These homegrown systems were typically run on early-generation mainframe computers and were tailor-made by "in-house" staff, producing considerable variation from school to school. ERPs are different from legacy systems in two important ways. First, a legacy system is highly dependent upon the expertise of the few people who built it, and their knowledge is valuable because it is idiosyncratic: it is embedded, local, and organization-specific. By contrast, a vendor's integrated information system is "standard" enough to serve the needs of a number of client organizations. Although ERP systems offer a variety of modules to be chosen by each client in different ways, PeopleSoft and its competitors depersonalize and delocalize the expertise and practices surrounding IT—and this inevitably makes a vendor's system less flexible for the adopting school. Second, legacy systems comprised a number of discrete databases spread throughout campus, and these functioned more or less autonomously. By contrast, ERPs centralize the collection, processing and storage of organizational data for a variety of purposes, and predictably managerial authority over these systems is also centralized.

In practice, ERPs in higher education do three things for a university (Brady, Monk, and Wagner 2001). First, they consolidate data management into a single system. Second, they integrate information systems in a way that makes discrete sub-units more dependent upon each other. Third, they digitize and automate the processes of data collection and retrieval—processes that once produced long paper-trails now appear as fleeting traces on a website designed by PeopleSoft. Each of these changes translates into significant cost savings and enhanced performance for universities and colleges, at least in theory. Legacy systems were becoming increasingly expensive to maintain, as the skilled programmers who designed them faded into retirement and as Web-based systems replaced mainframes. Costly repairs and constant updates now became PeopleSoft's burden. Moreover, the integrated and consolidated character of ERPs saves money by requiring data to be entered only once (previously, the same data might be entered repeatedly, in different functional domains). The likelihood of errors in data-entry is reduced if the data are entered only once, and there are other improvements in performance: because ERPs are capable of providing "real-time" data, university administrators can (in principle) make wiser decisions based on the best possible information.

Outsourcing IT is an especially effective way for universities and colleges to save money because of its back-office character. PeopleSoft itself does not appear on university websites, many of its tasks are invisible to students and their tuition-paying parents, and its influences seem remote from core activities of classroom teaching or research. In a sense, universities can "quietly" save money by adopting an ERP—avoiding the outcries that would surely attend an increase in tuition or a decrease in financial aid. But the advantages and efficiencies of PeopleSoft will stay quiet only if the new system works; if the "fit" between the newly adopted integrated information system and ongoing routine organizational practices goes bad,

PeopleSoft can become a lightning rod for rancor and complaint. And there are several good theoretical grounds for assuming that the adoption of an ERP will inevitably become a misfit.

One Source of Transfer Troubles: Transaction Costs

According to new institutional economics, transfer troubles in outsourcing result from cost-benefit decision-making by the two organizations involved in the outsourcing of IT: a university and a vendor. Both organizations pursue the economies they need in order to survive—vendors (like PeopleSoft) seek profits and market share, universities seek cost savings and enhanced performance. In this arrangement, universities must somehow deal with "transaction costs," which Williamson (1985) describes as the economic equivalent of friction in physical systems. Friction occurs when a process or product purchased from an external vendor is introduced into an organization—but where the resulting fit is less than perfect. The organization must then pay to fix these imperfections, i.e., transaction costs. Williamson's model is designed to explain the organization's choice to outsource (or not) in terms of anticipated transaction costs.[1] Transaction costs are one important source of "transfer troubles" facing organizations that outsource, but (as we shall see) they are not the only source.

Vendors face the following challenge: How can the many idiosyncratic information practices among a population of universities be translated into a single standardized system that will be "close enough" to extant practices? To be sure, PeopleSoft could build individualized information systems for each university in order to satisfy the unique needs of each and every potential client—but that would eliminate the significant economies of scale gained by the design of a single system that can be sold more or less off-the-shelf to a large number of universities and colleges. That is, the number of clients for an ERP must be large enough to cover the significant upstream costs of design—plus profits. The risk would be that the standardized package is so much at variance with routine practices at different universities that none of them would be enticed to purchase the system. To reduce that risk, PeopleSoft chose seven "beta universities" and modeled its new information system on the existing data management practices at these schools. By mimicking practices at the "beta universities," and by removing or translating lingering idiosyncrasies, PeopleSoft hoped to arrive at software that would be "close enough" to the local practices at most universities and colleges throughout the country. But inevitably, the effort to generalize specific procedures from supposedly exemplary prototypes will fail to produce a system that matches the needs of any other potential client. PeopleSoft must hope to convince colleges and universities that its "off-the-shelf" base model is good enough at least to start with. If PeopleSoft were forced to build unique systems for each of its clients, the bottom-line price would likely become so high (design-work and programming costs are expensive) that no school would gain worthwhile savings over their legacy system. PeopleSoft ultimately produced a somewhat standard system good enough for most universities, but a perfect fit for none of them. Remaining gaps are filled-in by PeopleSoft's willingness to allow any university to add functionalities to the standard base by customizing—in exchange, of course, for a not-so-modest fee.

Universities and colleges face a different cost-benefit quandary. The goal is to off-load their information systems onto a system built and maintained by a private vendor, at a cost less than what they are paying to keep the legacy system operational. The cheapest product for them to buy would be the "vanilla" PeopleSoft, an integrated information system with absolutely no customizations or add-ons. Such a purchase would seem to guarantee the greatest cost savings—except, of course, for costs associated with the bad fits that emerge after the system is implemented. Either the university can pay up front for costly customizations that would make the transfer smooth, or they can pay down the line to correct misalignments (for example, retraining staff to do routine tasks in a completely new way). Financially strapped public universities like Indiana are more likely to come down on the side of lower initial cost—buying the standard model with only the absolutely essential local modifications, hoping that the later costs of bad fits can be minimized or spread out over time.

A bad fit results from PeopleSoft's need to standardize their product to cover the costs of design, and the clients' need to save money by not insisting on expensive customizations that could in principle create a perfect copy of their unique data management practices. This situation will be familiar to those who study the effects of transaction costs on firms' decisions to outsource. If the organization is unwilling or unable to pay up front for a sufficiently customized IT system, it must be willing (later on) to absorb the inefficiencies created by the bad fit between their extant practices and the exigencies imposed by the new standardized package. Williamson (1981, 1985) classically described these inefficiencies as "transaction costs"—costs beyond the actual price that are associated with drag or friction inherent in the transfer process itself. If costs are exorbitant, the organization might rationally decline the opportunity to outsource or find a cheaper vendor. Our only reason for rehearsing the transaction costs model—where organizations make choices to in-source or outsource based on available information—is to suggest that this narrowly economistic orientation fails to capture all of what brings about transfer troubles as universities and colleges adopt ERPs. The literature in science studies suggests what might be missing: transfer troubles that emerge from the processes of articulating and translating an organization's routines.

Replication in Science: More Transfer Troubles

A working premise among scientists is that all experimental findings are, in principle, replicable. Confidence and trust in scientific knowledge is grounded on the assumption that if competent scientists were to repeat the same experiment in the same laboratory conditions but distant from where it was first conducted, the observed results would be identical. However, the replication of experiments is better thought of as an institutionalized myth of science rather than as an accurate sociological description of routine practice. The assumption that experimental claims are replicated is a key determinant of the public credibility of scientific knowledge and of scientists' cultural authority, even if—on the ground—few experiments are ever replicated, and even when they are, the decisiveness of the replication in confirming or falsifying the original claim is fraught with complications (Collins 1975).

Most experiments are never replicated by anybody. Merton's studies of the reward and evaluation system of science suggest that there are few career payoffs to be won by doing an experiment that somebody else has already done—Nobel Prizes (and lesser recognitions, along with symbolic and material resources) are awarded to those scientists accorded priority in making a revolutionary scientific discovery (Merton 1996). To be sure, some experimental replications are attempted, and occasionally their results are reported in the scientific literature—but typically only when the original findings are dramatically at variance with common understandings of reality or when suspicions are raised about the technical ability or moral propriety of the scientists who first performed the experiment. In 1989, Pons and Fleischmann's "cold fusion" claim generated countless replication attempts, for all of these reasons—and that episode begins to suggest exactly why experimental replication (even in those rare cases where it is attempted at all) is something less than a straightforward and immediately convincing adjudication of the initial claim (Gieryn 1999, chapter 4).

Transfer troubles in scientific replication, as Collins (1975: 206) first pointed out, arise from complexities involved in deciding exactly what constitutes an "exact copy of the original." Did the replication in fact use exactly the same specimens (or whatever raw data), exactly the same instruments and equipment, in exactly the same laboratory conditions? Perhaps most importantly, did the replication follow exactly the same step-by-step methodological procedures as the original? Collins presents a conundrum: how can anyone know whether a replication has been competently performed in a way faithful to the initial experiment? Plainly, if the results from the replication match those found in the original experiment, scientists are inclined to say that the copy was a good one—and that the claims should be sustained. But if different results are found, then it is not easy to decide if the original experiment is simply wrong in its depiction of Nature or if the replication attempts are incompetently and inaccurately performed. As Collins (1975: 219–220) puts it, "scientists' actions may then be seen as negotiations about which set of experiments in the field should be counted as the set of competent experiments." Pons and Fleischmann tried to "save their phenomenon" in the face of failed replication attempts by suggesting that something in the ambient environment at their Utah lab made cold fusion happen only there.

Collins's analysis of replication in science enables us to think about transfer troubles as resulting from something more intransigent than calculated transaction costs (as in the economistic frame). The adoption of PeopleSoft in organizations of higher education involves a transfer of practices and procedures from their initial organizational setting (the university) to the ERP vendor (where those practices and procedures are more or less fit into standardized packaged software), and then they are transferred back again to the client-university (where goodness of fit is discovered during implementation). To be sure, some of the bad fit that inevitably occurs when a university brings in PeopleSoft to manage its information system results from the market necessity of the vendor to sell a standardized product and the reluctance of the client to pay huge sums for customization. But there is a deeper source of transfer troubles, inherent in the problematic notion of a "working copy," which is apparent in Collins's description of two models that he uses to describe the process of transfer in scientific replications.

In an "algorithmic" model, the knowledge and skill required to perform an experiment (or, by extension, to run the IT system of a university) can be expressed as a sequence of precise commands—like those found in a computer program. These "instructions" would ideally incorporate all of the information needed to replicate the experiment—and present it in an "unambiguous" way (Collins 1975: 206). However, based on his long-standing studies of scientists who work on gravitational radiation in physics, Collins (2004) suggests that it is impossible to capture in a finite set of instructions all of the knowledge that the replicating scientist might need to know in order to produce an exact "working copy" of the original experiment. He proposes an "enculturation" model to describe better how knowledge gets transferred from one experimental group to another—and we think this model also adds to the understanding of what happens when universities outsource their data management systems.

The knowledge and skill needed to do an experiment is deeply embedded in the local culture of a group of scientists—specifically, it is "embodied in their practices and discourse and could not be explicated" (Collins 2004: 608). Much of what goes into experimental work is "tacit knowledge," a term that Collins borrows from Michael Polanyi (1958) and then redefines this way: "knowledge or abilities that can be passed between scientists by personal contact but cannot be ... passed on in formulas, diagrams, or verbal descriptions and instructions for action" (Collins 2004: 609). By participating in the life of a specific laboratory group, a scientist gets enculturated into a distinctive way of seeing and acting—a "disposition," for Bourdieu (2004)—that enables the performance of experimental operations. That group's local knowledge cannot be articulated fully and unambiguously as a set of rules or procedures—so that any account of experimental protocols will inevitably be incomplete, which in turn leads to transfer troubles when a replication is tried by a scientist who lacks the necessary tacit skills. Experiments are so thoroughly entrenched in the local culture of a laboratory that their performance cannot be extracted from that initial context and successfully replicated elsewhere—unless accompanied by scientists who shared first-hand the tacit understandings and skills of the original group.

When a university enters into a contract with PeopleSoft to outsource its IT systems, it begins by preparing an algorithmic list of the "functionalities" they need from the modular but standardized ERP. University administrators and technical support staff must describe their extant data management processes and systems in enough detail so that both parties can see how closely (or not) the "standard issue" PeopleSoft approximates the procedures already in place. When discrepancies become apparent, the university must decide whether or not it wants to pay for customizations designed to preserve some "idiosyncratic" features of the legacy systems. Not all customizations will be affordable, which means—following

Williamson—the university will later absorb transactions costs inherent in the imperfect fit between PeopleSoft's packages and previously established routines.

But when Collins's analysis of experimental replications in science is extended to IT outsourcing, an even deeper source of transfer troubles is expected. Every university has its own organizational culture, and its employees share tacit knowledge about their embodied practice that is impossible to retrieve and articulate precisely and unambiguously—as an algorithmic set of rules and procedures then compared to PeopleSoft's standardized package. In effect, as the university debates about which costly customizations to buy, they are able to consider only a truncated subset of all the possible differences between their legacy system and PeopleSoft. Transfer troubles arise not just from an economistic decision to delay the costs of outsourcing until after an imperfect "replication" is implemented, but also from the unavoidable and unrecognized inability of university officials and staff to formalize exactly and completely their practices *status ante quo*. Universities will be able to anticipate only those transfer troubles that result from their choice not to purchase some expensive customizations; they will be surprised by additional instances of "bad fit" emerging from deeply sedimented implicit knowledge—which is exactly what happened to Indiana University as it transferred the "I-Core" program to PeopleSoft.

Case Study: The I-Core "Fit Gap"

The Integrated Core ("I-Core") is the signature feature of the undergraduate training program in the Kelley School of Business at Indiana University. I-Core is an integrated set of courses taken by business majors during their junior year. Four classes covering basic areas of business expertise—finance (F370), marketing (M370), operations (P370), management (J370), along with an integrated discussion section (I370)—are all taken during the same semester. The pedagogic goal is for students to be able to make substantive connections among issues raised in these courses. Instructors in the course on finance, for example, make explicit links to what is being discussed in the courses on marketing, operations, and management.

Saving I-Core with "Block Enrollment"

As Indiana University contemplated the outsourcing of its data management systems to PeopleSoft, it was obvious to all that the vaunted I-Core was one functionality that needed to be maintained at any cost. That is, the university was willing to pay extra for a customization of the standardized PeopleSoft package that would enable I-Core to continue through the IT transition. Specifically, the Kelley School of Business needed a system that would allow their students to "co-register" for the five courses that make up I-Core—so that admission into one of those courses automatically involved admission to the other four. This "co-registration" facility was not available on PeopleSoft's standardized package of services.

To preserve I-Core, Indiana University and PeopleSoft devised a plan for "block enrollment," in which the five courses are conceptually tied together into a "pseudo-block." Under PeopleSoft, students entering I-Core enroll in a single "pseudo-course" labeled BE370 (Block Enrollment 370), rather than registering for separate courses with distinctive numbers (F370, M370, etc.). In effect, all I-Core courses share the same number for purposes of registration—BE370—but in actuality, BE370 is just a fictive place holder that secures a student's seat during the registration and enrollment process. Only on the very last day of registration are I-Core students then "decomposed" from BE370 into the five constituent courses that make up the program. Block enrollment is a significant change from the now-retired legacy system, where students registered separately for each of the five courses.

Because I-Core is a crown jewel of the undergraduate business program at Indiana University, everybody agreed that it must be allowed to continue in spite of the different course registration procedures introduced by PeopleSoft. "The university thought it [block enrollment] would work," said one official, "and so did I." Administrators at the Kelley School "told them [the transition team] what we wanted … at the beginning, there was a Christmas list of all the things we needed"—and the ability to "co-register" students into the five integrated courses was at the top of the list. PeopleSoft's block-enrollment function appeared to be the right customization for the job: it would, in principle, allow students to get seats in all five I-Core classes during the same semester, in a way that was hoped to be just as smooth as in the tried and true legacy system.

Authorization Complications

Unfortunately, the transfer of I-Core registration procedures to PeopleSoft's block enrollment system introduced several unanticipated troubles—that became apparent only when the ERP went "online." Not just any Indiana University student may be admitted to I-Core: only those students who are "authorized" to enroll may actually get seats, and this authorization process made the transfer of I-Core less than smooth. To understand those troubles, we need to say a little more about how Indiana University organized its course registration procedures under the legacy system.

Before PeopleSoft, Indiana University worked with a hierarchical system involving both "courses" and "sections." Courses were given names like Introduction to Sociology and unique numbers that included a letter indicating its department or school—in this instance, S100. Some courses—but not all—have multiple sections, and each of these was assigned a four-digit code number (in any given semester, the Sociology Department might offer as many as twelve different sections of S100, on different days and times, in different rooms, with different instructors and substantive content). Importantly, when registering for classes, students signed up for a specific section rather than for the course as such. Under the legacy system, granting authorizations for I-Core was a laborious but doable process: the integrated discussion course had three sections, while the other four courses had two sections each. In practice, students could mix and match as many as 23 different combinations of sections that would get them seats in the five different courses that make up I-Core. In a typical fall semester, 700 students are seeking authorization for I-Core: the challenge was to get these students distributed among the different sections in a way that did not overbook the capacity of assigned classrooms and that did not introduce overlapping time-conflicts for the student. We were told that "it took one whole day for one administrative assistant" to grant authorizations for I-Core courses—but it got done.

PeopleSoft has introduced a different terminology as well as different procedures. Now, what used to be called "sections" are called "classes"—and each class is assigned a unique five-digit number. In the legacy system, individual sections were (from a data management perspective) grouped under a designated course—so, for example, when a student was authorized for section 3865, they were automatically authorized for S100. In PeopleSoft, individual classes now functioned as autonomous units (again, from a data management perspective), and the system did not automatically link a specific class number to the umbrella course number. This seemingly trivial difference—one that went largely unnoticed as the transfer of I-Core was planned out—introduced massive complications that resulted specifically from the Block Enrollment innovation that was supposed to be a quick fix. Recall that I-Core students register for BE370—a dummy course number that supposedly would secure seats in all five courses. Now, because the block is only "conceptually" related to the specific courses, and because individual classes cannot be mapped directly onto the actual five course numbers (which, in effect, disappear from the authorization and registration process until the very last day when "decomposition" occurs), authorization must be granted for the block *and* for the separate classes. This has greatly enlarged the work-load: "Now, authorization takes more than a week for three assistants." Transfer troubles indeed.

More Complications: Timing and Room Assignments

I-Core courses are integrated not only in their substance, but temporally and spatially too. The Kelley School of Business wishes to organize I-Core class times back to back, so that students move directly from finance at 2:30 p.m. to management at 4:00 p.m. These classes are 75 minutes long, and there is little downtime between them—a virtue, because students get the feeling of being thoroughly immersed in their studies. Moreover, lessons from the early class may be picked up for further consideration in the later one. Because there is so little time between classes, instructors also prefer to locate the sequential classes in rooms that are adjacent or at least nearby—this proximity also reinforces the "integrated" nature of I-Core.

The sequential scheduling of classes in rooms close to each other was routinely accomplished under the legacy system—largely because it was relatively easy throughout the authorization and registration process to attach specific sections to specific times and rooms. In the old system, students signed up for a "real" section, with a designated time and room (not, as with PeopleSoft, for a pseudo-block). It was ordinarily a simple matter for Kelley School administrators to change class times and rooms during the registration process (in response, say, to a professor's request for back-to-back classes to be closer together): a student's schedule could instantly be updated because each section was a real and autonomous entity.

These adjustments of class times and room assignments quickly became unmanageable once PeopleSoft went live. Because of the Block Enrollment system, it was impossible to know whether the time or room assigned to a class could be changed—because, until the very end of registration, it was impossible to know which students were in which specific classes. The Block Enrollment system was the only way, under PeopleSoft, that students could be co-registered for the five composite course-set in I-Core. Yet this necessary "solution" created a huge new problem: any requested change in time or room could introduce impossible overlaps where a student was now assigned to more than one class on the same day at the same time—and these potential overlaps were not easy to detect or avoid with

PeopleSoft. The I-Core schedule of courses remained in flux throughout the registration process, frustrating coordinators and confusing students. In response to this instability, a student might provisionally be assigned to three different blocks with three different sets of room assignments and starting times—and the situation would be resolved only just before the semester would begin. One I-Core administrator felt like an "air traffic controller of students; they keep coming in and we need to land them somewhere, somehow." Blocks, they claim, are a "pain to arrange, a pain to decompose and a pain to track."

Conclusion

A rather large "fit gap" emerged when the Kelley School of Business tried to transfer the registration procedures for its I-Core program from Indiana University's legacy system to PeopleSoft. These troubles did not arise from the university's unwillingness to spend money to customize the off-the-shelf ERP in a way that would avoid the costly problems we have just described. Importantly, the new complications arising from PeopleSoft were not anticipated as university officials decided which customizations they needed to buy. Given the lofty stature of the Kelley School of Business on the Bloomington campus, and given the profound importance of I-Core for that School's reputation, Indiana University would have been happy to pay for a customized solution to the authorization and scheduling problems— if only they could have seen them coming. The university did not knowingly choose to absorb downstream "transaction costs"; they did not know, in advance, that the Block approach would necessitate so many expensive expenditures for increased staff labor and (eventually) re-training.

Transfer troubles result from something more than the willingness of an organization to incur transactions costs associated with outsourcing. Our analysis shifts the focus from market pressures to the inherent cognitive and communication problems associated with the transfer of practices among organizational contexts. Universities can calculate transaction costs, but only if they are capable of describing their extant practices in sufficient and accurate detail. Outsourcing IT is like the replication of a scientific experiment in that it is impossible to create an algorithm that effectively captures everything involved in the original experimental protocols—or (by extension) in the previously established data management system. The old routine procedures for handling authorizations for I-Core classes—and for assigning their times and rooms—were so deeply embedded in a local organizational culture that they were beneath explicit recognition and articulation. Those who ran the legacy system possessed an abundance of tacit knowledge and skill that they were unable to transfer into a set of formalized rules that PeopleSoft could—at a price—incorporate. One spokesperson for Indiana University lamented the disappearance of its legacy system: "it's the little things … that's what I miss … but you don't know till they're already gone."

Acknowledgments

The research on which this chapter is based is funded by a National Science Foundation Dissertation Improvement Grant SES #0551802. One primary source of information is 42 interviews conducted at large public universities. In view of the detailed subject matter of this chapter, it is especially appropriate to thank the Kelley School of Business, Indiana University, and those individuals interviewed specifically for this book chapter. We are also grateful to those who have commented on earlier drafts of this paper and presentations prepared on the subject matter, especially Tim Bartley, Fabio Rojas, and Brian Steensland.

Note

1. Williamson's model assumes that a vendor's product is either standardized (general asset specificity) or customized (firm-specific) when it goes to market, and that it remains fixed while on the market. Moreover, the model assumes that the purchasing organization has fixed needs, either for a standardized and general product or for something highly customized and tailor-made. As it happens, the adoptions of ERPs in higher education have not conformed exactly to the assumptions in Williamson's model. Neither the products made available by vendors nor the level of generality/ specificity needed by the client is a fixed and stable entity. Rather, universities and PeopleSoft *negotiate the substance* of the outsourced product—moving it along a hypothetical gradient in which one pole would be "exact equivalence to the university's legacy system" and the other pole would be "PeopleSoft's baseline standard package of services." We describe in the text how and why PeopleSoft and universities negotiate the range and scope of customized modules, but the important point is this: from the standpoint of identifying "transfer troubles" resulting from outsourcing, there is essentially no difference between an organization choosing from a menu of fixed outsourced products and an organization negotiating with a vendor over the substance of the product and its cost. At the end of the day, no matter how much negotiation and customization might have gone on upstream, the organization must still make a decision to buy the product or not—and, thus, to absorb or not the transaction costs that Williamson has discussed.

References

Barnett, R. 2000. *Realizing the University in an Age of Supercomplexity*. Open University Press.

Biemiller, L. 2005. The life of a campus: Books, buses, and chicken strips. *Chronicle of Higher Education*, January 28: B15–B19.

Blom, S., and S. Beckley. 2005. Six major challenges facing student health programs. *Chronicle of Higher Education*, January 28: B25–B26.

Bourdieu, P. 2004. *Science of Science and Reflexivity*. University of Chicago Press.

Brady, J., E. Monk, and B. Wagner. 2001. *Enterprise Resource Planning*. Course Technology.

Breneman, D., J. Finney, and B. Roherty. 1997. *Shaping the Future: Higher Education Finance in the 1990s*. California Higher Education Policy Center.

Collins, H. 1975. The seven sexes: A study in the sociology of a phenomenon, or the replication of experiments in physics. *Sociology* 9: 205–224.

Collins, H. 2004. *Gravity's Shadow*. University of Chicago Press.

Davies, P. 2005. Outsourcing can make sense, but proceed with caution. *Chronicle of Higher Education*, January 28: B20–B22.

Etzkowitz, H. 1990. The capitalization of knowledge. *Theory and Society* 19: 107–121.

Gieryn, T. 1999. *Cultural Boundaries of Science*. University of Chicago Press.

Goldstein, P., D. Kempner, and S. Rush. 1993. *Contract Management of Self-Operation: A Decision-Making Guide for Higher Education*. APPA.

Gose, B. 2005. The companies that colleges keep. *Chronicle of Higher Education*, January 28: B1–B12.

Klaus, H., M. Rosemann, and G. Gable. 2000. What is ERP? *Information Systems Frontiers* 2: 141–162.

Kumar, K., and J. van Hillegersberg. 2000. ERP experiences and evolution. *Communications of the ACM* 43: 23–26.

Kvavik, R., et al. 2002. *The Promise and Performance of Enterprise Systems for Higher Education*. Educause Center for Applied Research.

Langenwalter, G. 1999. *Enterprise Resources Planning and Beyond: Integrating Your Entire Organization*. St. Lucie Press.

Lund, H. 1997. "Outsourcing" in commonwealth universities. CHEMS Paper 18. http://www.acu.ac.uk.

Merton, R. 1996. *On Social Structure and Science*. University of Chicago Press.

Norris, G., J. Hurley, J. Dunleavy, J. Balls, and K. Hartley. 2000. *E-Business and ERP: Transforming the Enterprise*. Wiley.

O'Leary, D. 2000. *Enterprise Resource Planning Systems: Systems, Life Cycle, Electronic Commerce, and Risk*. Cambridge University Press.

Polanyi, M. 1958. *Personal Knowledge*. University of Chicago Press.

PricewaterhouseCoopers. 1999. *Technology Forecast*.

Ross, J. 1998. The ERP Revolution: Surviving versus Thriving. Working paper 307, MIT Sloan School of Management.

Shrum, W., and Y. Shenhav. 1995. Science and technology in less developed countries. In *Handbook of Science and Technology Studies*, edited by S. Jasanoff et al. Sage.

Slaughter, S., and G. Rhoades. 2004. *Academic Capitalism and the New Economy: Markets, State, and Higher Education*. Johns Hopkins University Press.

Willcocks, L., and R. Sykes. 2000. Enterprise Resource Planning: The role of the CIO and IT function in ERP. *Communications of the ACM* 43: 33–38.

Williamson, O. 1981. The economics of organization: The transaction cost approach. *American Journal of Sociology* 87: 548–577.

Williamson, O. 1985. *The Economic Institutions of Capitalism: Firms, Markets, Relational Contracts*. Free Press.

About the Authors

Daniel Beunza is an assistant professor of management at the Columbia Business School and the acting director of Columbia University's Center on Organizational Innovation. He studies how social relations and technology shape value in the capital markets.

Michel Callon is a professor of sociology at the École des Mines de Paris and a researcher at the Centre de Sociologie de l'Innovation. He is working on the relations between economic markets and technical democracy and is completing research with Volona Rabeharisoa on French and European patients' organizations.

Shay David is a Ph.D. candidate in Cornell University's Science and Technology Studies Department and a Fellow at Yale Law School's Information Society Project. He is working on a study of reputation economies and open information networks.

Barbara Grimpe holds a teaching and research position at the University of Konstanz. She is finishing a doctoral dissertation on the public debt management system DMFAS and its use in developing countries.

David Hatherly is an emeritus professor of accounting at the University of Edinburgh. His research interests include the application of judgment in auditing and accounting.

Karin Knorr Cetina teaches at the University of Konstanz and at the University of Chicago. She is working on a book on global foreign exchange markets.

David Leung is a doctoral student at the University of Edinburgh. He is researching ethnoaccountancy, which focuses on the ethnographic approach to accounting research.

Christian Licoppe is a professor of sociology of information and communication technologies in the Department of Social Science at the École

Nationale d'Études des Télécomunications. He is the author of *La formation de la pratique scientifique* (La Découverte, 1995) and of many studies on the uses of information and communication technologies.

Donald MacKenzie is a professor of sociology at the University of Edinburgh. His current research, which involves developing a "material sociology of markets," focuses on financial markets and on new markets in permits for greenhouse-gas emissions.

Philip Mirowski is Carl Koch Professor of Economics and the History and Philosophy of Science at the University of Notre Dame. He has two books forthcoming from the Harvard University Press: an edited volume (with Dieter Plehwe), *The Making of the Neoliberal Thought Collective*, and a book on the modern economics of science, *SciMart*.

Fabian Muniesa is a researcher at the Centre de Sociologie de l'Innovation at the École des Mines de Paris. He has published several articles on the social studies of finance. His research interests include economic experiments and the anthropology of calculation.

Edward Nik-Khah is an assistant professor of economics at Roanoke College. His recent work has examined the relationship between the auction theory project in economics and neoliberalism. He is working on a study of the development of the University of Chicago business program.

Trevor Pinch is a professor of Science and Technology Studies and of sociology at Cornell University. He has published numerous books and articles on the sociology of science and technology. His latest research concerns the use of vintage technologies in music and the sociology of online music communities.

Elizabeth Popp Berman is an assistant professor of sociology at the University at Albany. She is writing a book on the emergence of market-oriented institutions within academic science, and is also studying how economic growth became taken for granted as a public good.

Alex Preda teaches sociology at the University of Edinburgh. He is doing research on anonymous electronic markets.

David Stark is Arthur Lehman Professor of Sociology and International Affairs at Columbia University and an external faculty member of the Santa Fe Institute. He is the author of a number of articles in the *American Journal of Sociology* and elsewhere.

Richard Swedberg is a professor of sociology at Cornell University. He is the author of a number of works on economic sociology, including *Max Weber and the Idea of Economic Sociology*.

Index

Inside Technology
edited by Wiebe E. Bijker, W. Bernard Carlson, and Trevor Pinch

H. M. Collins, *Artificial Experts: Social Knowledge and Intelligent Machines*

Paul N. Edwards, *The Closed World: Computers and the Politics of Discourse in Cold War America*

Herbert Gottweis, *Governing Molecules: The Discursive Politics of Genetic Engineering in Europe and the United States*

Joshua M. Greenberg, *From Betamax to Blockbuster: Video Stores and the Invention of Movies on Video*

Kristen Haring, *Ham Radio's Technical Culture*

Gabrielle Hecht, *The Radiance of France: Nuclear Power and National Identity after World War II*

Kathryn Henderson, *On Line and On Paper: Visual Representations, Visual Culture, and Computer Graphics in Design Engineering*

Christopher R. Henke, *Cultivating Science, Harvesting Power: Science and Industrial Agriculture in California*

Christine Hine, *Systematics as Cyberscience: Computers, Change, and Continuity in Science*

Anique Hommels, *Unbuilding Cities: Obduracy in Urban Sociotechnical Change*

Deborah G. Johnson and Jameson W. Wetmore, editors, *Technology and Society: Building Our Sociotechnical Future*

David Kaiser, editor, *Pedagogy and the Practice of Science: Historical and Contemporary Perspectives*

Peter Keating and Alberto Cambrosio, *Biomedical Platforms: Reproducing the Normal and the Pathological in Late-Twentieth-Century Medicine*

Eda Kranakis, *Constructing a Bridge: An Exploration of Engineering Culture, Design, and Research in Nineteenth-Century France and America*

Christophe Lécuyer, *Making Silicon Valley: Innovation and the Growth of High Tech, 1930–1970*

Pamela E. Mack, *Viewing the Earth: The Social Construction of the Landsat Satellite System*

Donald MacKenzie, *Inventing Accuracy: A Historical Sociology of Nuclear Missile Guidance*

Donald MacKenzie, *Knowing Machines: Essays on Technical Change*

Donald MacKenzie, *Mechanizing Proof: Computing, Risk, and Trust*

Donald MacKenzie, *An Engine, Not a Camera: How Financial Models Shape Markets*

Maggie Mort, *Building the Trident Network: A Study of the Enrollment of People, Knowledge, and Machines*

Peter D. Norton, *Fighting Traffic: The Dawn of the Motor Age in the American City*

Helga Nowotny, *Insatiable Curiosity: Innovation in a Fragile Future*

Nelly Oudshoorn and Trevor Pinch, editors, *How Users Matter: The Co-Construction of Users and Technology*

Shobita Parthasarathy, *Building Genetic Medicine: Breast Cancer, Technology, and the Comparative Politics of Health Care*

Trevor Pinch and Richard Swedberg, editors, *Living in a Material World: Economic Sociology Meets Science and Technology Studies*

Paul Rosen, *Framing Production: Technology, Culture, and Change in the British Bicycle Industry*

Susanne K. Schmidt and Raymund Werle, *Coordinating Technology: Studies in the International Standardization of Telecommunications*

Wesley Shrum, Joel Genuth, and Ivan Chompalov, *Structures of Scientific Collaboration*

Charis Thompson, *Making Parents: The Ontological Choreography of Reproductive Technology*

Dominique Vinck, editor, *Everyday Engineering: An Ethnography of Design and Innovation*